CHEMTRAILS EXPOSED
A New Manhattan Project

by
Peter A. Kirby

Second edition

fourth printing

ISBN-13: 9798643048084

THANKS

My grade school English teacher Raymond Conti, my favorite high school English teacher Elizabeth Cornell, the Marin County Civic Center Free Library, the University of California Northern Regional Library Facility, the University of California at Berkeley, the UC Berkeley Newspapers and Microforms library, Ginny Silcox whose assistance was invaluable, Mauro Oliveira, Mark McCandlish, Michael Murphy, Catherine Frompovich (1938-2020), Environmental Voices, Allan Buckmann, Francis Mangels, Patrick Roddie, Arizona Sky Watch, Dane Wigington at *GeoengineeringWatch.org*, Rusty at *GlobalSkyWatch.com*, *Chemtrails911.com*, Jim Lee at *ClimateViewer.com*, Harold Saive at *ChemtrailsPlanet.net*, Mike Adams at *NaturalNews.com*, Roxy Lopez at *TheTruthDenied.com*, *ChemSky.org*, Marvin Herndon at *NuclearPlanet.com*, *ChemtrailsProject.com*, *ChemtrailsProjectUK.com*, *ChemtrailSafety.com*, *CheckTheEvidence.com*, *StopSprayingCalifornia.com*, Elana Freeland, Mark M. Rich, James Roger Fleming, George Noory at *Coast to Coast AM*, Clyde Lewis at *Ground Zero radio*, William Thomas, Ross at *ConsciousLifeNews.com*, *DavidIcke.com*, Shepard and Alex at *IntelliHub.com*, Aaron and Melissa Dykes at *TheDailySheeple.com* and *TruthstreamMedia.com*, David Zublick at *DarkOutpost.com*, Professor Chossudovsky at *GlobalResearch.ca*, Sean at *SGTReport.com*, Kerry at *ProjectCamelotPortal.com*, John B. Wells at *CaravanToMidnight.com*, Sarah Westall at *SarahWestall.com*, James Corbett at the *CorbettReport.com*, Michael at *ActivistPost.com*, and anybody else who has eyes to see and a brain to understand.

This book is dedicated to the memory of my father
Patrick A. Kirby (1926-2005).

The fight continues. This time, we put the New World Order out of business PERMANENTLY.

"Human understanding is a distinctly individual activity. An individual ponders and, through tedious efforts, places seemingly independent ideas and observations into a logical order in the mind so that causal relationships become evident and new understanding emerges."

-Dr. James Marvin Herndon, PhD

TABLE OF CONTENTS

Chapter 1
CHEMTRAILS EXPOSED

According to all relevant United States federal government organizations, chemtrails do not exist and are a conspiracy theory. They say the white lines in the sky stretching from horizon to horizon are normal jet contrails. They add that these emissions are nothing but harmless water vapor. So, they are saying that this water vapor can first appear as white lines high in the sky then, over the course of hours, expand as it floats down to earth; first creating clouds and finally a haze. Folks, that's not how water vapor behaves. That's how particulate matter behaves.

Contrails (not to be confused with chemtrails) are visible jet airplane engine emissions resulting from the combustion of unadulterated fuel. Most of the time, jet airplane exhausts are invisible, but when unadulterated jet engine exhausts are introduced into an atmosphere of high humidity (over 70%) and low temperature (below -76 degrees F), a contrail forms. Contrails then quickly (under 10 seconds) disappear.

Chemtrails are a different animal. Chemtrails are jet airplane emissions which do not quickly dissipate and are not necessarily the product of engine exhausts. They are seen in the sky above our heads on a daily basis. They are the lines in the sky. Your subconscious knows them already. The word 'chemtrail' is an amalgamation of the words 'chemical' and 'contrail.' You could call chemtrails 'chemical contrails' or 'chemical trails.' The earliest mention of the word 'chemtrail' the author could find appears as the title of a 1990 U.S. Air Force Academy chemistry manual.

The word 'chemtrail' is recognized by the Oxford dictionary. It is defined as, "A visible trail left in the sky by an aircraft and believed by some to consist of chemical or biological agents released as part of a covert operation."

The term 'persistent contrail' is often used to describe chemtrails, but this term is a politically correct oxymoron which the author uses only to condemn. The term creates confusion. Contrails do not persist. Chemtrails do. The earliest use of the term 'persistent contrail' known to the author appears in the 1970 proceedings of the Second National Conference on Weather Modification.

Lab tests of rainwater samples taken around the world have shown that chemtrails commonly consist of aluminum, barium and strontium. Further, as we will see in chapter 6, a world famous PhD scientist has come forward with rock-solid evidence that the substance commonly being sprayed is coal fly ash - a toxic waste.

Oh no they wouldn't!

Oh yes they would. Dr. Leonard Cole's 1988 book *Clouds of Secrecy* documents our military intentionally exposing unsuspecting United States citizens to chemicals and biologicals hundreds of times over the past 60 years. Andrew Goliszek's illuminating book *In the Name of Science* provides another examination of our government's indiscretions. Going back many decades, this book documents megatons of chemical and biological agents ranging from VX nerve gas to zinc cadmium sulfide to bacillus globigii to radioactive iodine to many other substances released onto unsuspecting American citizens in thousands of open air tests conducted all over the country. Many of these experiments involved the agents being released from airplanes.

Governments spray us with toxic chemicals openly. What about cloud seeding with silver iodide? What about the spraying of malathion in the 1980s to save California from the mediterranean fruit fly, or the more recent light brown apple moth sprayings? Did everybody consent to these actions? In some areas of the Country, city works trucks will routinely drive down the street spraying herbicides all over the sidewalk and anyone who might be unfortunate enough to be walking down it. In Florida, large aircraft will drop mosquito spray over populations without prior consent.

Observations

A typical chemtrail first appears as a white line high in the sky emitted by a large jet airplane. These lines become diffuse as they float down to earth. Heavy spraying creates a thick haze over vast areas. Chemtrails often form Xs and sometimes grid patterns. Chemtrails are also sometimes sprayed in a circular fashion. Although seemingly endless amounts of photos and videos are posted online, you can observe these phenomena with your own eyes. One just needs to look up.

Although trolls online try to drag people into highly technical, obscure, and deceptive arguments about the formation of contrails, simple applications of deductive reasoning show that we are being sprayed. When two jet airplanes fly at a similar altitude, why is it that one will emit a persistent trail while the other does not? The fuels are not different. Jet fuel, like gasoline, is standardized. How does a trail appear behind one engine of an airplane, but not the other engine? How does a plane fly along emitting a trail that persists, then abruptly stop emitting a trail, then abruptly start again? It is safe to assume that the engine is not being turned off and on. Why is it that one day the sky will be full of lines in every direction and then the next day there will be none, even though the weather is identical? Why do the lines in the sky often not follow common flight paths? Your author has witnessed all these things, and you can too.

When they spray above natural clouds, we get a milky sky. Chemtrails sprayed above clouds descend and commingle with the clouds below to cause a loss of cloud definition.

Sun dog w/ chemtrail

Chemtrail spray can cause iridescence. Often when they spray, something called a 'sun dog' occurs. Also referred to as a 22 or 46 degree halo, a sun dog is a big, rainbow-like halo encircling the sun, although a sun dog's colors are not as bright as a rainbow's. You may also see little rainbow segments called 'chembows.' Lastly, you may see small, iridescent cumulus clouds. Spraying in general tends to make clouds iridescent; especially around the edges and when backlit by the sun. Chemtrails manifest themselves in many other ways not noted here. These are merely the most common and noticeable.

Coastal areas are seeing a depletion of marine layers. There is a district of San Francisco, CA (the author's home town) called the Sunset District. Until about 2005, this district had literally 5-10 days of sunny weather per year. Before 2005 the adjacent Pacific Ocean provided the Sunset District with an almost ever-present blanket of clouds and fog. Now in the Sunset District it is sunny most days. Due to the hygroscopic (moisture-absorbing) properties of the spray, chemtrail spraying has destroyed the Sunset District's marine layer. It is safe to assume that chemtrail spraying has caused similar effects along the rest of California's Pacific Coast as well as desiccation of the state's interior. What do you think that does to a native ecosystem?

San Francisco's summer fog is gone. San Francisco used to be world famous for its thick fog which usually occurred in the summer. When was the last time you heard about it? Until about 2005, many summer days in San Francisco were so foggy that one could not see for more than one half of a city block. One never hears about it anymore because now, due to the spraying, San Francisco *never* has that really thick fog that it used to have.

Personally, I can smell and taste chemtrails. Chemtrail spray has an ashy, metallic smell and taste. Others report a burning rubber/metallic type of smell. Also, chemtrail spray irritates my eyes and makes them water. After so much exposure, many others are apparently not sensitive enough to observe these things, or maybe they think it's just common industrial pollution.

At the time of this writing, chemtrails are appearing almost every day above San Francisco and San Rafael, CA; where the author currently lives. Judging by the first wave of reports, large-scale chemtrail spraying became nationally prevalent in the mid 1990's. Although chemtrails have been mostly documented in North Atlantic Treaty Organization (NATO) countries such as America, Japan, Australia, and those in Europe, this is a global operation. The number of chemtrail reports from all around the world has been steadily increasing and chemtrails can now be observed above all adequately industrialized regions of the planet. In fact, the more industrialized and populated a region is, the more chemtrails can be observed.

Geoengineering

The only commonly known reason for jet airplanes to be emitting sprays at altitude is for weather modification. Geoengineers have been publicly proposing just that.

Many people calling themselves geoengineers have appeared on many television and Internet broadcasts. They often propose spraying megatons of chemicals into the atmosphere utilizing jet airplanes. They say it can save us from global warming. They don't admit that it is currently happening, they just say it might be good and sometimes press for the adoption of such measures. Harvard University has a whole department dedicated to geoengineering.

Although geoengineering encompasses many schemes such as 'fertilizing' the ocean with iron filings or building giant terraforming machines to remove atmospheric carbon dioxide, the most famous and most promoted geoengineering scheme is something called Solar Radiation Management (SRM). Solar Radiation Management involves spraying fine particles into the lower stratosphere from jet airplanes. Amongst us common folk, the term 'geoengineering' has become synonymous with SRM. Proponents of all forms of geoengineering claim these activities will save us from the dreaded global warming and climate change.

Geoengineers say they want to alter the climate. Climate is synonymous with weather. Because geoengineers are publicly advocating for climate modification, they are advocating for weather modification. Just as climate is synonymous with weather, geoengineering is synonymous with weather modification.

The thesis of this book is that geoengineering, specifically the spraying of stratospheric aerosols, is a current, ongoing operation. It's not being done to save us all from global warming, though. It is mostly done as part of a global weather modification project. Just as the prominent geoengineer David Keith has claimed that he doesn't know what a chemtrail is, geoengineers are simply playing word games. This thesis will not and cannot be disproven. The New Manhattan Project exists. Chemtrails are real.

Although today's geoengineers often claim that the field only began in the late 1970s, geoengineering is actually a newer name for an old global weather modification project with roots going back to the 1940s and earlier. In fact, the history of weather modification, or 'weather control' as the Library of Congress calls it, is the history of geoengineering.

'Cloud seeding,' which has been done openly for a long time, is a pre-cursor to today's geoengineering programs. Government agencies officially oversee the cloud seeding industry and there are many companies involved. They have an industry association. They engage in conventional cloud seeding activities where permitted aircraft usually dump silver iodide on us to either make it rain or suppress hail. But that's old technology. As this book will explain, today's geoengineers spray us with completely different materials and, in addition, use electromagnetic energy to manipulate the particles once they are dispersed.

You see, what today's scientific establishment calls 'SRM geoengineering' is actually code for the greatest scientific endeavor in Human history. This scientific project involves the dispersion of tiny particles from aircraft which are then manipulated *en masse* by electromagnetic energy generated by ground-based antennas as well as satellites. In this way, along with the electromagnetic manipulation of the ionosphere and other techniques, the weather can be comprehensively modified and/or controlled. It is chiefly this use of electromagnetic energy which distinguishes the New Manhattan Project from conventional weather modification activities.

Perhaps a former executive adviser for aerospace & defense at Booz Allen Hamilton by the name of Matt Andersson put it best when he said, "Few in the civil sector fully understand that geoengineering is primarily a military science and has nothing to do with either cooling the planet or lowering carbon emissions. While seemingly fantastical, weather has been weaponized. At least four countries - U.S., Russia, China, and Israel - possess the technology and organization to regularly alter weather and geologic events for various military and black operations, which are tied to secondary objectives, including demographic, energy and agricultural resource management."

The term 'geoengineering' was introduced in a 1977 paper by Cesare Marchetti titled "On Geoengineering and the CO2 Problem." It appeared in the premiere edition of a publication called *Climatic Change*.

They couldn't!

Although many people unfamiliar with the literature and the science argue that weather modification and control are physical impossibilities, many of those 'in-the-know' contend otherwise. Rudimentary weather modification is a <u>fact</u> and a substantial body of credible evidence indicates that weather can be comprehensively controlled. Technical arguments regarding the physics of weather modification and control could be made here, but *prima facie* evidence as well as summaries posited by scientists and public policy experts are noted here instead because these are more easily understandable.

First, let's take a look at rudimentary forms of weather modification. It is a scientific fact that if unsheathed wires are strung over an area in a relatively tight grid pattern and then electrified, ground fog will condense on the wires and the area will be cleared of fog. Is that not an example of weather modification?

Although many have questioned the efficacy of seeding clouds with materials such as silver iodide, as we will learn in chapter 8, study after study has shown otherwise. Do you think power companies would have been spending so much money on dispersing silver iodide all these years if it didn't work? They have been doing it since the late 1940s and they are still doing it today. Our federal government is on record as having spent tens of billions of dollars on conventional weather modification as well. As the pioneering weather modifier Bernard Vonnegut wrote, "Cloud seeding works and works every time."

The Western world's most highly powered scientific organizations have been deeply involved in the atmospheric sciences and weather modification since early on. As we will see time and time again throughout the pages of this book, the Defense Advanced Research Projects Agency, Stanford Research International, the Rand Corporation, National laboratories, the National Science Foundation, the National Oceanic and Atmospheric Administration, the National Aeronautics and Space Administration (NASA), Naval research, Air Force research, General Electric, Raytheon, the American Meteorological Society, the Massachusetts Institute of Technology, and many more have all contributed greatly.

There are at least hundreds of applicable patents and major government reports going back over 100 years. There are easily thousands of papers - many from some of the biggest names in science. Nobel Prize for chemistry winner Irving Langmuir devoted most of his later work to weather modification. Edward Teller is widely recognized for his work in the field. Former United States Science Czar John Holdren, University of Calgary professor David Keith, Stanford University professor and Carnegie Institution member Ken Caldeira, and Microsoft's Bill Gates are currently very prominent proponents.

Now let's take a look at the reality of more comprehensive weather control.

~ ~ ~

Before he died in 1957, the famous Manhattan Project scientist John von Neumann said that, "Intervention in atmospheric and climatic matters on any desired scale" was only decades away. Von Neumann was also quoted as saying, "Our knowledge of the dynamics and the controlling processes in the atmosphere is rapidly approaching a level that will make it possible, in a few decades, to intervene in atmospheric and climatic matters. It will probably unfold on a scale difficult to imagine at present. There is little doubt one could intervene on any desired scale and ultimately achieve rather fantastic effects." We will have more about von Neumann in the next chapter.

~ ~ ~

Beginning in 1957 as a senator, then later as president, Lyndon B. Johnson (1908-1973) spoke in favor of global weather control many times. Johnson was also the political driving force behind the creation of NASA. In 1958 he stated:

"The testimony of the scientists is this: Control of space means control of the world, far more certainly, far more totally than any control that has ever or could ever be achieved by weapons, or by troops of occupation. From space, the masters of infinity would have the power to control the earth's weather, to cause draught and flood, to change the tides and raise the levels of the sea, to divert the Gulf Stream and change temperate climates to frigid... If, out in space, there is the ultimate position - from which total control of the earth may be exercised - then our national goal and the goal of all free men must be to win and hold that position."

~ ~ ~

Dr. Joseph Kaplan (1902-1991), chairman of the International Geophysical Year, said, "Control of earth's weather and temperature is within the realm of practicability now."

~ ~ ~

High-profile weather modifier Archie Kahan pointed towards future large-scale weather control by stating, "Changes in the atmospheric circulation produced by modifying the radiation balance of the atmosphere and control of major storms are possibilities held forth for the future."

~ ~ ~

Distinguished Rear Admiral Luis de Florez (1889-1962) was best known for something called Project Whirlwind. He also chaired a special science advisory board of the Central Intelligence Agency (CIA). De Florez's CIA committee met three or four times each year in Washington to talk about how new technologies could help CIA missions. De Florez wrote of comprehensive weather control many times.

Luis de Florez

It was as chairman of this CIA science committee, and after discussions with Thomas F. Malone of The Travelers insurance company, that de Florez sent a 1960 memo to General Charles Cabell (1903-1971) which spoke to new, comprehensive climate control capabilities. The paper itself was produced by the Research Division of the Travelers Insurance Company and it stated, "It has now become necessary for us to recognize the realities and potentialities of modern science for what they are and what they can mean for the possibilities of climate control." As our story unfolds, we will see the pervasive role of the CIA in all of this.

In a 1961 article he wrote for *Aerospace Engineering*, de Florez stated, "By subjecting the problem of weather control to an attack equal in magnitude and quality to that which brought about our great discoveries in the fields of flight, nuclear power, chemistry, medicine, and other disciplines, we can expect results of equal, if not greater, value." He explains that man, "can and will achieve control of weather if he but seek it with all his might." De Florez then mentions the so-called 'butterfly effect;' a phenomenon central to the New Manhattan Project and explained in the next chapter. He writes, "There is reason to believe that we can find triggering mechanisms which could make or dissipate atmospheric disturbances." Later in the piece, he advocates for further investigation of these atmospheric triggering mechanisms. De Florez finishes with, "Perhaps it is a matter for the United Nations to consider." By the end of the same month of this article's publication, as we will see next, President John F. Kennedy was speaking before the UN in favor of weather control. The aforementioned Thomas Malone of The Travelers insurance company advised Kennedy on his speech.

~ ~ ~

On September 25, 1961 President John F. Kennedy (1917-1963) addressed the United Nations proposing a global system of weather control. He said, "We shall propose further cooperative efforts between all the nations in weather prediction and eventually in weather control."

~ ~ ~

In his paper "How to Wreck the Environment," LBJ's presidential science advisor Dr. Gordon J.F. MacDonald (1929-2002) wrote of man controlling weather. As a chapter in the 1968 book *Unless Peace Comes*, MacDonald writes:

"*Operations producing such conditions might be carried out covertly, since nature's great irregularity permits storms, floods, draughts, earthquakes and tidal waves to be viewed as unusual but not unexpected. Such a 'secret war' need never be declared or even known by the affected populations. It could go on for years with only the security forces involved being aware of it. The years of draught and storm would be attributed to unkindly nature, and only after a nation was thoroughly drained would an armed takeover be attempted."*

As this book unfolds, we will have much more about the infamous Dr. MacDonald.

~ ~ ~

During House weather modification hearings in 1977, the former U.S. ambassador to the North Atlantic Treaty Organization (NATO) Harlan Cleveland (1918-2008) advocated for more national and international weather modification efforts. He said:

"The review here is an informal general theory of what kind of changes can be wrought in the atmospheric environment by what the experts call brute force, by seeding of various kinds in different circumstances - at the moment with chemical agents, perhaps in the future with forms of electromagnetic energy - and by altering the lower layers of the atmosphere."

Harlan Cleveland was the chairman of a government group called the Weather Modification Advisory Board. The Weather Modification Advisory Board submitted a paper to these proceedings titled "A U.S. Policy to Enhance the Atmospheric Environment." In this paper, they reiterate Mr. Cleveland's assertion, writing that Earth's atmosphere may be manipulated by, "Introducing perturbation energies to redirect the atmosphere's 'natural' energies. The seeding of different clouds in different ways, with chemical agents (and perhaps, in the future, with some form of electromagnetic energy),..."

Harlan Cleveland was chairman of the Weather Modification Advisory Board which was created in January of 1977 to fulfill the directives of the 1976 National Weather Modification Act. The Weather Modification Advisory Board consisted of 17 citizens ordered to study the then current state of weather modification technology and to produce a report recommending national weather modification policy and research programs. The Weather Modification Advisory Board finished their report titled "Weather Modification: Programs,

Problems, Policy, and Potential" in the spring of 1978. Among many other findings, the 746-page report stated that, "It will soon be possible to influence the weather more reliably and in a much greater variety of ways."

We will have much more about Dr. Cleveland in chapter 14.

~ ~ ~

On April 28, 1997 at the University of Georgia, U.S. Secretary of Defense William S. Cohen said:

"Others are engaging even in an eco-type of terrorism whereby they can alter the climate, set off earthquakes, volcanoes remotely through the use of electromagnetic waves... So there are plenty of ingenious minds out there that are at work finding ways in which they can wreak terror upon other nations. It's real, and that's the reason why we have to intensify our efforts, and that's why this is so important."

~ ~ ~

Throughout the pages of this book, we will see time and time again many people in-the-know telling us about the realities of a new generation of weather modification technologies capable of changing the weather in ways previously thought impossible.

A new Manhattan Project

Have you heard of the original Manhattan Project? It was the gigantic, super-secret government research and development project which produced the two atomic bombs dropped on Japan in 1945. It was a massive undertaking employing over 125,000 people. It was all kept secret largely through a bureaucratic process known as compartmentalization.

Over the years, many weather modifiers have compared weather control to the power of atomic bombs. Weather modification-pioneering General Electric scientist and Nobel Prize winner Irving Langmuir often mused about it. The 'Father of the Hydrogen Bomb' and Manhattan Project scientist, Edward Teller suggested using atomic bombs detonated in the sky as a means to modify the weather many times. In fact, Edward Teller advocated for weather control in all of his post-war years. In 1958, high-altitude atmospheric detonations of nuclear bombs were actually carried out in Operations Argus, Starfish, and Newsreel (as part of Operation Hardtack) as part of experiments designed to map Earth's natural magnetic energy.

Ross Gunn, Donald Hornig, Vannevar Bush, Bill Nierenberg, Horace Byers, Bernard Vonnegut, Marvin Wilkening, and John von Neumann are some other famous scientists who worked on the original Manhattan Project and then later contributed to the atmospheric sciences and/or weather modification. It is this way because both atomic bombs and weather control are subjects for the physical sciences. They are physics problems. This is a new Manhattan Project in weather control brought to us by the producers of the original Manhattan Project.

Meteorology and atomic energy have another fundamental connection. Meteorology is necessary, for the health and safety of all those affected, to track the atmospheric movements of radioactive particles. Both in the use of radioactive materials for energy and for weapons, varying amounts of radioactive material are released into the atmosphere either on purpose or by mistake. In any case, it takes a knowledge of the movements of the atmosphere in order to keep track of these harmful particles.

The scientific era of weather modification began with General Electric in 1946 (ch 3), only one year after the detonation of the first atomic bombs. Five years later, in 1951, reputable scientists who brought us this initial foray into weather control were testifying before Congress - enumerating the similarities between weather modification and atomic bombs. During 1951 testimony before the United States Senate, Chauncey Guy Suits (1905-1991), the director of research at General Electric Laboratories said:

"*It is a fact that has been repeatedly demonstrated that under suitable circumstances one may with 1 pound of dry ice cause a thundercloud to precipitate a heavy rainstorm. In a typical case the energy of condensation which has been released is equivalent in magnitude to the energy of several atomic bombs. There are so many points of similarity between the release of atomic energy and the release of weather energy that it is well to consider them in detail. The similarities are - and I quote from a letter I wrote to you, Senator*

Anderson, dated November 22, 1950: 1. Large amounts of energy are involved. The energy release (in the form of heat or condensation) in a small thunderstorm equals the energy of several atomic bombs. 2. A chain reaction is an important basic mechanism in many meteorological phenomena and in atomic reactions. This permits a small initiating force to generate large-scale effects. 3. The national defense and the economic possibilities are vital aspects of both problems. 4. Both problems transcend State and National boundaries in their influence and importance, and ultimately will involve international agreements. 5. Extensive research is required to fully develop the economic and military applications of both forms of energy."

Chauncey Guy Suits

General Electric (GE), who we will soon recognize as the original developers of the New Manhattan Project, also developed the first peaceful application of atomic energy. In the mid 1950s, GE began using a prototype nuclear submarine reactor to generate electrical power at West Milton, New York.

~ ~ ~

In 1954, one of President Eisenhower's science advisors, Howard T. Orville (1901-1960) wrote in an article titled "Weather Made to Order?," "And before we can hope to control the weather, we must learn what causes weather. To gain this knowledge would probably require an effort as large as the Manhattan Project for the development of atomic energy."

~ ~ ~

In 1961, the aforementioned Rear Admiral Luis de Florez advocated that the US government should, "Start now to make control of weather equal in scope to the Manhattan District Project which produced the first A-bomb." He also wrote, "We must realize that weather control or modification, like nuclear energy, solves world problems." He continues, "Certainly the end results, weather control, justify the same consideration as that which was given to nuclear development and it should be treated as such. We have led the world in nuclear power, why not lead the world in meteorology and its application?" The last paragraph of the piece makes his point more explicitly. De Florez writes, "From the standpoint of importance to our present and future existence, we might well consider the control of [sic] modification of weather in the same category of importance as that we attached to the Manhattan Project which yielded the first atomic bomb."

~ ~ ~

The executive summary of the seminal 1996 document "Owning the Weather in 2025" reads, "A high-risk, high-reward endeavor, weather-modification offers a dilemma not unlike the splitting of the atom." We will have much more about this Air Force document in chapter 3.

~ ~ ~

The House Committee on Government Reform held a hearing on September 21, 2006 titled *Climate Change Technology: Do We Need a 'Manhattan Project' for the Environment?*

~ ~ ~

Geoengineers themselves are calling this National effort in weather modification a new Manhattan Project. In late 2009 and early 2010, the U.S. Congress heard detailed testimony from top geoengineers. They called the hearings *Geoengineering: Parts I, II, and III*. These geoengineering hearings referenced the Manhattan Project *three times*. Geoengineer Philip Rasch, in written testimony, provided the best example:

"*In my opinion before a nation (or the world) ever decided to deploy a full scale geoengineering project to try to compensate for warming by greenhouse gases it would require an enormous activity, equivalent to that presently occurring within the modeling and assessment activities associated with the Intergovernmental Panel on Climate Change (IPCC) activity, or a Manhattan Project, or both. It would involve hundreds or thousands of scientists and engineers and require the involvement of politicians, ethicists, social scientists, and possibly the military. These issues are outside of my area of expertise. Early 'back of the envelope' calculations estimated costs of a few billion dollars per year for full deployment of a stratospheric aerosol strategy (see for example, Crutzen, (2006) or Robock et al (2009b)).*"

Documented reports

Although the story deserves much more coverage, local professional news reporters around the nation have investigated and/or made objective mentions of the chemtrail phenomenon many times. A plethora of professional and objective newspaper, magazine, and Internet-based reports about chemtrails have been published. National news outlets almost exclusively ridicule those aware of the situation and say that nothing is going on.

Despite big media's disinformation, there is a global grassroots political movement against chemtrails and geoengineering. There are hundreds of thousands (probably millions) of independent eyewitness accounts on the Internet. Many millions of people are already fully chemtrail aware. People from all walks of life around the world have come forward and spoken out against chemtrails. Airline pilots, police officers, U.S. Marines, air traffic controllers, a senior Air Traffic Control manager, Air Force tanker crews, career scientists and many others are speaking out.

The rest of this section is comprised of excerpts from the 2004 book *Chemtrails Confirmed* by William Thomas, who is an award winning Canadian journalist. Thomas' writing and photography have appeared in more than 50 publications in eight countries, including translations into French, Dutch and Japanese. His editorial commentaries have appeared in *The Globe and Mail*, *The Toronto Star*, *The Vancouver Sun* and *Times-Colonist* newspapers. A frequent radio talk-show guest, Thomas has also appeared on the CBC and New Zealand's national television. He currently lives and works among the Gulf Islands of British Columbia. Thomas first popularized the term 'chemtrails.' He writes:

"It was nearly noon when S.T. Brendt awoke and entered the kitchen of her country home in Parsonfield, Maine. As she poured her first cup of coffee, the late night reporter for WMWV radio could not guess that her life was minutes away from drastic change.

"Her partner Lou Aubuchont was already up, puzzling over what he had seen in the sky a half hour before. The fat puffy plumes arching up over the horizon were unlike any contrail he had ever seen, even during his hitch in the Navy.

"Lou got up and looked. What kind of clouds run exactly side-by-side in a straight line? He wondered. It's just too perfect to happen naturally. When he said he wasn't sure, S.T. stopped smiling and went outside.

"Looking up towards the southeast over West Pond, she spotted the first jet. A second jet was laying billowing white banners to the north. Both aircraft appeared to be over 30,000 ft. Turning her gaze due west, Brendt saw two more lines extending over the horizon. She called Lou. Within 45 minutes the couple counted 30 jets.

"This isn't right, S.T. thought. We just don't have that kind of air traffic here. While Lou kept counting, she went inside and started calling airports. One official she reached was guarded but friendly. He had relatives in West Pond.

"The Air Traffic Control manager told Brendt her sighting was 'unusual.' His radars showed nine commercial jets during the same 45-minuite span. From her location, he said, she should have been able to see one plane.

"And the other 29? The FAA official confided off the record that he had been ordered 'by higher civil authority' to re-route inbound

European airliners away from a 'military exercise' in the area. 'Of course they wouldn't give me any of the particulars and I don't ask,' he explained. 'I just do my job.'

"Excited and puzzled by this information, S.T. and Lou got into their car and headed down Route 160. Looking in any direction they could see 5 or 6 jets flying over 30,000 ft. Never in the dozen years they'd lived in rural Maine, had they seen so much aerial activity. As a former U.S. Navy intelligence courier, Aubuchont was used to large-scale military exercises. But he told S.T. he had never seen anything this big.

"'It looked like an invasion,' he later recounted. Another driver almost went off the road as he leaned over his dashboard trying to look up. As they passed, he acknowledged them with a nod.

"As far as they could see stretched line after line. Two giant grids were especially blatant. In stead of dissipating like normal contrails, these sky trails grew wider and wider and began to merge. Looking towards the sun, Aubuchont saw what looked like 'an oil and water mixture' reflecting a prismatic band of colors.

"At approx. 3:55 they headed home to Parsonfield. They (the jets) were still up there. What's worse is that these grids were now merging to the point of greying their beautiful skies... By 5:30 their beautiful day had turned dingy and hazy like air pollution and the sunset was dirty. Lou remembers seeing the last jets at about 5:15 pm leaving chemtrails. They were spaced further apart than the earlier jets.

"Richard Dean called back. After receiving S.T.'s message, the assistant WMWV news director had gone outside with other news staff and counted 370 lines in skies usually devoid of aerial activity. The most jets they could see at any one time was 17."

~ ~ ~

"Dave Dickie's World Landscapes company performs contract landscape work for the City of Edmonton. 'Some contracts require us to utilize the services of environmental labs for soil tests,' says Dickie. 'Recent soils analysis have come back with a high EC [electrical conductivity] rating 4:l (toxic) and we've had some soil sources rejected of course they did not meet specifications.'

"In an interview with me on Nov. 23, 2002, Dickie explained that city landscape crews were finding widespread nutrient deficiency soils could cause severe problems for plant life - including trees.

"'Wait,' I interrupted. 'Aluminum sucks nutrients from the soil.'

"'No question' answered this soil expert. Moreover, added Dickie when measuring the electrical conductivity in Edmonton soil samples, 'city specifications call for a reading no higher than 1.'

"Dickie's crews are now finding readings from 4.6 as high as 7.

"The 'chlorosis' condition resulting from this drastically high electrical conductivity in soil was impacting their landscaping business, Dickie explained. 'We were not able to determine the cause of the EC, and many reasons are possible.'

"Presuming that unusual metal content in the soil could be causing the high readings, Dickie obtained samples of a fresh snowfall in sterile containers and took them to NorWest Labs in Edmonton. As explained, 'Our most recent snowfall was tested for aluminum and barium and we were not surprised with the results. You've said it all and this just substantiates some of your claims.'

"In Nov. 2002, lab tests of snow samples collected by the city of Edmonton, Alberta between Nov. 8 - 12, confirmed elevated levels of aluminum and barium. NorWest Labs lab report #336566 dated Nov. 14, 2002 found: aluminum levels at 0.148F milligrams/litre and barium levels of 0.006 milligrams/litre.

"Because aluminum is ubiquitous in the environment, and its chemistry depends on soil pH and mineralogical composition, it is difficult to provide generalized estimates of natural background concentrations.

"But according to Dickie, the NorWest Lab techs told him, 'That's interesting. Elevated levels of aluminum and barium are not usually found in Alberta precipitation.'

"'It may not prove that the aluminum came from atmospheric programs,' Dickie admits. 'However we are going to sample precipitation from various areas within a 40 mile radius of the City of Edmonton to determine aluminum/barium within precipitation.'

"Dickie says it's simple to test for aluminum and barium in soil samples. Labs typically charge about $15 for these tests. I suggested he add quartz to the list of chemtrail fallout components to check for.

In Espanola, quartz predominated rainfall samples, which also showed hazardous levels of aluminum.

"Though it must be emphasized that neither Dickie nor NorWest Labs are making any claims regarding these early test results, the correlation of known chemtrail chemistry with Edmonton's soil samples is compelling.

"This was hot. But imagine my shock when Dickie told me that he regularly visits Air Traffic Control at the Edmonton municipal airport and watches the chemplanes making repeated passes over the city!

"'I've been a plane spotter all my life,' Dickie explained. Blessed with good friends at work in the tower, he has watched radar-identified KC-135s 'on many occasions.'

"Last Father's Day (2002), Dickie and an excited group of 12 year-olds watched two sorties by two KC-135s. Petro 011 and Petro 012 were tracked by radar as HA (High Altitude) targets flying at 34,000 and 36,000 feet - 'one to the south, and one to the north of the city.'

"Both USAF tankers had flown south out of Alaska. As Dickie, the kids and the controllers watched, the big jets began making patterns over Edmonton - 'circuits' the controllers called it. The Stratotankers were working alone in 'commanded airspace' from which all other aircraft were excluded.

"And they were leaving chemtrails.

"'The signature is significant,' commented one radar operator, referring to a trail clearly visible on the scope extending for miles behind the KC-135. In contrast, a JAL flight on the display left no contrail.

"Going outside, Dickie and several controllers scanned clear blue skies. They easily located the KC-135 leaving its characteristic white-plume 'signature.' Visibility was outstanding. They also clearly saw the JAL airliner at a similar flight level. It left no contrail at all.

"On other occasions, Dickie has watched KC-135s on Edmonton radar leaving lingering trails as low as 18,000 feet.

"'We see these guys up here a lot,' Dickie says radar techs told him. The tanker flights originate in Alaska, grid the Edmonton area, and continue on into the States."

~ ~ ~

"The following unedited transcript is a recorded message from a Dec. 8, 2000 call by a Canadian aviation authority from the Victoria International Airport to a local resident.

"Stewart was responding to a call the previous day demanding to know why intense aerial activity had left lingering X's, circles and grid-like plumes over the British Columbia capitol on Dec. 6 and 7.

"'Mark, it's Terry Stewart, Victoria Airport Authority. Just calling you back from your comment. From what I gather, it's a military exercise; U.S. and Canadian air force exercise that's goin' on. They wouldn't give me any specifics on it. Hope that helps your interest. Very odd. Thanks a lot. Bye bye now.'"

All this and much more can be found in William Thomas' groundbreaking book *Chemtrails Confirmed.* Also check out his website *WillThomasOnline.net.*

Hard scientific evidence

Francis Mangels has a Bachelor of Science in Forestry from the International School of Forestry at Missoula, Montana. He spent 35 years with the U.S. Forest Service as a wildlife biologist and worked several years with the USDA Soil Conservation Service as a soil conservationist. Today he lives in Mt. Shasta, CA and works as a master gardener.

He took a sample of water from his backyard rain gauge on Feb. 1, 2009 and submitted it to Basic Laboratory of Redding, CA on Feb. 2, 2009. This sample showed aluminum at a level of 1010 micrograms per liter (μg/l). This same sample also showed barium at a level of 8 μg/l. Using the same sampling method and laboratory, he took a sample on Oct. 14, 2009 which showed aluminum at a level of 611 μg/l. The barium should not be there in any amount. He says that chemtrails have been known to consist of both barium carbonate and barium oxide. Barium carbonate is used in rat poison. The normal level of aluminum in rainwater is .5 μg/l. These samples show levels of aluminum at 2020 times and 1222 times the normal levels. There is no heavy industry in the Mt. Shasta area. There is no reason, other than chemtrails, for this stuff to be showing up at these levels. Similar test results from others have corroborated these findings hundreds, possibly thousands of times.

~ ~ ~

ChemtrailsProjectUK.com presents rainwater sample test results from Europe - mostly from the UK. There, one will find test after test showing elevated levels of all kinds of substances that are not supposed to be showing up in our rainwater.

GlobalSkyWatch.com is also another tremendous resource. There, one will find a plethora of scientific data supporting the chemtrail hypothesis not only in rainwater sample test results, but also other test results such as those of dust and soil.

Dane Wigington over at *GeoengineeringWatch.org* has compiled quite an impressive collection of supporting scientific data as well.

~ ~ ~

The mission statement of the California Air Resources Board (CARB) is, "To promote and protect public health, welfare and ecological resources through the effective and efficient reduction of air pollutants while recognizing and considering the effects on the economy of the state." This organization is not serving its purpose.

Data produced by the CARB shows elevated levels of chemtrail toxins. Between 1990 and 2002, CARB ambient air statewide average data shows elevated and increasing levels of aluminum and barium. From 1990 to 2002, aluminum was detected in the range of 1500 to 2000 nanograms per cubic meter. Barium, which between 1990 and 2002 consistently trended upwards, reached a peak of 50.8 nanograms per cubic meter in 2002.

The CARB classifies aluminum and barium as toxic compounds. The CARB website says, "For toxics compounds, there is generally no threshold concentration below which the air is healthy. For toxics compounds, the greater the quantified health risk, the more unhealthy the air is." In other words, *any* aluminum or barium is unhealthy. There *are no safe levels* except zero. Remember, these are *statewide averages*. God forbid you might be living in an area that increased the average.

You may ask why we are only referencing data up to 2002. This is 2020. Where is the missing data? The answer is that data from between

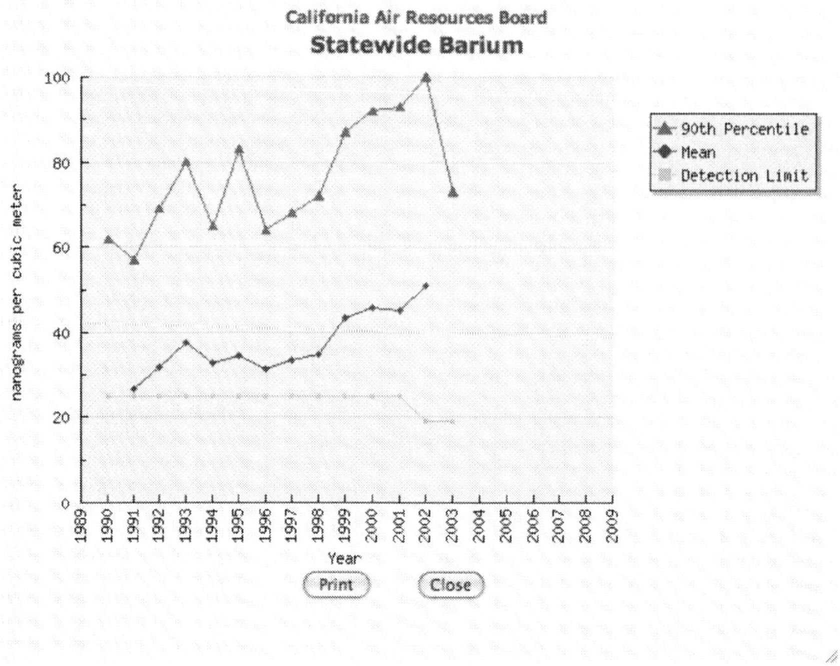
California Air Resources Board
Statewide Barium
nanograms per cubic meter
Year
90th Percentile
Mean
Detection Limit
Print
Close

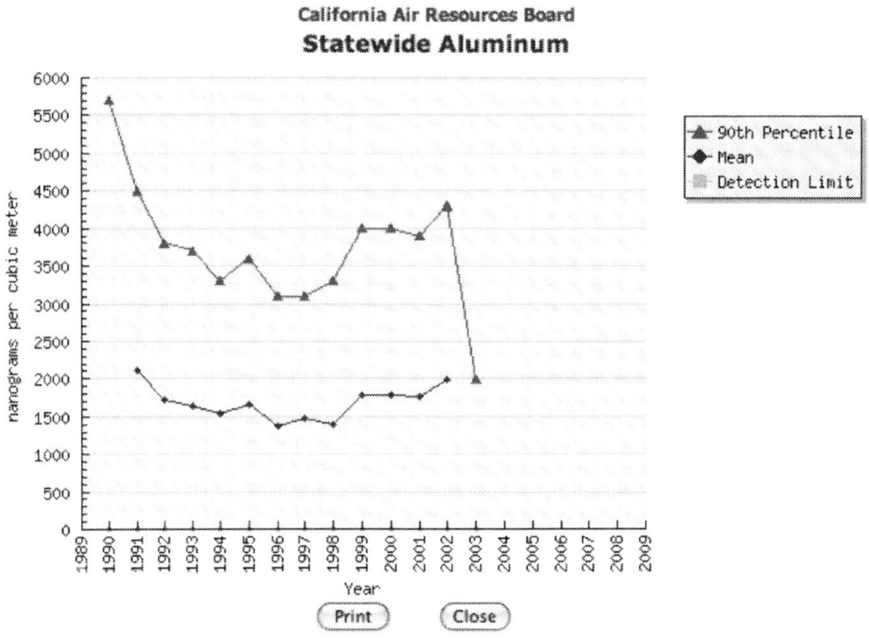
California Air Resources Board
Statewide Aluminum
nanograms per cubic meter
Year
90th Percentile
Mean
Detection Limit
Print
Close

1990 and 2002 is the only data which the CARB has widely distributed. As far as statewide averages for ambient aluminum and barium are concerned, these years are the only years for which their website and their "California Ambient Air Quality Data" DVD showed results. Their Public Information Officer Dimitri Stanich curiously *refused to answer questions about the missing data*. He referred me to documents which did not address the issue.

After discussions with staff, Mike Miguel, the chief of the Quality Management Branch of the Monitoring and Laboratory Division, wrote me an email dated Aug. 22, 2011 stating, "It is my understanding that the toxics air monitoring network (samples collected in Summa canisters) stoped [sic] analyzing for these compounds due to the low concentrations. However, the PM2.5 network does analyze for these compounds and that data was provided in the analyses and CD."

A statewide average of barium at 50.8 nanograms per cubic meter and aluminum at 2000 are low concentrations? Any levels of detectable aluminum or barium have been classified as unhealthy. The concentrations were trending upwards, yet they stopped analyzing for these compounds? This author has scoured their website, written letters and made many phone calls to the CARB and has not heard of or seen this missing data presented in any CD.

Thankfully, other people have been asking for this missing data as well. The organization known as Environmental Voices requested the missing data and on September 15, 2010, they got it.

Amazingly, after data showing many years of elevated and increasing levels of aluminum and barium, this newly produced data showed *much lower* levels. That's good news, right? We want to believe that everything is as it has always been. The problem is that the newly released data contradicts the previously released data.

Let us look at data for the year 2002 both new and old. 2002 is a year for which the CARB widely distributed data *and* it is a year for which they have provided data to only a select few researchers such as your author due to the efforts of Environmental Voices. We will refer to the widely distributed data as the 'old' data and the thinly distributed data as the 'new' data. The old data says that in 2002 the statewide average for ambient aluminum was 1980 nanograms per cubic meter. The new data says that the statewide average in 2002 was

67.5 nanograms per cubic meter. The new data also says that statewide average aluminum concentrations generally remained at this level through 2009.

As far as barium goes, the old data says that the statewide ambient air average barium concentration for 2002 was 50.8 nanograms per cubic meter. The new data says it was 27.5 nanograms per cubic meter. The new data says that statewide average barium concentrations only trended lower from 2002 to 2009.

Are certain people at the CARB trying to hide something? Why does their Public Information Officer, whose job is to answer questions from the public, refuse to answer questions about missing data? Why does he obfuscate the truth by referring me to documents that don't answer the question? Why does one of their division chiefs' response not make sense? How is it that data released to only a select few magically reduces the problem? Why didn't they post this new data on their website? All their answers so far only leave us with more questions.

Whistleblowers

Charles E. Jones III is a retired US Air Force brigadier general who served continuously in the Air Force and Air Force Reserve from 1954 to 1986. He has written a short piece acknowledging the reality of chemtrails. Jones writes:

"When people look up into the blue and see white trails paralleling and criss-crossing high in the sky little do they know that they are not seeing aircraft engine contrails, but instead they are witnessing a man made climate engineering crisis facing all air breathing humans and animals on planet Earth. These white aircraft spray trails consist of scientifically verifiable spraying of aluminum particles and other toxic heavy metals, polymers, and chemical components. Toxic atmospheric aerosols used to alter weather patterns, creating droughts in some regions, deluges and floods in other locations and even extreme cold under other conditions."

~ ~ ~

On the 16th of August, 2018, Dane Wigington of *GeoengineeringWatch.org* released a *YouTube* video of an interview with a retired U.S. Air Force Major General named Richard H. Roellig. The video is titled "Geoengineering: An Interview With A US Air Force General." Major General Roellig expressed that he is aware of and concerned about the current, ongoing large-scale uncontrolled climate geoengineering experiments.

~ ~ ~

On the 21st of August 2018, Dane posted a *YouTube* video of an interview with a former employee of the Environmental Protection Agency (EPA) named Michael Davis. The video is titled "Interview: EPA Scientist Sounds The Alarm On Geoengineering Contamination." Mr. Davis is an expert who has firsthand knowledge of the hard scientific evidence proving that we are being sprayed with massively destructive substances that work synergistically with the other toxic waste they call 'fluoride' in our drinking water to make us die and ruin our environment.

In the video Davis expounds upon the abject cowardice and selfishness of the legions of 'chair-warmers' at the EPA who are willing to watch the world burn so that they may receive a big paycheck. He was an Environmental Engineer for nearly 16 years in the National Pollution Discharge Elimination Systems Programs Branch of the Water Division in Region 5, Chicago of the USEPA. He was fired for blowing the whistle.

Complementing Davis' revelations, Wigington breaks protocol and names names of people in important positions of power who are fully aware of the geoengineering situation, but do nothing about it. In the interview the tension in their voices and how carefully they choose their words is palpable. They are both fully aware of just how crucial this information is.

Earlier, in February of 2017, Dane posted a statement by Davis as part of an article on *Geoengineeringwatch.com* titled "US EPA Scientist Fired For Trying To Tell The Truth About Climate Engineering And Fluoridated Water" in which Davis refers to the 'criminality' of the EPA and how the organization is 'a complete sham' that is 'corrupt to the core' and 'completely hijacked by multinational corporations.'

Admissions

In a video posted on *Globalskywatch.com*, an airline pilot calls chemtrails 'a necessary evil.'
http://globalskywatch.com/chemtrails/ubbthreads.php?ubb=showflat&Number=7126&an=294#Post7126

~ ~ ~

On July 8 of 2015, at a big climate conference in Paris, France, high-profile climate guru Tim Lenton admitted that geoengineering is an ongoing program TWICE. Audience member Dr. Colin Pritchard asked:

"My question is again for Tim. Colin Pritchard, Edinburgh University. Hi, Tim. Thank you very much for your very cogent explanation. I would basically agree with you on geoengineering – except, may I infer that you prefer an enormous global-scale uncontrolled experiment in geoengineering as opposed to a small-scale uncontrolled one? At the moment we are in the former. And it seems to be a little bizarre to prefer the former to the latter."

Lenton replied:

"I'm certainly not preferring carrying on with ***our current uncontrolled experiment.*** *And I'm not – what's the right word – I'm not monolithically set against things that are being discussed under the banner of geoengineering. So it's quite a nuance... I think that's quite a nuance discussion to have, perhaps over lunch, because it really depends on the options you're considering. So you've got some things which would be reflective roofs and road surfaces that are very practical, local adaptation options against urban heat islands that, if you did on a large enough scale, could have some measurable effect on regional climate and I think are very sensible. So we have to just be... I think we have to be nuanced on specific proposals, specific technologies. But I think we can perhaps all agree that certainly* ***none of us want to continue the current uncontrolled experiment.***"

In conclusion

This introductory chapter alone provides much convincing evidence for the existence of a new Manhattan Project in weather modification and for the assertion that the chemtrails we so often see are a part of it. For the researcher, the problem is not that of a lack of evidence, rather the problem is that of too much - with it all existing in a multitude of different disciplines. Fear not, though. Your intrepid author has been on the case for about ten years now and he has brought home the bacon. Can't you hear it sizzling? The rest of this book pieces together the history and current state of the biggest scientific effort in history.

References

"Chemtrails, Chemistry 131 Manual, Fall 1990" a study guide by the United States Air Force Academy, Department of Chemistry, 1990

"Pollution of the Upper Troposphere by Soot from Jet Aircraft and Its Relation to Cirrus Clouds" a paper by Freeman F. Hall, Jr., produced by the Douglas Advanced Research Laboratories, McDonnell Douglas Corporation as it appeared in the proceedings of the Second National Conference on Weather Modification of the American Meteorological Society, sponsored by the Atmospheric Sciences Section of the National Science Foundation, 1970

Clouds of Secrecy a book by Dr. Leonard A. Cole, published by Rowman & Littlefield, 1988

In the Name of Science a book by Andrew Goliszek, published by St. Martin's Press, 2003

Under an Ionized Sky: From Chemtrails to Space Fence Lockdown a book by Elana Freeland, published by Feral House, 2018)

"On Geoengineering and the CO2 Problem" a paper by Cesare Marchetti, published in Climatic Change, v1 n1, 1977

"When Will We Change the Weather?" an article by Bernard Vonnegut, 1967

"The Pathological History of Weather and Climate Modification" a paper by James Roger Fleming, published by the University of California Press, 2006

"Don't Like the Weather? Change it; The Weird Science of Weather Modification Makes a Comeback" an article by Bennett Drake, published by the Boston Globe, July 3, 2005

"Weather - Take It or Make It" an article by Luis de Florez, published in *Aerospace Engineering*, September, 1961

Memorandum for General Charles P. Cabell from Luis de Florez, November 22, 1960

The National Aeronautics and Space Administration a book by Richard Hirsch and Joseph Trento, published by Praeger Publishers, 1973

Fixing the Sky a book by James Roger Fleming, published by Columbia University Press, 2010

"Weather Modification: Panacea or Pipedream?" an article by Archie M. Kahan, published by the *SRI Journal*, second quarter, 1959

Jerome C. Hunsaker and the Rise of American Aeronautics a book by William F. Trimble, published by the Smithsonian Institution Press, 2002

"How to Wreck the Environment" a paper by Dr. Gordon J.F. MacDonald, as published in the book *Unless Peace Comes* edited by Nigel Calder, published by The Viking Press, 1968

Weather modification Hearing Before the Subcommittee on the Environment and the Atmosphere of the Committee on Science and Technology, U.S. House of Representatives, Ninety-fifth Congress, Wednesday, October 26, 1977

"Weather Control" by Marc Leepson, published by Congressional Quarterly, Inc., 1980

"Weather Modification: Programs, Problems, Policy, and Potential" a report by the Congressional Research Service, republished by the University Press of the Pacific, 1978

"United States Nuclear Tests: July 1945 through September 1992" a report by the U. S. Department of Energy, 2000

"Meteorology and Atomic Energy" a report prepared by the United States Department of Commerce, Weather Bureau for the United States Atomic Energy Commission, 1955

Hearing before the Subcommittees of the Committees on Interior and Insular Affairs, Interstate and Foreign Commerce and Agriculture and Forestry, United States Senate, Eighty-second Congress, first session, 1951 as it appeared in "Weather Modifications," published by the General Electric Research Laboratory, 1951

The Travelers: 100 Years a book by The Travelers Insurance Companies, published by The Travelers, 1964

"Weather Made to Order?" an article by Howard T. Orville as it appeared in the May 28, 1954 edition of *Collier's Magazine*

"Climate Change Technology Research: Do We Need a Manhattan Project for the Environment?" a hearing before the United States House Committee on Government Reform chaired by Tom Davis (R-VA), Sept. 21, 2006

Geoengineering: Parts I, II, and III hearings before the Committee on Science and Technology, House of Representatives, November 5, 2009, February 4, 2010, and March 18, 2010

Chemtrails Confirmed a book by William Thomas, Bridger House Publishers, 2004

What in the World Are They Spraying? a documentary film by Michael Murphy, Paul Wittenberger, and G. Edward Griffin, produced by Truth Media Productions, 2010

Geoengineering: Chronicles of Indictment: Exposing the Global Climate Engineering Cover-up a book by Dane Wigington, published by Geoengineering Watch Publishing and Media LLC, 2017

"Geoengineering: An Interview With A US Air Force General" an article by Dane Wigington, published by Geoengineeringwatch.org, August 16, 2018

"US EPA Scientist Fired For Trying To Tell The Truth About Climate Engineering And Fluoridated Water" an article by Dane Wigington, published on *Geoengineeringwatch.org*, Feb. 7, 2017

"Interview: EPA Scientist Sounds The Alarm On Geoengineering Contamination" an article by Dane Wigington, published by *GeoengineeringWatch.org*, August 21, 2018

"CRACKED! Top Climate Scientist Admits to Ongoing Geoengineering" an article by James Hodgskiss, published by *ChemtrailsProjectUK.com*, October 17, 2015

Chapter 2
TRULY a NEW MANHATTAN PROJECT

"The nation that first learns to plot the paths of air masses accurately and learns to control the time and place of precipitation will dominate the globe."
-General George Kenney addressing MIT's graduating class of 1947

In the preceding chapter, we saw the initial, compelling evidence for today's global, second-generation weather modification program being a new Manhattan Project. In this chapter we will take a long look at the people, institutions, and technologies that link the original to the new. In this chapter, we will dive deeper into the reasons why all of this is truly a new Manhattan Project.

The Massachusetts Institute of Technology Radiation Laboratory

During World War II, Massachusetts Institute of Technology Radiation Laboratory (Radiation Laboratory/Rad Lab) scientists developed early examples of technologies which have since gone on to become large parts of today's New Manhattan Project (NMP). Specifically, the Rad Lab made important early advancements in the areas of ionospheric heaters and air traffic control. These technologies involve the creative use of electromagnetic energy. The creative use of electromagnetic energy applied to the WWII effort was what the Rad Lab was all about.

The Massachusetts Institute of Technology (MIT) Rad Lab operated between 1940 and 1945 and it was big. Before the end of the war, the Rad Lab was employing nearly 4,000 people. The MIT Rad Lab was underwritten by a half-million dollars of John D. Rockefeller, Jr. money. By its last year, the Rad Lab reached a budget of about $125,000 per day, or close to $4 million per month. In 1945 the MIT Rad Lab had sixty-nine different academic institutions represented on its staff. It produced 150 radar systems. By the end of the war, the Rad Lab had delivered to the armed services 3 billion dollars worth of radar equipment. That's $40 billion in today's dollars; about $10 billion more than the cost of the original Manhattan Project.

Many scientists who worked on the original Manhattan Project also worked out of the Rad Lab. Famous Manhattan Project scientists

Kenneth Bainbridge (1904-1996), Jerrold Zacharias (1905-1986), and Robert Bacher (1905-2004) were all members of the Rad Lab Steering Committee.

A typical day at the Rad Lab

The MIT Rad Lab grew out of something called the National Defense Research Committee (NDRC) Microwave Committee. This Microwave Committee was part of an NDRC division headed by the famous scientist and president of MIT Karl Taylor Compton (1887-1954). Karl Compton also served on the Interim Committee which advised President Truman as to the use of the newly created atomic bomb. His brother was Manhattan Project scientist Arthur Compton (1892-1962). We will have more about the NDRC shortly.

At the behest of Karl Compton, the NDRC Microwave Committee was run by a man by the name of Alfred Lee Loomis (1887-1975). Alfred Loomis was also later a Rad Lab co-founder. The official NDRC historian during the second world war, James Phinney Baxter wrote an excellent book titled *Scientists Against Time* which efficiently describes the Microwave Committee's origins. Baxter writes:

"When the National Defense Research Committee was established, the first suggestions from the armed services of fields for NDRC investigation included basic research at ultra-high frequencies and studies of pulse transmission. The Air Corps was looking for solutions to the problem of fog and haze and was interested in the possibility of bombing through the overcast. [Karl] Compton, Chief of Division D, promptly established a section to study the applications of microwaves (radio waves 10 centimeters or less in length) to detection devices. This Section D-1, which later became Division 14 of NDRC, was headed by Alfred L. Loomis, a New York lawyer and a pioneer in the field of microwaves. It included scientists and engineers drawn from the American universities and industrial concerns which had done the most to develop microwave techniques. The Microwave Committee, as it was commonly called, continued with remarkably few changes in personnel throughout the war."

The MIT Rad Lab then evolved from the NDRC Microwave Committee. In mid-1940 a group of British scientists and officials, led by a man named Henry Tizard (1885-1959), arranged for their revolutionary short-wave radar prototype to be delivered to America. This was known as 'The Tizard Mission.' In exchange for this technology, the U.S. promised development and production. The British needed radar as part of their war effort against Germany and the Axis powers, and the U.S. wanted it too. In late 1940, the Microwave Committee unanimously voted to establish development and production facilities at the Massachusetts Institute of Technology. Research began at what was soon called the Radiation Laboratory on November 10, 1940, under an NDRC contract with MIT. The famous namesake of Lawrence Livermore National Labs and Manhattan Project physicist, Ernest O. Lawrence (1901-1958) recruited physicists in the early days, including the director of the laboratory, Dr. Lee DuBridge (1901-1994). Dr. E. G. 'Taffy' Bowen (1911-1991) of the Tizard Mission joined the staff as British Liaison Officer.

The Rad Lab developed and produced radar and radar-like systems and contracted for radar set (transmitter and receiver) production with five industrial concerns: Raytheon, General Electric, Radio Corporation of America (RCA), Westinghouse, and Philco & Sperry. MIT has a long and rich tradition of co-operating with industry.

The MIT Rad Lab is most well known for developing the airplane-mounted radar which allowed Allied fighters to see submerged German submarines called U-boats. Before the advent of this plane-mounted radar, the German U-boats were sinking Allied ships with relative impunity. Radar produced at the Rad Lab allowed our boys to blow their U-boats to smithereens, regain maritime supremacy, and ensure the safe passage of Allied shipping. The Rad Lab developed and produced scads of other technologies that dramatically helped our soldiers win, including some extremely effective submarine radar for the Pacific theatre. As it has been written and said many times: the atomic bomb only ended the war, radar won it.

Of particular note to our discussion, the Rad Lab had a meteorology department and that department will be the target of future investigations.

When the war ended in 1945, the Rad Lab was closed down and people who had been working there mostly found employment elsewhere in the vast, newly created military/industrial/academic complex. Once depopulated, the MIT Rad Lab was officially terminated on the last day of 1945, only to be bureaucratically reborn the next day (the first day of 1946) as something called the Basic Research Division. Six months later, the Basic Research Division re-joined MIT as a subdivision of the newly established Research Laboratory of Electronics (RLE). On the campus of MIT, both the Rad Lab and the later Research Laboratory of Electronics shared building space with the Office of Naval Research (ONR).

LORAN/SS Loran

Today, electromagnetic energy generated from ground-based antennas called 'ionospheric heaters' as well as their offspring is used to manipulate the atmosphere and thereby modify the weather. As noted in the first chapter, this use of electromagnetic energy is what distinguishes the New Manhattan Project from the conventional cloud-seeding industry. The biggest early developments of technologies which have since resulted in today's ionospheric heaters took place at the MIT Rad Lab. Ionospheric heaters have evolved from something developed at the Rad Lab called the Long Range Navigation (LORAN) system.

During the Rad Lab years with LORAN, Big Science got serious about bouncing radio waves off of the ionosphere. Before LORAN, the most significant contributions in this area were made by Nikola Tesla (1856-1943) and Guglielmo Marconi (1874-1937), and then later by scientists Gregory Breit (1899-1981) and Merle Tuve (1901-1982). Merle Tuve was the director of the terrestrial magnetism division of the Carnegie Institution. The LORAN concept was first proposed by the aforementioned Alfred Lee Loomis.

LORAN developments were originally produced under something called the Loran Group and later under the MIT Radiation Laboratory Navigation Group. The Loran Group started as a subcommittee of the National Defense Research Committee (NDRC) Microwave Committee not part of the MIT Rad Lab. In the early 1940s, the Loran Group was founded and administrated by a man named Melville Eastham (1885-1964). Famous scientists like Donald G. Fink (1911-1996), J. Curry Street (1906-1989), and Julius Adams Stratton (1901-1994) were members of the Loran Group and made significant contributions. By early summer of 1941, the LORAN project became the Rad Lab's when the MIT Radiation Laboratory Navigation Group took over the LORAN project. Fink and Street went on to run the project after Mr. Eastham's retirement.

The LORAN system involved sending electromagnetic signals from multiple ground-based transmitters to receiving sets aboard airplanes and boats in order to determine the position and bearing of the ship - a type of triangulation. LORAN allowed the ship's navigator to determine his position and bearing and thus proceed accordingly.

LORAN systems did not always bounce radio waves off the ionosphere. They did so only at night. When the electromagnetic signals were simply sent roughly parallel, above the surface of the Earth, it was called 'ground-wave' LORAN. This is what was done during daytime hours. At night, LORAN systems and their operators bounced the transmissions off the ionosphere using something called 'Sky-wave Synchronized LORAN,' or 'SS Loran' for short. SS LORAN had much longer ranges than the ground-waves. Ground-wave transmissions were used during the day and SS Loran at night because the ionosphere is adequately reflective at night and not so during the day due to atmospheric turbulence caused by heat from the sun. Ham radio operators have operated upon this principle for many decades. In the early summer of 1941, successful SS Loran field experiments were conducted between Montauk, New York and as far west as Springfield, Missouri - a distance of about 1,200 miles.

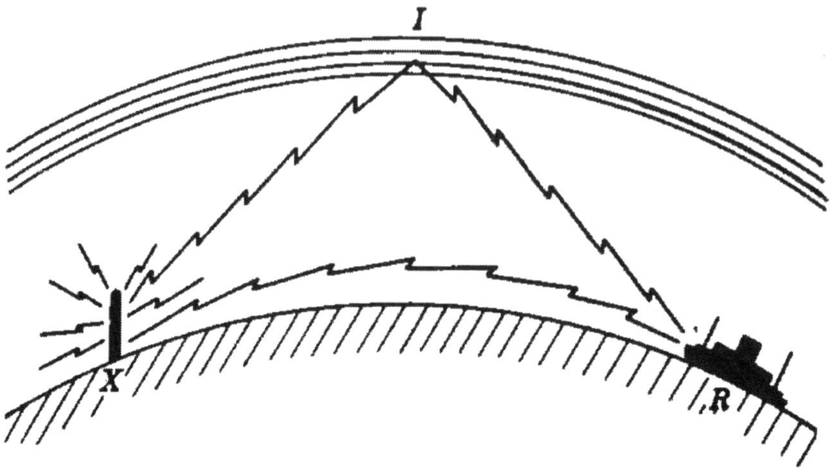

FIG. 3·4.—Ground waves and sky waves.

Ground-wave Loran and SS Loran

By mid-1942, LORAN field experiments were going very well and hence the project was enjoying lots of funding, access, and other help. Earlier in the year, the LORAN project had acquired a Naval Liaison Officer by the name of Captain L.M. Harding. By this time, the scientists of the Radiation Laboratory Navigation Group were successfully sending SS Loran signals from Fenwick Island, Delaware to the Azores Islands - a distance of over 4,000 miles. The British and Canadian governments and militaries were co-operating extensively.

Also in mid-1942, the development of the airborne SS Loran receivers, which had previously been conducted by General Electric, was turned over to the Aircraft Radio Laboratory of the Signal Corps at Wright Field. Although the first LORAN receiver sets were produced by the Rad Lab itself, the receiver sets were ultimately mass produced by the Radio Corporation of America (RCA).

In 1943 LORAN went fully operational in the North Atlantic. On Jan. 1, 1943 the U.S. Coast Guard (USCG) took over the aforementioned Fenwick and Montauk stations. In June of that year, transmitting stations in: Labrador, Newfoundland, and Greenland were also turned over to the USCG. In order to provide constant military navigation services, the USCG (in co-operation with the Navy and the Rad Lab) then began operating these stations 24 hours a day. By this time, LORAN was commonly able to transmit to a ship up to 700 miles away during daylight hours and up to 1,400 miles away at night. By 1945, the Canadians and the British along with the Americans (Allied Powers) were operating a slew of transmitting stations which provided LORAN coverage from the North Atlantic down to the Caribbean. Also by the end of WWII, LORAN systems were operational all over the northern and western Pacific Ocean and the Far East with coverage from Alaska to Japan to the China-Burma-India theater and on down into the East Indies. By the end of the war, LORAN covered three tenths of the Earth's surface.

The development of SS LORAN led to the later development of over-the-horizon (OTH) radar. Today's ionospheric heaters, which play a key role in modifying today's weather as part of the New Manhattan Project, are the direct descendants of OTH radar, and therefore, ionospheric heaters are the direct descendants of SS Loran. Amazingly enough, OTH radar transmitters bounce a signal off the ionosphere, then that signal hits a target which sends an echo back off the ionosphere again and back to the transmitter. In this way, we are able to remotely sense (see) objects thousands of miles away. OTH radar technologies were first developed in the 1950s. We left Kansas a long time ago.

Air traffic control

"I see a manless Air Force . . . [that] is going to be built around scientists - around mechanically minded fellows."

-Vannevar Bush

Today's New Manhattan Project air traffic control systems are the descendants of two different Rad Lab activities. One of these was a project called Ground-Controlled Approach and the other was research and development in the area of highly automated missile batteries.

Ground-Controlled Approach (GCA) systems allowed pilots to safely land their plane in zero visibility conditions and suggested the future development of radar-enhanced remote control. These truck-mounted systems used radar to determine an airship's position and bearing from far away. Once position and bearing had been determined, a GCA operator would give specific voice commands to the pilot over the two-way radio and talk the pilot down. It worked very well and was put into military use by the end of the war to save many lives. Manhattan Project scientist, Nobel Prize winner, and Rad Lab Steering Committee member Luis Alvarez (1911-1988) led the GCA group. Alfred Loomis was also instrumental in its development.

Luis Alvarez

The GCA system suggested the possibility of a new generation of remotely controlled aircraft systems. In the early 1940s, autonomous aircraft were only relatively new. Believe it or not, as early as 1918 autonomous bombers were being flown. The next logical step for the GCA was to use electronic signals which could control the plane's navigational systems, rather than voice commands and the actions of the pilot. In this way, GCA promised complete control of unmanned aircraft. Today's New Manhattan Project most probably utilizes unmanned aircraft extensively.

Another area of study at the MIT Rad Lab that most probably contributed to the air traffic control systems of today's NMP is that of automated missile batteries. During the Rad Lab years, our military was fantasizing about sweeping guided bombs across the enemy like they had been sprayed out of a fire hose. At that time, such remote launch and guidance of missiles had much room for improvement, though. Although our military's initial uptake of these types of systems was slow, our August, 1943 discovery of the Germans using radio-controlled missiles prompted our American military to get a move on. After 1943, a plethora of related programs were initiated. Technologies developed at the Rad Lab contributed greatly to this body of work.

These early developments of automated missile batteries later led to something called the SAGE system. The SAGE system was developed at MIT's Lincoln Laboratory. The MIT Lincoln Laboratory was founded by key figures in the MIT Rad Lab and the subsequent Research Laboratory of Electronics. Later still, in 1958 about 500 Lincoln Laboratory Employees left to form the MITRE Corporation. The MITRE Corporation is most probably the day-to-day manager of the scientific portions of today's NMP. Luis Alvarez went on to become a trustee of the MITRE Corporation. He also became a senior advisor to the secretive JASON Group which is also strongly implicated in the New Manhattan Project.

Vannevar Bush: architect and founder of the military/industrial/ academic complex

The aforementioned Vannevar Bush (1890-1974) was the man most responsible for the formation of today's military/industrial/ academic complex. Bob Dylan's song "Masters of War" is about him. We all know about Eisenhower's famous coining of the term 'military/ industrial complex.' The former President was referring to the combined power of giant organizations enabled by the bureaucratic framework established by Vannevar Bush and his cronies. Bush was the head of the WWII-era National Defense Research Committee (NDRC) as well as the head of the concurrent Office of Scientific Research and Development (OSRD). While overseeing massive growth for all parties involved, Bush brought our military, corporations, and universities together to work on enormous projects such as the original Manhattan Project and the MIT Rad Lab. Without congressional approval, Bush and his cronies created the first American 'black' military budgets and projects. Surprisingly, no, he was not biologically related to the Bush political crime family.

Dr. Bush was born in Everett, Massachusetts on March 11, 1890. He graduated from Tufts College in 1913 with degrees of B.S. and M.S. In that same year of 1913, he went to work for General Electric as a 'test man.' Bush got his PhD from MIT in 1916. He joined the MIT faculty in 1919. He became a Professor of Electrical Engineering at MIT in 1923 and Vice President as well as Dean of Engineering in 1932. Working with J.P. Morgan and others, Bush co-founded Raytheon in 1924. By 1926, Bush was a Raytheon consultant and he owned 3.6 percent of the company's outstanding stock. In the early 1930s, Bush's stature at MIT was greatly enhanced with the arrival of MIT President Karl Compton. Bush was then named Dean of the School of Engineering and a member of the MIT Corporation; the university's board of directors. After the war, Bush became Chairman of the MIT Corporation.

Vannevar Bush

At the start of 1939, before America's involvement in the war, Bush became president of the powerful Carnegie Institution in Washington D.C.. It was around this time that Bush began thinking that consolidating powers of the government, the military, and the private sector might match the efficiencies of the Nazi war machine then terrorizing Europe. With key support from such men as: Secretary of War Henry 'Skull & Bones' Stimson (1867-1950), the President of the National Academy of Sciences Frank Jewett (1879-1949), Manhattan Project chemist James B. Conant (1893-1978), Karl Compton, and Alfred Loomis, Bush went about doing just that.

By May of 1940, Bush had enlisted the help of a man named John Victory to draft legislative language calling for the creation of what he called the National Defense Research Committee (NDRC). The NDRC was to bring together America's militaries, corporations, and universities in an effort to more efficiently produce new weaponry. Bush called upon Frederic Delano (the President's uncle 1863-1953) to arrange an appointment with President Franklin D. Roosevelt. In early June, Delano's request landed Bush a meeting with the president's closest aide Harry Hopkins. Hopkins arranged for Bush to see the president. On June 12, at their meeting, Roosevelt endorsed the NDRC. With this endorsement, Roosevelt had promised Bush direct access to the White House, virtual immunity from congressional oversight, and his own line of funds. Without an act of congress, with the stroke of a pen, the military/industrial/academic complex was born. At a June 14 press conference, Roosevelt announced Bush's appointment as the head of the newly created NDRC. Compton, Jewett, and Conant quickly became his principal aides. Many other prominent experts (many of whom went on to be Manhattan Project and Rad Lab scientists) soon flocked to the organization. The NDRC leadership was stacked with MIT graduates.

Original NDRC members seated from left to right: Brigadier General George V. Strong, James B. Conant, Vannevar Bush, Richard C. Tolman, and Frank B. Jewett. Standing left to right: Karl T. Compton, Irvin Stewart, and Rear Admiral Harold G. Bowen, Sr.

The vast majority of Bush's funding decisions benefited his friends and colleagues. Bush himself later admitted he had pulled off, "an end run, a grab by which a small company of scientists and engineers, outside established channels, got hold of the authority and money for the program of developing new weapons." Even though the Carnegie Institution figured to be among the top recipients of NDRC contracts, Bush not only retained his presidency there, but the Carnegie Institution's headquarters became those of the NDRC. The Carnegie Institution ended up getting about $3 million in NDRC contracts. Scads of NDRC grants would also flow to his former employer MIT. In fact, MIT was the top academic recipient of NDRC contract dollars, raking in about $117 million. During the course of the war, Raytheon, whose stock Bush still owned, enjoyed a 60-fold growth in sales. Although Bush fretted about conflicts of interest and later tried to officially legalize his activities, President Roosevelt didn't initially see any of this as a problem.

By mid-1941, out of legal and funding concerns, President Roosevelt approved the creation of the Office of Scientific Research and Development (OSRD) to run the NDRC. The NDRC became the chief operating unit of the OSRD. Under the previous arrangement, the NDRC was funded by the executive branch. Under the OSRD, the NDRC became funded by congress. Bush remained in charge as he assumed the position of Director of the OSRD and Conant became NDRC Chairman. By 1944, Bush's OSRD was spending $3 million a week on 6,000 researchers at more than 300 industrial and university labs.

It is interesting to note that the NDRC did extensive work for the Office of Strategic Services (OSS). The NDRC served the Office of Strategic Services like the fictional Q Branch from the James Bond movies serves the British Secret Service; producing unorthodox weaponry and spying gear. This is significant because the OSS was the forerunner to today's Central Intelligence Agency (CIA) and, as we will see throughout the pages of this book, evidence indicates that the CIA is deeply involved in today's New Manhattan Project.

Creating the military/industrial/academic complex wasn't the only thing he did. In the historical weather modification literature, Vannevar Bush's name (pronounced *vuh-nee-ver* like *beaver*) comes up again and again. One might say that Vannevar was as busy as a beaver. Considering all that has been disclosed here, in combination with his well documented weather modification advocacy, Vannevar Bush is the putative founder of the New Manhattan Project. In late 1957, as an introduction to one of the most, if not *the* most cited weather modification document ever, he wrote:

"It is entirely possible, were he wise enough, that man could produce favorable effects, perhaps of enormous practical significance, transforming his environment to render it more salutary for his purposes. This is certainly a matter which should be studied assiduously and explored vigorously. The first steps are clear. In order to control meteorological matters at all we need to understand them better than we now do. When we understand fully we can at least predict weather with assurance for reasonable intervals in the future.

"With modern analytical devices, with a team of sound background and high skills, it is possible today to do a piece of work in this field which will render immediate benefits, and carry us far toward a more thorough understanding of ultimate possibilities. By all means let us get at it."

Yet another thing that Vannevar Bush accomplished was to ensure that technologists such as himself became vital to planning war strategy and operations. This was a radical idea at the time. As Bush's leading biographer writes, "In May 1942, the president endorsed the creation of a subcommittee to his new Joint Chiefs of Staff that would be devoted to new weaponry. Bush would chair the three-person advisory body, called the Joint Committee on New Weapons and Equipment, whose frank purpose was the 'education' of the military's top brass." G. Pascal Zachary goes on to write, "Bush was the first civilian outside the cabinet to ever formally have a line into the nation's military chiefs." This, along with all the other information detailed here, shows that Bush had the means, motive, opportunity, desire, and will to found the New Manhattan Project. Nobody else was even capable.

As if all this wasn't enough, Bush was also involved in the creation of huge post-war National organizations designed to enable state-funded and privately executed scientific research; thus continuing the legacy of the OSRD. These post-war organizations became primary vehicles for the development of America's weather modification programs. These organizations were and are the Office of Naval Research (ONR) and the later National Science Foundation (NSF).

To oversee nuclear weapons, Bush called for a world government. In his tome, *Endless Horizons*, Bush writes, "If the mechanism of world peace is available and is strong enough, peoples may be expected to relinquish something of their traditional nationalism to attain that equal footing through international organization." Bush goes on to suggest that the United Nations is the proper vehicle for world government.

The curious case of Edward Teller

Edward Teller (1908-2003) was one of the most famous scientists of the 20th century. He was famous because he was an extremely capable physicist and tirelessly promoted his projects. He worked as a senior scientist on the original Manhattan Project and went on to develop and promote the next generation of nuclear bombs known as *thermo*nuclear bombs (hydrogen bombs). He is also widely known for developing and promoting the Strategic Defense Initiative; a.k.a. Star Wars.

Edward Teller

We particularly concern ourselves with Mr. Teller because in the 1990s he famously co-authored a series of papers published by Lawrence Livermore National Laboratory proposing that the Earth and its biota be sprayed with stratospheric aluminum. We are concerned not only because on its face this proposition is extremely dangerous, we are concerned because, as we have seen, air and rainwater sample test results from around the world have shown aluminum to be a major constituent of today's common chemtrail spray. To top it all off, large-scale domestic chemtrail spraying operations began around the same time as when Teller's co-authored papers were published.

Teller never worked out of the MIT Rad Lab. He did, though, often theorize about how nuclear bombs detonated in the high atmosphere could influence the weather. He also advocated for, among many other things, using nuclear bomb blasts to reshape planetary landscapes. Unbelievably, Teller also wrote of periodically detonating underground nuclear explosions in order to generate a sort of man-made geothermal energy. Although most people today would probably call that 'completely freaking insane,' at the time, these types of proposed activities were called 'planetary engineering.' Planetary engineering is synonymous with today's geoengineering.

Teller wrote and spoke about weather modification countless times. In his memoirs, Teller writes of, "putting 1 billion small floating spheres into the atmosphere." We have noted how he wrote about spraying us with aluminum. In his book *The Legacy of Hiroshima* he suggests the solar radiation management (SRM) geoengineering thesis. In other words, he suggested spraying us with chemtrails many times. Teller served on the Panel on Weather and Climate Modification of the National Academy of Sciences where our good friend Gordon J.F. MacDonald was chairman.

Teller once testified before the Senate Military Preparedness Committee that if Russia was first to control the weather on a global scale, then the United States could be beaten without war. Teller also told the committee that he wouldn't be surprised if the Russians achieved global weather control in the relatively near future. Throughout his career, Teller used the Russian threat as a justification for more spending on weapons programs.

In the example reproduced next, we see how Teller notes atmospheric 'triggers.' He's writing about the notion that certain, relatively small atmospheric phenomena at certain points in space and time can create atmospheric chain reactions which eventually lead to large atmospheric phenomena such as storms. It is analogous to the so-called 'butterfly effect.' The butterfly effect is the notion that the flapping of a butterfly's wings in China can eventually cause a giant storm on Cape Cod. The thesis of atmospheric triggers is mentioned over and over again throughout the weather modification literature in the same way Teller mentions it here. In order to control the weather, Teller and his peers suggested that these triggers be artificially created and/or manipulated. The atmospheric trigger thesis is central to today's NMP operations. In fact, it was during the course of his work in computer-simulated weather models when American mathematician Edward Lorenz coined the phrase and wrote about it in his paper "Predictability: Does the Flap of a Butterfly's Wings in Brazil Set Off a Tornado in Texas?" Today's NMP supercomputer atmospheric modeling systems are designed to be able to identify these triggers and predict their outcomes. Knowing the probable outcomes is how the people running the NMP know when and where to perform atmospheric manipulations. The National laboratory Teller co-founded and directed (Lawrence Livermore National Labs) has been at the forefront of supercomputer atmospheric modeling since early on and continues there today. Teller writes:

"Before anything can be controlled, it first must be understood. We are just beginning to approach an understanding of weather. We know that very small causes can grow into very big effects. A slight disturbance of the air masses on the front separating the calm air of the poles from the steady westerly winds encircling the globe in temperate latitudes can trigger a whirlpool a thousand miles wide and can affect the weather over the United States for an entire week. We can and we should increase the number and range of our weather observations. We will use satellites and other means to keep track of clouds and winds. Then, using improved electronic computers, we shall be able to predict weather and trace the origin of each development back to its original trigger.

"When this high degree of meteorological understanding has been attained, we might be able to create triggers of our own and realize the age-old dream of actually doing something about the weather. We might ***spread a cloud of dust over a strategic location*** [author's emphasis] *or find some other way to upset the temperature balance between air masses. We might break droughts. We might regulate the precise location and time where a hurricane arises, thus predetermining the place where the destructive winds would dissipate.*

"Such new command over nature will give us responsibilities beyond our present ability to imagine. When rain will be the servant of man, man must be the master of himself. Control of clouds will bring either conflict or co-operation between nations. The prospect may seem terrifying, but in the long run this situation or one similar to it will surely arise. Science brings progress; progress creates power; power is coupled with responsibility. This responsibility we shall not escape."

As we can infer from his last paragraph, Teller was a global governance advocate. He openly professed it countless times as did many of his peers. Along with weather control, the threat of thermonuclear war has also been widely used as a catalyst for world government.

Edward Teller was also instrumental in the founding of Lawrence Livermore Labs and for many years served as its director. This is significant because Lawrence Livermore National Laboratory (LLNL) has an extensive history of developing atmospheric models and configuring supercomputers to potentially function as part of the New Manhattan Project. During Teller's directorship his direct superior was Army General Alfred D. Starbird (1912-1983).

Early on, Teller was a Rockefeller Fellow at the University of Copenhagen and later became close with Nelson Rockefeller. Perhaps the single most compelling bit of information linking Teller to the New Manhattan Project is found buried in a footnote to his memoirs. The footnote reveals that he was a longtime consultant to the aforementioned MITRE Corporation. The icing on the cake here is that Teller was an early, strong, and steadfast adopter of the theory of man-made climate change. The theory of man-made global warming and climate change has been used as a cudgel to get us to accept being sprayed with tens of thousands of megatons of toxic waste. The sum total of this evidence indicates that Edward Teller was instrumental in the development of the New Manhattan Project. In fact, he may have been the single most important scientist in its history.

The cult of John von Neumann

People saw the famous Manhattan Project physicist and mathematician John von Neumann (1903-1957) as a 'demigod.' The Nobel Prize winning physicist Hans Bethe wondered if von Neumann's brain indicated, "a species superior to that of man." Another Nobel Prize winner, Eugene Wigner (1902-1995) noted that von Neumann was not Human but a demigod who had, "made a detailed study of humans and could imitate them perfectly." His biographer Norman Macrae called him 'a prophet.'

These statements were justified by von Neumann's legendary scientific capabilities. Among other things, he was known for calculating in his head amazingly accurate answers to exceedingly complex mathematical equations. Given time, pencil, and paper, there was no problem he could not solve. He applied his brain power to whatever was needed at the time. Both the Army and the Navy treasured Johnny as an invaluable scientific resource. He was mild-mannered, affable and universally well liked.

In 1926 von Neumann was awarded a Rockefeller Foundation scholarship to study the foundations of mathematics at Göttingen University.

Johnny was part of the team who invented modern mathematical computer modeling. This field was invented and developed at Los Alamos in the context of the production of the world's first atomic bombs. Johnny and his team created mathematical models which ran on supercomputers that could predict outcomes of simulated nuclear detonations. Johnny later applied his knowledge of computer modeling attained during these experiments to atmospheric modeling.

Shortly after the war, in October of 1945, Vladimir Zworykin (1888-1982) and the Radio Corporation of America (RCA) published "Outline of Weather Proposal." This document called for geoengineering and a global weather control program. Von Neumann was a contributing author. In January of 1946, von Neumann and Zworykin went to the Weather Bureau and made their case for weather modification. They lobbied for the creation of a government-funded program in computerized atmospheric modeling at Princeton University's Institute for Advanced Study, and got it.

Princeton's Institute for Advanced Study (IAS) was founded in 1930. Like RCA's world headquarters, the IAS was located in Princeton, New Jersey. Von Neumann was one of the original five members. The other original members included Albert Einstein (1879-1955). In May of 1946, the Institute began their Meteorology Project. It was funded by the Office of Naval Research and was part of something called the Electronic Computer Project (ECP). Their Meteorology Project was designed to take the, "first steps towards influencing the weather by rational, human intervention." Johnny had informed Navy admirals about how computers could revolutionize meteorology and help attain weather control. Military men recognize an important weapons system when they see one, hence money for Johnny's Meteorology Project was never a problem. The ECP's Meteorology Project was presented to the meteorology community during a conference organized by von Neumann at Princeton in August of 1946.

Later, Jule Charney (1917-1981) and Norman Phillips (1923-2019) arrived at the IAS to help produce the first truly practicable atmospheric models. Faster computers filled in the other half of the equation. As time went on, the models got better and the computers got faster. By 1955 the U.S. Weather Bureau was issuing forecasts derived from computers. The basis for today's NMP supercomputer atmospheric modeling had been laid. Charney and Phillips later departed for MIT.

Von Neumann with an early IAS computer

Von Neumann was an early progenitor of the modern theory of man-made global warming and a geoengineering advocate. In the early 1950s, he speculated that the carbon dioxide released into the atmosphere from the burning of hydrocarbon fuels such as oil and gas could cause the Earth's average temperature to rise dramatically. Maybe he neglected to consider the fact that only a statistically insignificant 3% of the Earth's atmospheric carbon dioxide is man-made. Nevertheless, speculate he did. He also laid the foundation of today's Solar Radiation Management (SRM) Geoengineering thesis by theorizing that a layer of stratospheric aerosols, such as those created by a volcano, could lower the Earth's temperature by reflecting sunlight back into space. He was also an advocate for global weather control and global government. It's funny how all these things so often go hand in hand, isn't it? Von Neumann was one of Edward Teller's longtime friends and associates. He worked on the computer modeling for Teller's H-bomb.

On March 15, 1955, Johnny was sworn in as a member of the Atomic Energy Commission, the highest official position a scientist could reach in the U.S. In early May, his family moved to Washington. Three months later, in August of 1955, Johnny developed severe pains in his left shoulder. After surgery, bone cancer was diagnosed. By the end of 1955, his situation had deteriorated considerably. He was found to have several spinal cord lesions and began to have difficulty walking. By January 1956 he was in a wheelchair. John von Neumann died in Washington on February 8, 1957 and was buried in Princeton.

Alfred Lee Loomis: the father of psychotronic weaponry and major aspects of the New Manhattan Project

The most interesting has been saved for last. Alfred Lee Loomis (1887-1975) was a Wall Street lawyer and banker turned scientist. To gain an understanding of the man, one can read an excellent biography written by the granddaughter of the aforementioned James B. Conant. The book is titled *Tuxedo Park: A Wall Street Tycoon and the Secret Palace of Science That Changed the Course of World War II* by Jennet Conant. This book provides most of the information presented here.

As mentioned earlier, Loomis ran the Microwave Committee, was a Rad Lab co-founder, and first proposed the LORAN concept which eventually evolved into today's ionospheric heaters. In fact, the original work on LORAN began at his Tuxedo Park, New York laboratory. He was a Life Member of the MIT Corporation.

Alfred Lee Loomis is not to be confused with another important MIT Rad Lab scientist by the name of Francis Wheeler Loomis. The two were not biologically related, either. Alfred Loomis was, however, the first cousin and close friend of the aforementioned Secretary of War Henry 'Skull & Bones' Stimson.

Alfred Loomis owned several homes in an exclusive community north of New York City called Tuxedo Park. He converted one of these houses, called Tower House, into a laboratory. Calling it the Loomis Laboratory, there he worked with and entertained distinguished scientists from all over the world including: Enrico Fermi (1901-1954), Werner Heisenberg (1901-1976), Niels Bohr

(1885-1962), Ernest Lawrence (1901-1958), and Albert Einstein (1879-1955).

Alfred Lee Loomis

In 1930 Loomis became interested in brain waves. He was inspired by a German psychiatrist named Hans Berger (1873-1941). Loomis published his first paper on the subject in June of 1935. His paper went over experiments conducted at Loomis Laboratory noting, "...the very definite occurrence of trains of rhythmic potential changes as a result of sounds heard by a human subject during sleep." Using the electroencephalographic techniques he helped develop, Mr. Loomis noted different brain wave patterns and attributed these patterns to different states of sleep consciousness. Later, between 1937 and 1939, Loomis worked at Loomis Labs with a famous Harvard Medical School doctor by the name of Hallowell Davis (1896-1992) and his wife Pauline Davis. The trio solidified early brain wave research and continued development of the electroencephalogram. Today we have a much better knowledge of exactly how different electromagnetic frequencies, amplitudes, and waveshapes affect us.

All this about brain waves is significant because the ionospheric heaters and other machines used today as part of the New Manhattan

Project, which evolved from Loomis' LORAN, can produce such waveforms that can affect our moods, thoughts, and bodily functions. Equipment that can produce these types of electromagnetic signals is known as 'psychotronic weaponry.' Loomis is responsible for research helping to establish a basis for the second most probable NMP agenda (mind control) *and* the technology to deliver it!

Loomis was also interested in hydroelectric power. He was keenly interested in new technologies that would allow power generated in hydroelectric plants (dams) to be efficiently transmitted to far-off cities. As a Wall Street banker, he was very familiar with the value of these technologies. In fact, after the war, he made an enormous fortune financing public utilities. This is significant because the generation of hydroelectric power was later to become a major catalyst for the conventional weather modification industry.

Loomis was also a founding trustee of the Rand Corporation. As we will see, the Rand Corporation has many NMP connections. In fact, Loomis was persuaded to be involved in the formation of the Rand Corporation by a man named Rowan Gaither (1909-1961) who had been asked by the Air Force to organize the corporation. Later, in 1958, Rowan Gaither went on to become one of the five members of the MITRE Corporation's first Board of Trustees. At the time of MITRE's founding, the Rand Corporation (along with MIT and the Ford Foundation) was one of the groups disproportionally represented. Alfred Loomis was also a trustee of the Carnegie Institution of Washington.

After the war, Loomis became involved in funding and building astronomical observatories such as the the National Center for Atmospheric Research's High Altitude Observatory. This is significant because astronomy is relevant to the New Manhattan Project and the National Center for Atmospheric Research (NCAR) has many NMP connections.

Alfred Loomis' son Henry Loomis (1919-2008) majored in physics at Harvard, served in the Navy during WWII, worked at the University of California's Radiation Laboratory as well as at MIT, served as a special assistant to the Director of the Research and Development Board of the Secretary of Defense, became a 13-year member of the MITRE Corporation's board of directors beginning in 1976, and later became a member of the Council on Foreign Relations (CFR).

Conclusions

"Roosevelt called me into his office and said, 'What's going to happen to science after the war?' I said, 'It's going to fall flat on its face.' He said, 'What are we going to do about it?' And I told him, 'We better do something damn quick.'"
-Vannevar Bush

After the Axis powers were defeated, our military/industrial/academic complex wanted new enemies. New enemies justified its existence. Without a new enemy they were to be defunded. The initial enemy they claimed was the Russians. Another of the new enemies they eventually claimed was the weather - specifically the threat of catastrophic man-made global warming. Their prophet von Neumann had said so. They did not step aside gracefully. They jealously and dishonorably clung to power while declaring war on Mother Nature. After WWII, Vannevar Bush and his cronies spent their recently accumulated political capital building (among other things): a global weather modification project, a shadow United States government, and an authoritarian world government. Global warming was just the convenient lie they needed.

It is very interesting how this story and all the names listed here are new to most people. It was all mostly new to the author too. One might think that the story of the biggest scientific effort in Human history might have a little more recognition. One might think that this monolithic tale so important to our species would be known far and wide. Where is the Old Media? Where have they been all these years? Where are they today as fleets of jumbo jets routinely dump megatons of toxic waste into our atmosphere? They have been busy asserting that chemtrails do not exist and that anybody who suggests otherwise is just a nutty conspiracy theorist. It boggles the mind that a nobody such as your author has been able to uncover so much of this. Time proves this work correct.

Bush and his cronies riding on a monstrosity they created after the war

References

"$28,000,000 Urged to Support M.I.T." an article by Frank Kluckhohns, published by *The New York Times*, June 15, 1947

Scientists Against Time a book by James Phinney Baxter, published by the Massachusetts Institute of Technology Press, 1968

Tuxedo Park: A Wall Street Tycoon and the Secret Palace of Science that Changed the Course of World War II a book by Jennet Conant, published by Simon and Schuster, 2003

The Making of the Atomic Bomb a book by Richard Rhodes, published by Simon and Schuster, 2012

"Celebrating the History of Building 20" a report

LORAN: Long Range Navigation a book by J.A. Pierce, A.A. McKenzie, R.H. Woodward, and the Massachusetts Institute of

Technology Radiation Laboratory Series Board of Editors, published by McGraw Hill, v4, 1948

"An Introduction to Loran" a paper by Jack A. Pierce, published by the Institute of Electrical and Electronics Engineers, 1990

Alvarez: Adventures of a Physicist a book by Luis W. Alvarez, published by Basic Books, 1987

The Education of a College President a book by James R. Killian, Jr., published by the Massachusetts Institute of Technology Press, 1985

The Jasons: The Secret History of Science's Postwar Elite a book by Ann Finkbeiner, Pblished by Penguin Books, 2006

Endless Frontier: Vannevar Bush, Engineer of the American Century a book by G. Pascal Zachary, published by the Massachusetts Institute of Technology Press, 1999

Endless Horizons a book by Vannevar Bush, published by Public Affairs Press, 1946

M.I.T. in World War II Q.E.D. a book by John Burchard, published by John Wiley and Sons, 1948

"Final Report of the Advisory Committee on Weather Control" a report by the Advisory Committee on Weather Control, published by the University Press of the Pacific, 2003

"Global Warming and Ice Ages: Prospects for Physics-Based Modulation of Global Change" a paper by Edward Teller, Lowell Wood, and Roderick Hyde, published by Lawrence Livermore National Laboratory, 1997

"Long-Range Weather Prediction and Prevention of Climate Catastrophes: A Status Report" a paper by E. Teller, K. Caldeira, G. Canavan, B. Govindasamy, A. Grossman, R. Hyde, M. Ishikawa, A.

Ledebuhr, C. Leith, C. Molenkamp, J. Nuckolls, and L. Wood, published by Lawrence Livermore National Laboratory, 1999

"Active Climate Stabilization: Practical Physics-Based Approaches to Prevention of Climate Change" a paper by Edward Teller, Roderick Hyde, and Lowell Wood, published by Lawrence Livermore National Laboratory, 2002

The Legacy of Hiroshima a book by Edward Teller and Allen Brown, published by Doubleday and Company, 1962

"Weather as a Weapon" an article by Howard T. Orville as told to John Kord Lagemann, published by *Popular Science*, June 1958

"Weather and Climate Modification Problems and Prospects: Final Report of the Panel on Weather and Climate Modification" a report by the National Research Council, 1966

The Story of Western Science: From the Writings of Aristotle to the Big Bang Theory by Susan Wise Bauer, published by W.W. Norton & Company, 2015

Edward Teller: The Real Dr. Strangelove a book by Peter Goodchild, published by the Harvard University Press, 2004

Teller's War: The Top-Secret Story Behind the Star Wars Deception a book by William J. Broad, published by Simon and Schuster, 1992

Adventures in the Atomic Age: From Watts to Washington a book by Glenn T. Seaborg w/ Eric Seaborg, published by Farrar, Straus, and Giroux, 2001

John von Neumann: The Scientific Genius Who Pioneered the Modern Computer, Game Theory, Nuclear Deterrence, and Much More a book by Norman Macrae, published by the American Mathematical Society, 1999

The World as a Mathematical Game: John von Neumann and 20th Century Science" a book by Giorgio Israel and Ana Millán Gasca, published by Birkhäuser, 2009

Pieces of the Action: The Personal Record of Sixty Event-filled Years by the Distinguished Scientist Who Took an Active and Decisive Part in Shaping Them a book by Vannevar Bush, published by William Morrow and Company, 1970

MITRE The First Twenty Years: A History of the MITRE Corporation (1958-1978) a book by the MITRE Corporation, published by the MITRE Corporation, 1979

"Henry Loomis, 89; Led Voice of America" an article by William Grimes, published by *The New York Times*, Nov. 14, 2008

Paid notice: Deaths LOOMIS, HENRY, published by *The New York Times*, Nov. 14, 2008

"Henry Loomis, 89; Physicist Led VOA and Public Broadcasting" an article by Joe Holley, published by *The Washington Post*, Nov. 8, 2008

None Dare Call it Conspiracy a book by Gary Allen and Larry Abraham, published by Buccaneer Books, 1971

Chapter 3
ELECTRIFYING the ATMOSPHERE

The distinguishing aspect of the New Manhattan Project (NMP) is its extensive use of man-made atmospheric electricity. This extensive exploitation of man-made atmospheric electricity distinguishes the NMP from conventional weather modification practices and, in combination with the small, atmospheric particles dispersed from aircraft, allows for comprehensive weather control. In manipulating the electrical characteristics of the atmosphere, ionospheric heaters such as the High-frequency Active Auroral Research Program (HAARP) antenna in Alaska and many other similar devices play a central role. The man-made sources of electricity and electromagnetic waves utilized as part of the NMP can work either in combination with or against naturally occurring atmospheric phenomena in order to produce the desired results.

This chapter is a non-technical discussion of the modern weather modification tools available to today's geoengineers and the history thereof. This chapter is about the electrification and electromagnetic manipulation of our atmosphere as part of the New Manhattan Project.

Nikola Tesla

Nikola Tesla (1856-1943) probably first conceived of the plans to manipulate the weather with electromagnetic energy which eventually grew into today's New Manhattan Project. In the late 1800's, American inventor Nikola Tesla popularized the practical use of electromagnetic energy. He made sure that our power grid ran on alternating current (AC) rather than Edison's inferior direct current (DC). The Supreme Court found that U.S. patent #645,576 "System of Transmission of Electrical Energy" shows that he invented radio, not Marconi. Radar was first conceived and pioneered by Tesla. He invented autonomous vehicles along with wireless signal and power transmission. Today's Tesla automobile engines are partially based on Nikola Tesla designs.

Tesla's work originally inspired today's ionospheric heaters which use electromagnetic energy to cause atmospheric perturbations from great distances and play a defining role in the New Manhattan Project. Relevantly, Tesla also pioneered the use of certain types of electromagnetic energy called very-low frequency (VLF) and extremely-low frequency (ELF). These types of energy are known to be used in the NMP as they are within the frequency range naturally produced by the Earth and the Human brain. Bernard Eastlund (1938-2007), the inventor of one of today's most significant, powerful, and versatile electromagnetic energy generators, the High-frequency Active Auroral Research Project (HAARP) antenna array, attributed its fundamental technologies to Tesla. There are many ionospheric heaters all around the world today being used to manipulate the weather as part of the NMP.

Nikola Tesla

Tesla did a good deal of theorizing about weather control. In his autobiography he wrote about how, as a child, he hypothesized that man could control the weather with electrical forces. He wrote:

"The sun raises the water of the oceans and winds drive it to distant regions where it remains in a state of most delicate balance. If it were in our power to upset it when and wherever desired, this might[y] life sustaining stream could be at will controlled. We could irrigate arid deserts, create lakes and rivers, and provide motive power in unlimited amounts. This would be the most efficient way of

harnessing the sun to the uses of man. The consummation depended on our ability to develop electric forces of the order of those in nature."

Tesla went on later to actually reproduce lightning bolts and his biographer John O'Neill writes that Tesla intended to use his artificial lightning bolts as a means to produce rain.

When Tesla died, all of his effects (most notably his scientific papers and instruments) were confiscated by the United States government. This was á massive amount of stuff. We're talking about two truckloads as well as 80 barrels and bundles. Although Tesla was an American citizen, this confiscation was carried out by the Office of the Alien Property Custodian under the direction of one Walter C. Gorsuch, because Tesla's legal heir was not a citizen and these effects were of National security importance. Among these confiscated papers may have been the initial New Manhattan Project plans.

The timing was right. As noted above, Tesla died in 1943, then (as we will soon learn) just three years later, Bernard Vonnegut, who went on to conduct early NMP field experiments, and two other General Electric (GE) scientists were simultaneously kicking-off the scientific era of weather control and the New Manhattan Project. The three years plus between Tesla's death and GE's famous exploits may have provided time to review Tesla's material and make plans for the future.

Most curiously though, the scientific expert charged with reviewing Nikola Tesla's posthumously confiscated papers was a Massachusetts Institute of Technology Radiation Lab Steering Committee member and Assistant Director by the name of John G. Trump (1907-1985). For our military, John G. Trump translated the Tesla papers from 'Scientist' to English. Our military, in turn, decided which of Tesla's ideas were of military value and thus worthy of further exploration. John G. Trump went on to serve as the head of the British Branch of the MIT Radiation Laboratory. If you are wondering... yes, John G. Trump was the uncle of United States President Donald J. Trump. In fact, The Donald's middle name of 'John' was given to him in honor of his uncle.

Prior to examining Tesla's documents, in the late 1920s John Trump worked for General Electric. By 1936 he had a doctorate in electrical engineering from MIT and an appointment as an assistant professor there. Gwenda Blair, biographer of the Trump family writes that, by WWII, "John Trump, who led the MIT High Voltage Research

Laboratory, had already made a name for himself through his work with Robert J. Van de Graff on the first million-volt X-ray generator. Used for radiation therapy, this invention would prolong the lives of cancer patients around the world." In his later years, Trump was a physics professor at MIT.

John G. Trump

Presumably there to help get into any sealed containers, another man present at the posthumous examination of Tesla's documents was an Office of Strategic Services agent by the name of Willis De Vere George. George wrote a 1946 book called *Surreptitious Entry* in which he describes his exploits of breaking into secure locations and stealing things for the United States government. George performed such operations for the Treasury Department, the Office of Naval Intelligence, and the US Army Office of Strategic Services (OSS). Working for the Office of Naval Intelligence in New York City during WWII, George and his 'little squad of government burglars' conducted operations somewhere in the neighborhood of 100 times as part of domestic counter-intelligence activities waged against the Axis powers.

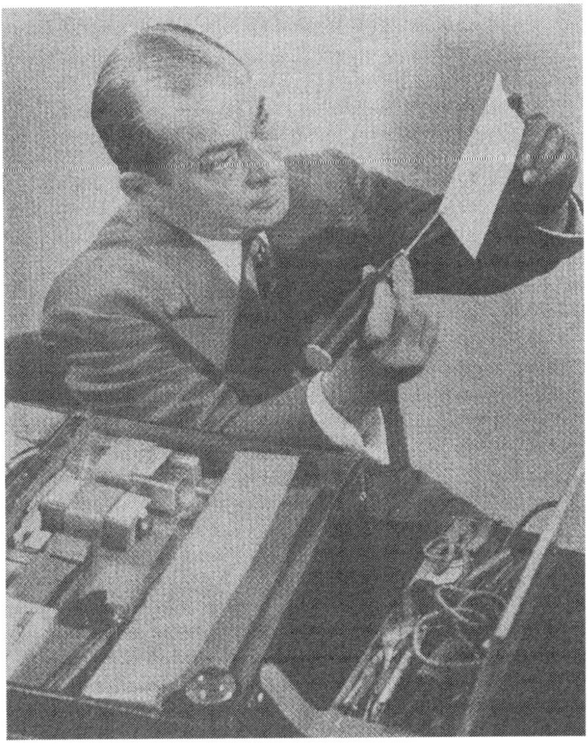

Willis De Vere George

Following service in the First World War, George worked as a Wall Street stockbroker. In 1927 he became a member of the New York Stock Exchange. He quit Wall Street in 1931. Working for the United States Treasury Department in Havana, Cuba, George conducted himself as a U.S. Government terrorist engaged in: blowing up the boats of alcohol smugglers, breaking and entering, extra-legally arresting people, threats and intimidation, and more. George successfully thwarted a kidnapping of Irénée du Pont of the Wilmington du Ponts by warning him and taking him to a secure location. Through du Pont, George was later rewarded for this by none other than John D. Rockefeller, Sr. with one of his trademark dimes, which George thereafter kept with him as a good luck charm.

The von Karman Reports

In May of 1946, less than one year after the world's first atomic bombs were used in combat to end World War II, the United States Army Air Force released a restricted series of reports titled *Toward New Horizons*. Commonly referred to as the *Von Karman Reports*, this series of reports featured a constellation of leading American weather modifiers and experts in applied electromagnetic energy as contributors such as: Lee A. DuBridge (1901-1994), Irving P. Krick (1906-1996), George E. Valley, Jr. (1913-1999), and the aforementioned Vladimir Zworykin.

The *Von Karman Reports* are similar to the *Air Force 2025* set of documents published exactly 50 years later, which contain the infamous "Owning the Weather in 2025" report and outlined the New Manhattan Project multiple times. Like *Air Force 2025*, *Toward New Horizons* spoke to a broad overhaul of U.S. military air operations.

One of the *Von Karman Reports* was titled "War and Weather." "War and Weather" speaks to a second generation, domestic weather modification effort apparently involving the electrification of the atmosphere. The authors of "War and Weather" repeatedly stress the importance of their mission, writing, "The importance of pursuing research projects in peacetime, directed at fulfilling the future requirements of the military weather services, cannot be overemphasized." They go on to make many recommendations as to how this should be accomplished. As we will see in the next section, by the end of 1946 (the year of "War and Weather's" publication), General Electric, the Air Force and the Navy were rolling out the modern era of weather modification.

Generally consistent with what Edward Teller stated, "War and Weather" notes that, "Atomic energy may be used to control certain weather phenomena." What makes this quote particularly interesting and different from others is the fact that the authors write generically of 'atomic energy' rather than the detonation of nuclear bombs specifically as having an effect upon weather. As noted in the previous chapter, when writing of using atomic energy to modify the weather, Teller commonly wrote only of atomic bombs being detonated in the atmosphere. The detonation of atomic bombs is only but one application of atomic energy. The term 'atomic energy' used here loosely applies to anything having to do with atomic physics. Heck, the term 'atomic energy' can loosely apply to simple electricity (which is technically sub-atomic). The authors' use of the generic term 'atomic energy' here opens the door to a whole range of possible applications of atomic energy as a way to control the weather - including the use of electromagnetic energy which is what is used today to control the weather as part of the New Manhattan Project.

In fact, on the last page of "War and Weather," under the heading 'Atomic Energy Applied to Meteorology,' the authors write, "The controlled use of atomic energy by meteorologists may result in the synthesis of certain weather phenomena or forced local release of atmospheric instability ... Furthermore, it is conceivable that **the peacetime applications of such concepts may be more significant than their military uses** [author's emphasis]." When these authors write of 'forced local release of atmospheric instability,' once again they are writing about the so-called 'butterfly effect.' Again, this is the notion that a relatively small amount of directed energy can have a tremendous later effect upon the weather. As explained earlier, this is a thesis central to the peacetime (i.e. domestic) New Manhattan Project and will be mentioned time and time again throughout the pages of this book.

The namesake of the *Von Karman Reports* was a Hungarian/American scientist by the name of Theodore von Kármán. Immediately following the end of hostilities in the European WWII theatre, von Kármán led a delegation of American scientists surveying captured German military sites in order to assess the status of their military's scientific progress - and they found a lot. Von Kármán and his fellow scientists produced a report about their findings titled "Where We Stand." "Where We Stand" was another report included in the *Von Karman Reports*.

We concern ourselves with the "Where We Stand" report here as well because, once again, the authors of this report write about a second generation weather control program utilizing atomic energy. As he and his co-authors supply some general guidelines for such a program, Von Kármán advocates that research in meteorology should be 'vigorously' pursued. The authors write:

"The conditioning of weather over large territories has not been seriously considered in the past, however, the progress of meteorological science and the possibility of introducing in the air large amounts of energy by nuclear methods, might bring this aim into the realm of possibility. For example, the amount of energy required for forced local release of atmospheric instability in the case of convective storms and for the dissipation of fog should be within the limits of available energy from atomic sources. The general problem consists essentially of three parts: (1) exact knowledge of the weather parameters in the domain in which we want to produce changes, including both instantaneous values and their tendency of variation; (2) methods of computing the future weather, as dependent on the presence or absence of available control measures; and (3) means of applying the controls, such as adding energy in certain regions, modifying the reflection coefficient of certain areas, etc. It seems possible, with the aid of electronic computers, to produce a model of a certain region of the earth's surface and the existing weather situation, which can be used not only for fast weather prediction, but also for direct rapid experimentation, on a mode scale, with various control methods."

Theodore von Kármán (1881-1963) was born in Budapest on May 11, 1881. He graduated in 1902 from the Royal Technical University of Budapest with highest honors as a mechanical engineer. From his graduation until 1906, he taught at his alma mater. Von Kármán left Budapest in 1906 to study at the University of Göttingen. Von Kármán received his PhD from Göttingen in 1908. He was appointed as a Göttingen privatdozent following his graduation and remained so until 1912 when he organized an aerodynamics institute at the Technical University of Aachen. He became Professor of Aerodynamics and Mechanics and Director of this Aachen Aeronautical Institute. During WWI, von Kármán served as Director of Research of the Austro-Hungarian Aviation Corps, returning to Aachen with the end of the war. Back at Aachen, both von Kármán and the Aerodynamics Institute grew to eminence and attained a world-wide reputation.

Theodore von Kármán

Von Kármán first came to the United States in 1926 on the invitation of Robert Millikan (1868-1953), President of the California Institute of Technology, and of Harry Guggenheim (1890-1971), President of the Daniel Guggenheim Fund for the Promotion of Aeronautics. At that time, Guggenheim had recently established the Daniel Guggenheim Fund for the Promotion of Aeronautics to speed the development of civil aviation. Guggenheim had also provided funding for the establishment of a Daniel Guggenheim School of Aeronautics at the California Institute of Technology (Caltech) where von Kármán became a consultant. Von Kármán eventually became director of the Daniel Guggenheim Aeronautical Laboratory at the California Institute of Technology. At Caltech, von Kármán dealt with theoretical aspects of meteorology. Von Kármán became a U.S. citizen in 1936.

In 1944 von Kármán's career entered a new phase when Army Air Force General H. H. Arnold (1886-1950) asked him to organize and chair a Scientific Advisory Group to study the use of science in warfare by the European nations and to interpret the significance of new developments in rockets, guided missiles, and jet propulsion for the future of the U.S. Air Force. This Army Air Force Scientific Advisory Group produced the *Von Karman Reports* which were requested by General Arnold. As part of Operation Lusty, von Kármán and the Scientific Advisory Group traveled the world surveying the latest in aeronautics. In his duties as head of the Scientific Advisory Group, von Kármán reported to General Arnold and von Kármán was Arnold's chief scientific adviser. The Scientific Advisory Group sent what they found directly to Wright Field.

The Army Air Force Scientific Advisory Group, which was designed to carry out the objectives set forth in the *Von Karman Reports*, was succeeded by the U.S. Air Force Scientific Advisory Board, of which von Kármán was Chairman until 1954 and Chairman Emeritus until his death in 1963. With the 1946 retirement of General Arnold, the newly formed Scientific Advisory Board (SAB) became executively managed by General Curtis LeMay (1906-1990). The Air Force SAB between 1955 and 1995 went on to produce over 350 studies, some pertaining to aspects of the New Manhattan Project such as remotely piloted vehicles and directed energy weapons.

In mid-1949 the Ridenour Committee of the SAB met for the first time. This SAB Ridenour Committee spent two months touring and evaluating Air Force facilities in order to get a picture of what the SAB should do in order to implement the dictums of *Toward New Horizons*. In September of 1949, the Ridenour Committee report was submitted to von Kármán and he passed it along to General Hoyt Vandenberg (1899-1954). In light of the findings of the Ridenour Committee as well as those of another, independent committee known as the Anderson Committee, the Air Force Air Research and Development Command (ARDC) was established on January 23 of 1950 in order to reorganize the Air Force.

Another organization springing to life from *Toward New Horizons* was the Air Force Air Engineering and Development Center (AEDC). The AEDC was created to better pursue the ends outlined in *Toward*

New Horizons. It was later named after General Arnold as the Arnold Engineering and Development Center.

Perhaps most significantly, in light of *Toward New Horizons'* multiple assertions that atomic energy might be used to modify the weather, the SAB established their Nuclear Weapons Panel in March of 1953. The SAB's Nuclear Weapons Panel reunited the two most notable Manhattan Project scientists who were known to have done significant work in weather modification and the atmospheric sciences: John von Neumann and Edward Teller. This *ad hoc*, so-called 'Von Neumann Committee' was also joined by many other top American nuclear physicists.

As we can see, Theodore von Kármán had a tremendous influence upon America's post-war scientific activities, and more pertinently, he had an influence upon the development of technologies that have since been incorporated into the New Manhattan Project. Von Kármán, while touring Europe in the days following the end of WWII hostilities, persuaded many former Nazi scientists to emigrate to the U.S. Von Kármán persuaded these scientists to come work in America because, as previously noted, it was during his post-war European travels that he learned the extent of the new technological achievements produced by Nazi Germany and he wanted their help in further efforts. As alluded to earlier, it was these European travels that initiated the production of the *Von Karman Reports*. During his European travels immediately following WWII, von Kármán learned about Germany's inventions in the areas of remotely controlled aerial vehicles, target-seeking missiles, and electronic communications. As we have seen or will see, all of these areas have relevance to the New Manhattan Project.

~ ~ ~

Another early person of interest in our investigation is a man by the name of Theodore F. 'Teddy' Walkowicz. Walkowicz was a consultant to the production of the "War and Weather" report and the lead author of another report in the series titled "Future Airborne Armies."

In the same year that our good friend Gordon J.F. 'How to Wreck the Environment' MacDonald became a trustee of the aforementioned

MITRE Corporation, 1968, Dr. T.F. Walkowicz from Rockefeller Family and Associates became a trustee of the MITRE Corporation as well. Dr. Walkowicz was a major in the Army Air Corps and the boss of something called the National Aviation and Technology Corporation of New York, NY. He stayed on the MITRE Corporation board of directors for at least 10 years.

Air Force Major Theodore Walkowicz was a WWII veteran. He also had a doctorate from the Massachusetts Institute of Technology and was selected by von Kármán to work on the *Toward New Horizons* reports. Walkowicz and von Kármán became close associates and friends. Working with his friend Colonel Bernard Schriever (1910-2005), through the Air Force Scientific Advisory Board, where Walkowicz served as Secretary from 1948 to 1950, Walkowicz brought new technologies off the drawing board, into production, and onto the battlefield.

Dr. Theodore Walkowicz

Following the Korean War, Walkowicz left the services and went to work for Laurance Rockefeller (1910-2004) as part of his venture capital team. Rockefeller had first heard of Walkowicz from a Princeton aeronautical engineer named Courtland Perkins (1912-2008). After leaving the air force, Walkowicz continued to hold his security clearances, continued to work with the Air Force Scientific Advisory Board, and continued his friendship with Colonel (later General) Bernard Schriever. As a civilian venture capital advisor,

Walkowicz focused Rockefeller's interest on rocketry and intelligence systems. Other members of Rockefeller's staff included Edward Teller and media magnate Henry Luce (1898-1967).

Walkowicz wrote a paper for the Rockefeller Brothers Fund Special Study Project titled "Survival in an Age of Technological Contest." The paper proposed streamlining the military R&D process by turning the military into a science-driven organization and emphasized America's military dominance of space. The paper highlighted the fact that, to a large extent, Walkowicz had left the government for Rockefeller's ventures because he saw Rockefeller's efforts as much more effective in attaining the goals set forth in this paper.

The scientific era begins

The scientific era of weather modification began famously in November of 1946 with a trio of scientists from General Electric Laboratories: Irving Langmuir, Vincent Schaefer (1906-1993), and Bernard Vonnegut. Leading the group was the world famous Nobel Peace Prize winning scientist Irving Langmuir (1881-1957). This trio popularized the fact that, under certain circumstances, dumping substances from airplanes into clouds can cause precipitation. The initial experiments of late 1946 used dry ice while experiments conducted in early 1947 pioneered the use of silver iodide. Also invented were silver iodide generation equipment and many other weather related scientific instruments. This trio's work here was done in co-operation with the Office of Naval Research, the Army Signal Corps and the Air Force. These efforts were known as Project Cirrus. General Electric (GE) employees were merely advisors and were to, "refrain from asserting any control or direction over the flight program. The GE Research Laboratory responsibility [was] confined strictly to laboratory work and reports." Although others had previously dumped stuff out of airplanes in attempts to modify the weather, the GE scientists practiced a rigorous scientific method previously unseen in the field. Clouds were seeded near GE's headquarters in Schenectady, NY and in New Mexico where White Sands and the Nazi scientists working with GE were.

Langmuir, Schaefer, & Vonnegut

Following the famous scientific weather modification efforts of the GE Labs trio, the public's imagination was apparently sparked and, beginning a year later in 1947, a government regulated weather modification industry began to flourish. To this day, the conventional, government regulated weather modification industry expels silver iodide from airplanes usually in order to make it rain or suppress hail.

However, as previously noted, the conventional weather modification industry is distinct from the New Manhattan Project and therefore is not the focus of this chapter or book. The New Manhattan Project employs electromagnetic energy to manipulate dispersed particles while conventional weather modifiers do not. Different substances are sprayed and conventional weather modification efforts are conducted on a regional basis while the New Manhattan Project is global.

As we are about to see, not only did the GE Labs trio start the conventional weather modification industry and the scientific era of weather modification, they also kicked off the New Manhattan Project. In the next section we will see how one of these three scientists went on to lead the Project. Before we do, though, let's take a quick look at the first industrial scientist to win a Nobel Prize, Irving Langmuir.

Less than a month after Germany's surrender, Langmuir was recruited to begin conducting work for the government in the area of weather modification. Two Army officers came to his brother's apartment in New York City and asked him to participate in this new assignment. Langmuir was one of the first scientists to conduct experiments on ionized gases which he called 'plasmas.' In doing so, he invented the field of plasma physics. This is relevant to today's NMP because, as part of the Project, electromagnetic energy is currently being used to manipulate ionospheric plasmas.

Bernard Vonnegut

One member of the GE Labs trio, Bernard Vonnegut (1914-1997) went on to pioneer weather modification research involving the use of artificial electric charges and atmospheric aerosols. His work in this area was largely performed under government contracts outsourced to a research and development firm called Arthur D. Little Inc. Irving Langmuir's earlier work provided much of the foundation for Vonnegut's and as Vonnegut went forward with the NMP's standard, Langmuir continued to influence him greatly.

Vonnegut graduated from MIT in 1936. For his graduate thesis "A Freezing Point Apparatus," he designed a device to measure the exact point at which water with other substances dissolved in it will freeze. This work was relevant to cloud nucleation, and therefore, weather modification. He got his PhD in physical chemistry from MIT in 1939. He joined GE Labs in 1945.

During WWII, before he became director of the MIT meteorology department's aircraft de-icing project, he worked at the MIT Chemical Engineering Department where he developed smokes for the U.S. government's Chemical Warfare Service. Vonnegut got Manhattan District clearance so he could use radioactive tracers to measure

German gas mask smoke penetrations. It was during this time when he met Irving Langmuir who was working on smokes as well. These MIT smoke programs were carried out with the participation of General Electric.

Bernard Vonnegut

Incidentally, General Electric and MIT have historically had a very close professional relationship. Willis R. Whitney, the founder of GE Labs, was a member of the 1890 MIT graduating class and later an MIT professor. The man who asked Whitney to organize GE Labs, Elihu Thomson served as MIT's acting president from 1920 to 1923. Karl Compton, who served as both MIT President and President of the MIT Corporation, was a consultant to the GE Research Lab in the 1920s. There are many other examples.

After Vonnegut's WWII jobs with Langmuir, he worked for the Air Force Meteorology Department where he got closer to both Langmuir and his assistant Vincent Schaefer while discussing aircraft deicing. He joined GE in the Fall of 1945 and didn't leave until July of 1952.

Space charge experiment

Beginning in 1955, Bernard Vonnegut, Arthur D. Little *et al.* conducted experiments involving unsheathed stainless steel wires miles long strung from the tops of telephone poles. These wires were connected to a strong DC power supply and thus created a discharged corona. The coronal discharge's effect upon aerosols (ambient and man-made) and the clouds above was monitored and analyzed. Through the late 1960s, these experiments were carried out in Massachusetts, New Hampshire, Texas, Illinois, and New Mexico. These types of experiments are referred to as 'space charge' experiments. The U.S. Signal Corps and the U.S. Coast Guard provided support. Others performed similar experiments.

The earliest recorded instances of electricity being intentionally used to modify particles in the atmosphere can be found in the 1884 experiments of Sir Oliver Lodge (1851-1940). The 1918 U.S. patent #1,279,823 "Process and Apparatus for Causing Precipitation by Coalescence of Aqueous Particles Contained in the Atmosphere" by J.G. Balsillie built upon Lodge's work. Using this knowledge as a basis, Mr. Vonnegut resumed Lodge's and Balsillie's work - this time with massive funding and modernized scientific equipment.

As noted in the previous chapter, the 1958 *Final Report of the Advisory Committee on Weather Control* is one of the most, if not *the* most referenced weather modification document ever. This report contained an article by Bernard Vonnegut, Vincent Schaefer, J. S. Barrows, and Paul MacCready titled "The Future." In it, they outline an atmosphere saturated with 'chemicals' and 'altering' atmospheric electrical variables. This is the earliest explicit description of the New Manhattan Project yet found by the author. It reads:

"When the nature of thunderstorm electrification is understood it may prove possible to control this process by the introduction of chemicals into the atmosphere or by altering electrical variables. Such variables might be atmospheric conductivity, field, and space charge, or perhaps the corona giving properties of the earth's surface.

"When we become sufficiently sophisticated concerning the dynamics of the atmosphere it is possible that weather may be controlled by the large scale release of chemical or more probably thermonuclear heat energy."

Vonnegut worked on space charge experiments with his longtime colleague Charles B. Moore (1920-2010). Vonnegut met Moore at Arthur D. Little. Moore got his bachelor's degree in chemical engineering from the Georgia Institute of Technology in Atlanta. During WWII he had done graduate level work in the Army Air Corps and was chief weather equipment officer for the China-Burma-India theatre.

Vonnegut and Moore's first space charge experiments were sponsored by the U.S. Army Signal Corps and were carried out on the roofs of buildings in West Cambridge, Massachusetts in 1955. A fine wire (.025 cm diameter) 200 m long was connected to a high-voltage (about 25,000 V) power supply in order to produce negative coronal discharge. The downwind fluctuations in atmospheric electricity were monitored on the ground as well as from a light airplane. During their monitoring activities from the airplane, they observed that power plants and other industrial installations using Cottrell precipitators produced large plumes of negative space charge. The experimenters ultimately found the heavily-populated, urban setting for their earliest experiments to be limiting. This setting limited the length of the wire and, hence, the amount of negative charge they could use. The setting also hampered their investigations by being too close to the ocean in that the ocean's salt water mists distorted their measurements. For these reasons, subsequent experiments were carried out at other locations.

The environs of Jacksboro, Texas, about 130 km WNW of Dallas, proved better for their experiments. The usual moist, onshore wind coming from the Gulf of Mexico kept the space charge plumes drifting over gently rolling hills. The relatively flat land also allowed for easy construction and maintenance of the line. Further, the location was suitable for the fact that the region's air is well saturated with natural condensation nuclei which carry electric charges well. Lastly, as is true for much of the central United States, the background concentration of naturally occurring space charge is quite low, which allowed for better detection of their artificially generated space charge. A local Texan by the name of James Cook found a location for the experiments and installed and maintained the space charge generating equipment.

The Jacksboro wire was held aloft on about 80 telescoping masts which were either lashed to existing fence posts or set in holes in the ground. Raising and lowering the telescoping masts allowed for easy line maintenance. The high-voltage (30 kV), DC power supply was capable of providing currents of up to about 2 mA. Like the experiments at West Cambridge, all the experiments at Jacksboro involved negative charge. As the electrical properties of the atmosphere were observed from the ground, a time-lapse camera simultaneously photographed the electric field meter and the entire sky. During some of the experiments, observations were also recorded from a light airplane.

In his 1959 patent titled "Generation of Electrical Fields," Vonnegut writes about creating clouds of charged particles. Vonnegut postulated that lightning strikes caused by potential differences between alternately charged particle clouds can cause precipitation.

Attracting and repelling atmospheric clouds of charged particles may be a method of controlling today's weather as part of the New Manhattan Project.

In his 1962 patent "Atmospheric Space Charge Modification," Vonnegut explicitly states that during space charge experiments, artificial particles are generated for ionization. He also states that atmospheric particles may be ionized by a 'radioactive source' or by 'electrostatic induction.' As he gives other details of his space charge experiments, he notes that, "In order to produce any significant results, the area covered (i.e. the area of the grid) should be at least about 1,000 square miles, and preferably in the order of 10,000 square miles."

In a 1965 article titled "Electricity in Volcanic Clouds," Vonnegut, Moore *et al.* write of experiments involving their measurements of atmospheric electricity downwind from an erupting volcano. They found that the particles emitted from the volcano were sufficiently ionized so that the situation presented itself as a large-scale, naturally-occurring space charge experiment.

In 1967, as part of an article titled "When Will We Change the Weather?," Vonnegut stated the SRM geoengineering thesis. He writes:

"Thinning supercooled clouds or breaking them up could increase the amount of solar energy that reaches the earth. There is an opposite effect that could also be beneficial. By seeding the clear, supersaturated atmosphere to produce an overcast of cirrus clouds, we may be able to reduce the flow of solar energy to the earth. In the daytime this would slow down the 'heat engine' that produces the convective clouds we associate with storms. At night, artificial creation of a cirrus overcast could slow down the cooling of the earth by reducing the radiation of heat back into space."

He continues:

"Weather scientists hope that lightning experiments now under way will lead to techniques for artificially modifying the electrical properties of the atmosphere, thereby influencing the physical processes in clouds... dispersing small metallic needles into a cloud from an airplane may be a way to reduce the charge of electricity in a thunderstorm."

Again in 1967, Vonnegut had a paper published that explicitly notes the creation of artificial, man-made atmospheric particles in the course of conducting space charge experiments. Vonnegut and his co-authors write of conducting atmospheric experiments using charged particle generators. They write that they, "incorporated an aerosol particle generator and a corona charging device similar to that used in Cottrell precipitators." The similarities between their experiments and what a Cottrell electrostatic precipitator does are noted many times. The preferred particle size was 1-.1 micron. The particles were generated by the vaporization of heavy crankcase oil.

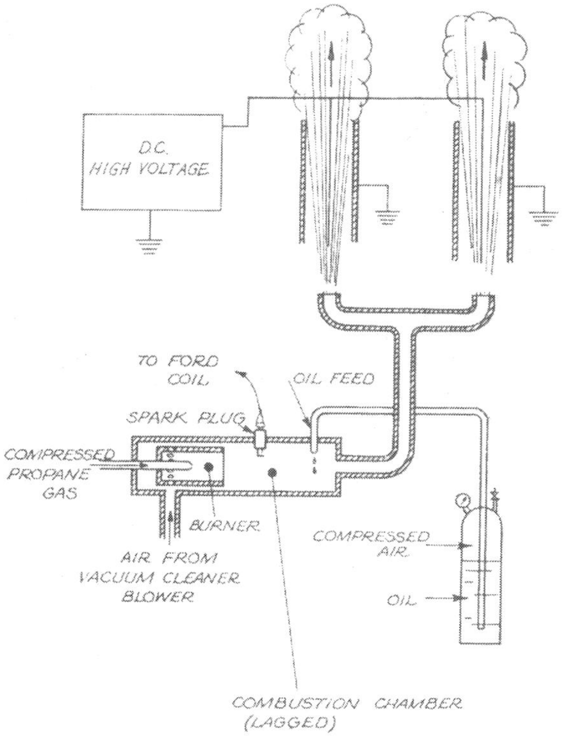

Fig. 1. Schematic diagram of small size charged aerosol generator used in early experiments.

Vonnegut's early charged particle generator

The first experiments were conducted at Scituate, Massachusetts where an offshore breeze carried the electrified smoke over the ocean. Atmospheric electrical field perturbations were then measured from a small boat crossing the plume.

Following these first experiments, a much larger particle generator was constructed and semipermanently installed on a pier in the former U.S. Naval Shipyard at Hingham, Massachusetts. Once again a small boat equipped with the appropriate devices, downwind from the generator was used for measuring atmospheric perturbations. The authors write that, when using this larger generator, "it was found that from 300-400m down wind there were large quantities of fairly large salt water drops supported by the electric fields, and the sensation going through the plume was very much like that of being in a mild rain storm."

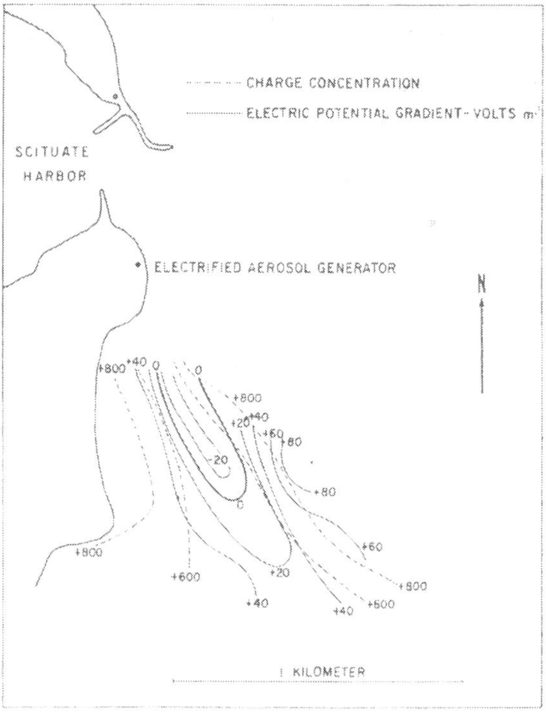

Fig. 3. Perturbations in the fine weather electric potential gradient and space charge concentration over the ocean produced down wind of small size charged aerosol generator with wind from northwest.

The authors conclude by writing, "Electrified aerosol generators appear to be a promising tool for meteorological experiments. Potential uses are production of locally highly concentrated space charge and high intensity electric fields in the free atmosphere." And,

"Because the technique makes it possible to suspend in the atmosphere by electrical forces large particles that would normally fall out quite rapidly, the technique may have practical uses in the application of agricultural sprays or powders." This mention of agricultural crop dusting suggests the spraying of substances from aircraft.

As part of the aforementioned Project Cirrus, Vonnegut spent many summers in New Mexico studying convective cirrus clouds. Convective cirrus clouds are the clouds that are formed from solar heating of surface moisture. The water vapor rises and forms cirrus clouds. These are the clouds that typically bring summertime rainfall in many areas of the Country, including New Mexico. While the headquarters of Project Cirrus remained in the northeast, during the course of the operation, hundreds of clouds in the Albuquerque and Socorro area were seeded. The Project Cirrus team had first arrived in New Mexico in October of 1948. Vonnegut and Moore spent their first summer in New Mexico when they joined New Mexico School of Mines (later called the New Mexico Institute of Mining and Technology/New Mexico Tech) Research & Development Division (R&DD) scientists in 1956. Since 1952, scientists from the R&DD had been measuring atmospheric electricity and using radar.

Under the auspices of the National Science Foundation, construction of the New Mexico Tech Irving Langmuir Laboratory of Atmospheric Physics began in the summer of 1962 on top of South Baldy Mountain. The lab was finished during the summer of 1963. The primary support for Langmuir Laboratory came from the Office of Naval Research and the National Science Foundation in the form of grants both to the lab and to individual research projects. In 1966 the NSF provided $117K for an expansion of the Lab.

From 1963 to 1969 the chairman of the Langmuir Laboratory Committee was a Manhattan Project scientist by the name of Marvin Wilkening. Wilkening had come to the School of Mines in 1948. Once at the School of Mines, he was taken under the tutelage of Dr. William Crozier. Wilkening worked in the field of tracking air movements by tracing naturally occurring atmospheric radionuclides. His work in this field of study lead him to his work with Vonnegut and Moore. A long-lasting personal and professional relationship resulted from Wilkening's collaborations with Vonnegut and Moore.

The first device built at Langmuir for the purpose of space charge experiments was something called 'Stirling's cloud machine.' Joe Chew, the author of *Storms Above the Desert* writes:

"The machine consisted of a barn-like structure with a 1,200-horsepower airplane engine and propeller in one end and a chimney on top. When a storm passed over the ridge between Baldy and the lab, the engine was fired up and lubricating oil was injected into its exhaust. This produced clouds of smoke that were blown past a grid of high-voltage electrodes. From out of the chimney came billows of greasy space charge headed for the cloud at a high velocity. (The Forest Service was not informed of all the oily details of how the machine worked.)"

Stirling's cloud machine

Stirling's cloud machine was named after a former Lawrence Livermore National Laboratory scientist by the name of Stirling Colgate (1925-2013) who, at the time, was the president of New Mexico Tech. Colgate served as president until 1974 when he left in the midst of a personal scandal.

Other space charge experiments carried out at the Langmuir Lab involved a little more than a mile of high-strength, half-inch-diameter steel cable strung between a tower near the Lab and a tower on the other side of a canyon. The wire could be electrified with up to 150,000 volts of DC power.

During these experiments, instrumented aircraft flew over the canyon and collected scientific data. For these activities, the only restricted airspace ever set up for civilian research was established above the Langmuir Lab. One of the aircraft used was called the Special Purpose Test Vehicle for Atmospheric Research, or SPTVAR I and it was provided by the Office of Naval Research. Another plane used in these experiments was assembled from various sources including the National Oceanic and Atmospheric Administration (NOAA), the Air Force and the Navy. Scientists at the Langmuir lab have also used a Sabreliner business jet from the National Center for Atmospheric Research (NCAR) to drop aluminum chaff.

NOAA and NCAR's participation here was not limited to airplanes. In 1984, NOAA and NCAR provided the Langmuir Laboratory with four sophisticated Doppler radars.

Of particular interest to our discussion, from 1978 to 1986, a scientist by the name of Marx Brook was the director of the New Mexico Tech R&DD. As Chew writes, in the early 1980s, "Brook helped to develop a new type of electrostatic precipitator to clean the fly ash out of smokestack emissions. By using an electrified aerosol of water droplets - in other words, applied cloud physics - this precipitator can scrub out tiny particles that conventional precipitators miss." As noted in chapter 1, today's common chemtrail spray has been found to be coal fly ash and the machinations of electrostatic precipitators (Cottrell precipitators) have everything to do with today's New Manhattan Project.

In 1965, after leaving Arthur D. Little, Vonnegut's colleague Charles Moore became a professor of atmospheric physics at New Mexico Tech. Throughout the 1970s and well into the 1980s, Charles Moore went on to have a large impact upon the work done at the Langmuir Lab.

In 1967 Vonnegut ended up taking a faculty position at the Atmospheric Sciences Research Center of the State University of New

York at Albany, becoming a professor at the school founded six years earlier by his colleague Vincent Schaefer.

If you are wondering... yes, Bernard Vonnegut was related to the novelist Kurt Vonnegut (1922-2007). They were brothers. In fact, Kurt also worked for GE as a publicist. In his creative writing, GE-like corporations are a recurring theme.

Historical NMP examples

Ten years after the early scientific breakthroughs of the GE Labs trio, fueled by high-level political rhetoric and apparent popular interest, the United States federal government began pouring hundreds of millions of dollars annually into basic atmospheric research. The bulk of the taxpayer money spent on this first large-scale movement into atmospheric research is documented in a series of semi-annual reports called the Interdepartmental Committee for Atmospheric Sciences (ICAS) reports. The ICAS reports focused on weather modification.

The ICAS reports published by the Federal Council for Science and Technology reveal that between 1958 and 1978, in dollars adjusted for 2020, our U.S. government spent roughly $27 billion dollars on basic atmospheric research. That's just for starters. On the rest of the New Manhattan Project, much more has been spent. In comparison, the total cost of the original Manhattan Project (adjusted for 2020 dollars) was somewhere in the neighborhood of $30B. What has been spent on the new Manhattan Project dwarfs what was spent on the original.

All this research was performed because if one is to control the weather, one must know how the atmosphere works. Or as geoengineer Dr. Clement J. Todd wrote in 1970, "Our ability to manage precipitation depends upon four factors: (1) understanding the physical processes of the atmosphere, (2) real-time knowledge of the weather we wish to manipulate, (3) devising the optimum treatment material and technique, and (4) delivery of that treatment to the cloud where and when we wish." Achieving these factors was understandably not cheap.

The majority of the vast expanses of historical literature pertaining to weather modification and the atmospheric sciences is geared towards conventional weather modification. However, both the New

Manhattan Project and conventional weather modification are supported by basic atmospheric research. So, buried in this body of literature, one may find glimpses of the New Manhattan Project. Although much of the pertinent literature undoubtedly remains classified, quite a bit of this information is available. The rest of this section recounts some of the available glimpses.

~ ~ ~

In 1958, the aforementioned chief White House advisor on weather modification, Captain Howard T. Orville, said the U.S. defense department was studying, "...ways to manipulate the charges of the earth and sky and so affect the weather..." by using an electronic beam to ionize or de-ionize the atmosphere over a given area. The same article, which appeared in *Popular Science*, suggested that weather might be changed on a global scale by, "Electronic bombardment of the ionosphere to alter its electrical charge."

~ ~ ~

The Department of Commerce Weather Bureau reported in 1960 that they were conducting a weather modification study in which, "Chemicals are introduced into the cloud which noticeably changes the surface tension of the droplets. Electrification effects are being observed by artificially electrifying the droplets and subjecting them to impressed electric fields."

~ ~ ~

Now let's get into the aforementioned ICAS reports. The now defunct Interdepartmental Committee for Atmospheric Sciences (ICAS) was created by the Federal Council for Science and Technology in 1958 in order to oversee and coordinate a wide range of basic atmospheric research originating from many previously disparate government offices. Their focus was weather modification. Reporting members of the ICAS included: the departments of Agriculture, Commerce, Defense, Interior, Transportation, and State as well as the Environmental Protection Agency, the Energy Research and

Development Administration, the National Aeronautics and Space Administration, and the National Science Foundation. These government agencies have been involved in weather modification all along.

As previously mentioned, the ICAS produced a series of semi-annual reports between 1958 and 1978. In these reports, ICAS member organizations' weather related scientific activities and expenditures were recounted. The ICAS reports' areas of study included: Earth's natural geomagnetic energy, different ways clouds form and different ways they precipitate, lightning, hurricanes and other extreme weather, inadvertent weather modification, intentional weather modification, and extra-planetary atmospheres. The ICAS is duly noted here because so much of the history of the New Manhattan Project is accounted for in the pages of their reports.

In 1966, the ICAS Select Panel on Weather Modification produced a document titled "Present and Future Plans of Federal Agencies in Weather-Climate Modification." On page 17 of this report, it reads:

"It is anticipated that there will be a few large-scale facilities funded for the testing of modification schemes. Typical schemes might be the suspension of a spray nozzle over a valley between two mountain peaks to produce cloud-sized droplets into which electrical charges can be introduced in either polarity, contaminants can be introduced, and the drop size spectrum can be adjusted to any reasonable distribution."

The Interdepartmental Committee on Atmospheric Sciences subsequently agreed to proceed with the development of a national weather modification program along the lines of this report.

In the 1969 ICAS report, under the heading of 'Cloud Electricity Modification,' it is written that the National Science Foundation is developing, "Means for injecting significant quantities of charge artificially into clouds..." Again in this 1969 report, on page 37 it describes the Army's intentions in the area of weather modification. It reads, "Studies will continue on upper atmospheric structure and dynamics, lasers and other electromagnetic propagation, and acoustic propagation. New approaches to atmospheric modification will be studied."

On page 42 of the 1971 ICAS special report "A National Program for Accelerating Progress in Weather Modification," the authors write of fog being cleared by airplanes releasing chemicals and 'electrical methods' of fog dissipation.

Throughout the ICAS reports, the electrical and electromagnetic manipulation of the atmosphere is mentioned so many times that to cite every example here would be quite tedious. Even beyond what the ICAS reported, the authors of these ICAS reports admit there were classified weather modification experiments underway all along as well.

~ ~ ~

In 1969 R.J. Blackwell was granted a U.S. patent for his "Method of Producing Precipitation from the Atmosphere and Apparatus Therefor." This patent involved producing rainfall with the use of an electrode attached to and extending from the rear of an airplane. As the airplane flies along, so the patent states, a charge can be injected into the atmosphere and cause rain to fall.

~ ~ ~

In 1970, P.H. Wyckoff and the National Science Foundation wrote that, "Ice nucleation experiments in the presence of an electrostatic field were carried out."

~ ~ ~

For the 1972 Third Conference on Weather Modification hosted by the American Meteorological Society, a paper was submitted by a Paul L. Smith, Jr. of the South Dakota School of Mines and Technology. In this paper, Mr. Smith writes of a wire under the influence of high-voltage DC current discharging corona, such as those used for space charge experiments. He proposes that such a wire could be used in combination with an airplane to modify the weather, stating, "...one possible method for seeding clouds electrically is to carry the charging device through a cloud on an airplane, leaving behind a wake of charged cloud droplets."

~ ~ ~

The 1987 U.S. patent #4,686,605 "Method and Apparatus for Altering a Region in the Earth's Atmosphere, Ionosphere and/or Magnetosphere" by Bernard Eastlund shows how stratospheric and tropospheric aerosols can be manipulated using electromagnetic energy in order to modify the weather. The ground-based antennas (known as ionospheric heaters as well as other devices) needed to produce the appropriate electromagnetic energy exist. We will have detailed analyses of ionospheric heaters and the New Manhattan Project shortly.

~ ~ ~

In 1996, our Air Force produced the previously mentioned document called "Weather as a Force Multiplier: Owning the Weather in 2025." The document was produced by the Department of Defense and written as ordered by the chief of staff of the Air Force, Ronald R. Fogleman. "Owning the Weather" was but one in a series of 39 documents speaking to a great overhaul of Air Force operations to be achieved by the year 2025. The larger set of documents is called *Air Force 2025*. "Owning the Weather" describes a system of weather modification combining atmospheric aerosols with electromagnetic energy. On page 2 the document reads:

"*Prior to the attack, which is coordinated with forecasted weather conditions, the UAVs begin cloud generation and seeding operations. UAVs* [unmanned aerial vehicles] *disperse a cirrus shield to deny enemy visual and infrared (IR) surveillance. Simultaneously, microwave heaters create localized scintillation to disrupt active sensing via synthetic aperture radar (SAR) systems such as the commercially available Canadian search and rescue satellite-aided tracking (SARSAT) that will be widely available in 2025. Other cloud seeding operations cause a developing thunderstorm to intensify over the target, severely limiting the enemy's capability to defend. The WFSE monitors the entire operation in real-time and notes the successful completion of another very important but routine weather-modification mission.*"

The document mostly speaks to military combat applications, but there are some very interesting quotes. Here's one, "In the United

States, weather-modification will likely become a part of national security policy with both domestic and international applications." Let's hear more about those 'domestic applications.' Another on page 34 reads, "The ability to modify the weather may be desirable both for economic and defense reasons." The economic and military motives of weather modification are well documented.

Also in 1996, as part of the same series containing "Owning the Weather," the Air Force produced a document entitled "An Operational Analysis for Air Force 2025" which briefly outlines something they call a, "...weather analysis and modification system." This system is described as employing both particulate seeding and microwave energy.

Shortly after its publication, Ronald R. Fogleman, the Air Force Chief of Staff who ordered the *Air Force 2025* series of documents, retired in 1997. Fogleman later became the chairman and CEO of Durango Aerospace, an international aviation consulting firm with a client list that has included Boeing, Northrop Grumman, and Raytheon. Since his retirement, Fogleman has been executively involved with a slew of aerospace firms. He has also been a board member and trustee of the MITRE Corporation.

Ronald R. Fogleman

~ ~ ~

The National Aeronautics and Space Administration (NASA) released a document in 2007 titled "Workshop Report on Managing Solar Radiation." The authors of this document advocate SRM geoengineering and write about atmospheric particles influenced by electromagnetic energy. The authors write:

"Alternatives to dielectrics have been suggested, such as metallic or resonant particles (see, for example, Teller, 1997). Metals interact with electromagnetic radiation strongly and might conceivably require much less particle mass than would non-conducting (dielectric) particles."

Aluminum oxide is basically a dielectric particle. A little later, it concludes:

"Several options exist or are conceivable for deploying the radiation reflecting materials into the stratosphere. These include naval artillery, high-altitude transport aircraft, and unpiloted vehicles."

~ ~ ~

Although apparently not part of the New Manhattan Project, a new type of weather modification method harkening back to the space charge experiments of old has been cropping up lately. It's a rainmaking operation that involves the generation of atmospheric space charge for the purpose of ionizing ambient atmospheric particles which can then float upwards into the lower atmosphere and form rainclouds. The firms that build-out and perform these types of operations have websites and they are looking for more work. Some of the names of these outfits producing these systems are: Meteo Systems, Advanced Synoptic Technologies, Australian Rain Technologies, Aquiess, Sciblue, Ionogenics, and Earthwise. Jim Lee over at *ClimateViewer.com* has been doing some good work on this subject, so please check in there for more information.

Ionospheric heaters

Central to today's electrification and electromagnetic manipulation of our atmosphere as part of the New Manhattan Project are the aforementioned ionospheric heaters. As the evidence presented here indicates, scientists figured out a long time ago how to use electromagnetic energy to modify the weather. Although people running these programs don't admit that ionospheric heaters are being used, or even *can* be used for such a purpose, it is well established.

As we saw in the previous chapter, the initial prototypes of what eventually became today's ionospheric heaters were the antennas used during and after WWII that performed SS LORAN functions. A progression ensued. In 1958 a scientist named Nicholas Christofilos (1916-1972) inspired a project which, over time, was called (in chronological order): Bassoon, Shelf, Sanguine, Seafarer, & ELF. It involved a system of buried and above ground wires laid over vast portions of Wisconsin and Michigan. What has been declassified about these projects says that it was all about naval communications with distant, submerged submarines. But from the HAARP documents, we know that antennas which produce ELF and similar waves can do a whole lot more than produce communication signals. Specifically and pertinently, we have seen that these types of antennas can be used for weather control. Project Bassoon/Shelf/Sanguine/Seafarer/ELF (Project B/S/S/S/ELF) operations continued through to 2004 and some of the operations remain classified.

The earliest known ionospheric heater was located in Pennsylvania. Rosalie Bertell, PhD writes:

"As early as 1966 researchers at Pennsylvania State University built and operated an ionospheric heater, using electromagnetic energy to stimulate or heat the bottom of the ionosphere. Because the device caused problems for pilots, it was removed to a more remote location, Platteville, Colorado. By 1974, similar research facilities were located at Arecibo, Puerto Rico, and in Armidale, New South Wales, Australia."

This first known ionospheric heater was later moved to Poker Flat, Alaska where it continued to be operated by Pennsylvania State University and Dr. Anthony Ferraro under the U.S. Navy.

In mid-July of 1968 construction began on Poker Flat Research Range, near Chatanika, Alaska, 30 miles north of Fairbanks. The name of the research range was taken from a Bret Harte short story called "The Outcasts of Poker Flat." The Penn State ionospheric heater at Poker Flat, and others that were to come to Alaska later, were mainly used in combination with the launching of atmospheric sounding rockets. The first atmospheric sounding rockets were successfully launched from Poker Flat in March of 1969. From humble beginnings, the Poker Flat range expanded throughout the 1970s and into the late 1980s.

Another early ionospheric heater came to Poker Flat in the early 1970s. The authors of the Poker Flat Research Range User's

Handbook write, "Installation of an 88-foot incoherent (Thomson) scatter radar, located three miles from the launch area, was completed by SRI [Stanford Research International] in spring 1971. Called the Chatanika Radar, they [SRI] supported many programs launched out of Poker Flat for 11 years."

In the mid-to-late 1960s, the Chatanika Radar was built and tested on the campus of Stanford University in Palo Alto, California by Stanford Research International. The Chatanika Radar was operated at Poker Flat from 1971 to 1982. The Department of Defense funded the move to Alaska and covered the initial operating costs. At Poker Flat, operations became fully funded by the National Science Foundation. For coordinated measurements, the Chatanika Radar was operated in concert with other OTH (over-the-horizon) radars around the world. It was shut down in March of 1982, disassembled and shipped to Sondrestrom, Greenland, then reassembled near a U.S. air base to begin a new phase of operations as part of the Sondrestrom Upper Atmospheric Research Facility.

The Chatanika radar

The powerful, early ionospheric heater at Arecibo, Puerto Rico mentioned earlier is especially significant. The Defense Advanced Research Projects Agency (DARPA), Stanford Research International, and the National Science Foundation (among many others) have all been very active there. The Caribbean location of the Arecibo ionospheric heater compliments the geographic location of the

HAARP facility well in that their respective locations provide excellent coverage of the contiguous United States from Washington to Florida.

The Arecibo ionospheric heater

Another early and significant ionospheric heater known as the HIPAS antenna was located in Alaska. HIPAS stands for High Power Auroral Stimulation. Construction of the HIPAS facility began in 1980 and was completed in 1986. Primary funding for the original construction was provided by the Office of Naval Research. The National Oceanic and Atmospheric Administration (NOAA) and Lawrence Livermore National Labs (LLNL) were involved as well. The HIPAS facility covered 120 acres. A University of California at Los Angeles (UCLA) professor by the name of Alfred Wong was the founder and director of HIPAS. Following construction in the late 1980s, HIPAS was one of the two most powerful such facilities in the world. HIPAS is said to have shut down in 2007, but may have continued to operate covertly.

HIPAS

HAARP

One of the world's most powerful and versatile electromagnetic energy-generating ionospheric heaters is part of something commonly known as HAARP. Located on a United States Air Force site near Gakona, Alaska, the High-frequency Active Auroral Research Project (HAARP) antenna is one of the world's largest, most powerful, and most functional ionospheric heaters and affords us our best view into this field of technology. Construction began in 1993. Today, the HAARP antenna can generate super high-powered beams of directed electromagnetic energy. The HAARP antenna is designed to shoot these energy beams hundreds of kilometers (the distance varies depending upon the operation) up into the sky; affecting Earth's ionosphere. In doing this, the HAARP antenna can perform a number of functions. For a long time, the HAARP antenna array was the most powerful and versatile ionospheric heater in the world, but it has apparently now been eclipsed recently by the European Incoherent

Scatter Scientific Association (EISCAT) facility located near Tromso, Norway.

The High-frequency Active Auroral Research Project

The first version of HAARP was completed in 1995. Over the years, there have been many upgrades. The British firm BAE Systems completed a major HAARP upgrade in 2007 making it into today's behemoth.

The known uses of the HAARP antenna are: weather modification, power beaming, earth tomography (mapping of our planet's interior), Star Wars-type defense capabilities, enhanced communications, communication disruptions, and mind control. For an in-depth discussion about what the HAARP antenna does and how it does it, you simply must read the 1995 book *Angels Don't Play this HAARP* by Dr. Nick Begich and journalist Jeane Manning. Although lesser ionospheric heaters do not generate energy beams as powerful or possess the same functionality as HAARP, similar facilities are located around the world. The HAARP website explains the differences between HAARP and other ionospheric heaters like this: "HAARP is unique to most existing facilities due to the combination of a research tool which provides electronic beam steering, wide frequency coverage and high effective radiated power collocated with a diverse suite of scientific observational instruments." HAARP can be remotely operated. In order to increase functionality and effectiveness, ionospheric heaters can be used in combination.

World ionospheric heaters

The HAARP program is now owned by the University of Alaska at Fairbanks and operated by their Geophysical Institute. But HAARP was originally ran by the Air Force, the Navy, and the Defense Advanced Research Projects Agency (DARPA). Here's more from the HAARP website, when it was under previous management:

"Technical expertise and procurement services as required for the management, administration and evaluation of the program are being provided cooperatively by the Air Force (Air Force Research Laboratory), the Navy (Office of Naval Research and Naval Research

Laboratory), and the Defense Advanced Research Projects Agency. Since the HAARP facility consists of many individual items of scientific equipment, both large and small, there is a considerable list of commercial, academic and government organizations which are contributing to the building of the facility by developing scientific diagnostic instrumentation and by providing guidance in the specification, design and development of the IRI [HAARP antenna]. *BAE Advanced Technologies (BAEAT) is the prime contractor for the design and construction of the IRI. Other organizations which have contributed to the program include the University of Alaska, Stanford University, Cornell University, University of Massachusetts, UCLA, MIT, Dartmouth University, Clemson University, Penn State University, University of Tulsa, University of Maryland, SRI International, Northwest Research Associates, Inc., and Geospace, Inc."*

Bernard Eastlund

12 U.S. patents are commonly recognized as directly applicable. These are known as 'the HAARP patents.' Dr. Bernard Eastlund is

listed as the inventor on two of these patents and a co-inventor on another. Eastlund was the inventor of HAARP. Dr. Eastlund got his Bachelor's degree in physics from the Massachusetts Institute of Technology and based his work heavily upon Nikola Tesla's.

The 12 HAARP patents were all assigned to ARCO Power Technologies Incorporated (APTI); a subsidiary of Atlantic Richfield Company (ARCO). APTI also won the original contract to build HAARP. In 1994, APTI was sold to a company called E-Systems. E-Systems then changed APTI's name to Advanced Power Technologies Incorporated. Largely involved in communications and information systems, E-Systems got most of its business from and had extensive ties to the National Security Agency and the Central Intelligence Agency. In 1995, Raytheon acquired E-Systems. Raytheon, the defense contracting behemoth, now holds all 12 HAARP patents.

HAARP and weather modification

HAARP and similar devices like it can modify the weather. Bernard Eastlund made no secret of it. Let us reference a passage from Nick Begich and Jeane Manning's book *Angels Don't Play this HAARP*:

"Eastlund's enthusiasm for planetary-scale engineering came through just as clearly in an interview with Omni Magazine. While acknowledging that many of the uses of his invention are warlike, he also talked about 'more benign' uses. His view of benign included using the technology to reroute the high-altitude jet stream, which is a major player in shaping global weather. Another way to control the weather with his technology would be to build 'plumes of atmospheric particles to act as a lens or focusing device' for sunlight, he told Omni. With this, the people controlling the antennae could aim in such a way that the return beams would hit a certain part of the earth. With the heating ability, they could experiment until they could control wind patterns in a specific place.

"The Omni article explained. 'What this means, he says, is that by controlling local weather patterns one could, say, bring rain to Ethiopia or alter the summer storm pattern in the Caribbean.'"

In 1998, a paper written by Dr. Eastlund titled "Systems Considerations of Weather Modification Experiments Using High Power Electromagnetic Radiation" was published. Eastlund wrote:

"In the mid-1980's, antennas producing up to 10[to the twelfth power] watts using natural gas on the North Slope of Alaska were studied by ARCO and the U.S. Department of Defense for military applications in the ionosphere. Because of the similarity between the proposed antenna power and the energy turnover of some typical storm systems, applications for weather modification in the troposphere were proposed."

The document continues with technical details as to how this weather modification, employing the use of satellites and computerized atmospheric models, is achieved. As we can see from this passage, natural gas from Alaska's North Slope oil and gas reserves serves as the fuel that runs HAARP. The original owner of the HAARP patents, APTI was a subsidiary of the owner of trillions of cubic feet of North Slope natural gas, Atlantic Richfield (ARCO).

United States patent #4,712,155 "Method and Apparatus for Creating an Artificial Electron Cyclotron Heating Region of Plasma" is one of the aforementioned 12 HAARP patents and states multiple times that HAARP can be used for weather control. Dr. Eastlund is noted as the inventor. The patent states:

"Since there is evidence that wind currents in the stratosphere appear to be linked to certain weather patterns on earth, such winds can be produced with the present invention at selected locations and altitudes which can be used in establishing such weather patterns."

United States patent #4,686,605 "Method and Apparatus for Altering a Region in the Earth's Atmosphere, Ionosphere and/or Magnetosphere" is one of the 12 HAARP patents and provides another link between HAARP and weather control. Dr. Eastlund is credited as the inventor. It states, "Weather modification is possible by, for example, altering upper atmosphere wind patterns or altering solar absorption patterns by constructing one or more plumes of atmospheric particles which will act as a lens or focusing device."

The patent goes on:

"Also as alluded to earlier, molecular modifications of the atmosphere can take place so that positive environmental effects can be achieved. Besides actually changing the molecular composition of an atmospheric region, a particular molecule or molecules can be chosen for increased presence. For example, ozone, nitrogen, etc. concentrations in the atmosphere could be artificially increased.

Similarly, environmental enhancement could be achieved by causing the breakup of various chemical entities such as carbon dioxide, carbon monoxide, nitrous oxides, and the like. Transportation of entities can also be realized when advantage is taken of the drag effects caused by regions of the atmosphere moving up along diverging field lines. Small micron sized particles can be then transported, and, under certain circumstances and with the availability of sufficient energy, larger particles or objects could be similarly affected."

Chemtrails enhance the electrical connection between the lower and upper atmosphere. Although chemtrails are not sprayed as high up as the ionosphere, the electrical and pressure effects caused by ionospheric heaters such as HAARP can strongly influence chemtrail laden skies below. A strong physical connection between the ionosphere, stratosphere, and troposphere is well documented in the atmospheric sciences and weather modification literature.

Let us refer to *Angels Don't Play this HAARP*:

"*...there is a super-powerful electrical connection between the ionosphere and the part of the atmosphere where our weather comes onstage, the lower atmosphere. Furthermore, scientific theories describe how the electrical energetic levels of the atmosphere are connected to cloud processes.*"

In a 2005 *USA Today* article, Eastlund was quoted as saying, "The technology of artificial ionospheric heating could be as important for weather modification research as accelerators have been for particle physics."

HAARP 2.0

It is reported that smaller and mobile versions of HAARP exist. In fact, there are many, smaller HAARP-like facilities all around us. This type of technology currently comprises something called the NEXRAD system of weather radar. Let us refer again to *Angels Don't Play this HAARP*:

"'*Is it possible that the HAARP scientists could have miniaturized the technology so that they don't need such a large area of land and electrical power as called for in Eastlund's patents?' Manning asked him.*

"'It's entirely possible,' he [Eastlund] replied. 'They have had a lot of good engineers working on it for some time. I would hope they have improved it.'"

In 2005, scientists at the University of California Los Angeles and the High Power Auroral Stimulation (HIPAS) Observatory in Alaska reported a new generation of ionospheric heaters. This technology, known as 'Hertzian antennas,' can, "...match or exceed the peak power of the new HAARP system for a fraction of its cost." Though most implementations are not capable of producing all of the complex, phased signals produced by facilities like HAARP or the antenna array at Jicamarca, Peru. Named after the man who first proved electromagnetic energy, an effective Hertzian antenna can fit in a big back yard.

Have these relatively new Hertzian antennas been placed strategically around the country and the world in a coordinated effort to modify the weather? As noted earlier, devices such as these can be networked and synchronized to increase functionality and effectiveness. Today's weather modification operations most probably involve a large network of smaller antennas such as these. All these antennas could be controlled from a central location. This proposed scenario jibes with what the scientists at the HAARP facility are saying. They say that the HAARP ionospheric heater antenna array goes largely unused. If a network of smaller antennas across the country is handling the day-to-day dirty work, then the HAARP array may only be used occasionally or not at all.

This image was captured from Google Earth:

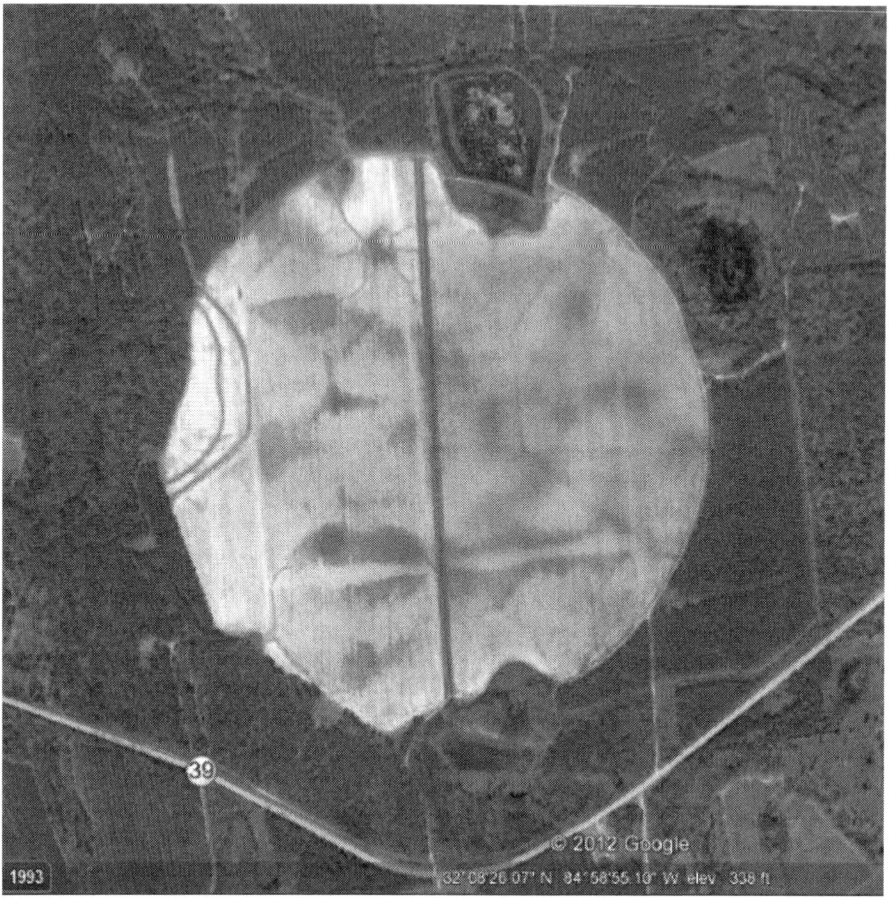

This, at first glance, appears to be a circle formed by a center pivot irrigation system. Such circles appear all over the country as this method of irrigation is common. But, as one can see, the edges of the circle are irregular and there appears to be diamond shaped structures under the soil. What accounts for these irregularities?

My science advisor, who is an expert in the field of electromagnetic energy, tells me that this may be an ELF antenna buried in the ground. Although she says this particular one is aiming inward rather than upward, an antenna such as this can produce the electromagnetic energy needed to modify the weather as part of the

New Manhattan Project. There may be hundreds of such antennas buried in a similar fashion across the country. My science advisor tells me that she has been finding these types of things on mountain tops as well.

There is precedent for this. The March, 1971 Interdepartmental Committee for Atmospheric Sciences (ICAS) report makes mention of our navy developing 'sub-surface transducers.' The aforementioned Project B/S/S/S/ELF employed sub-surface electromagnetic energy producing wires. In 1973, the United States Navy produced a document titled "Project Sanguine: Michigan site." This document details a planned program involving antennas emitting extremely low frequency (ELF) electromagnetic energy buried under the ground. The electromagnetic energy produced in this fashion was intended to be one-way transmissions to submerged submarines thousands of miles away. This was a proven technology. The system was designed to be able to survive a nuclear holocaust. U.S. patent #4,839,661 "Guided Wave Antenna System and Method" describes how ionospheric heater-like antennae may be buried about a meter deep and still remain highly functional.

Maybe it's time we get out there with electromagnetic energy detectors and find out what this thing is. The coordinates for the above image are 32 08'26.07" N 84 58'55.10" W. You can plug those coordinates into Google Earth. Those coordinates translate to a field alongside Highway 39 near Fort Benning, Georgia. The circle appears to be a little less than one mile in diameter.

A pattern of deception

When it comes to HAARP, our government has shown a pattern of denial, obfuscation and outright lies. Burying infrastructure to cover up a weather modification project would be consistent behavior. In many instances, the HAARP website and the military contradict each other and/or the authors of *Angels Don't Play this HAARP*.

The HAARP website claims that HAARP is not used for weather modification and the military has never admitted to these capabilities. My science advisor, the patents, the inventor of HAARP, Nick Begich, Jeane Manning, the European Parliament, and *their own documents* say HAARP *can* modify the weather.

Although the HAARP website has claimed that experiments are only being carried out in a relatively small portion of the ionosphere directly over the facility, the original HAARP executive summary says, "For broader military applications, the potential for significantly altering regions of the ionosphere at relatively great distances (1000 km or more) from a heater is very desirable."

The HAARP website claims that HAARP does not make holes in the ionosphere. The European Parliament, my science advisor, and the authors of *Angels Don't Play this HAARP* say that it does.

The HAARP website and the military deny that HAARP is a 'Star Wars' defense type weapon. The European Parliament and the authors of *Angels Don't Play this HAARP* insist it is; the patents support their position. *Angels* says:

"*In February, 1995, the Star Wars missile defense shield was supposed to be dead. The United States House of Representatives by a 218 to 212 voted to kill the program. Yet HAARP continues on while the motives of the military are hidden from the world.*"

The military and the HAARP website have both claimed that HAARP is not a classified project, but leaked documents show that military planners intended to keep the program under wraps.

The HAARP website contradicted itself about military involvement. In their self-description, they said that they are a military project, but in the FAQs, they said HAARP is, "...not designed to be an operational system for military purposes." All this while the military's executive summary said that HAARP is used to 'exploit' ionospheric processes for Department of Defense purposes.

Dr. Eastlund contradicted the official military position many times. Even though they have been exhaustively proven, our military denies connections between Eastlund, APTI, and HAARP. Eastlund himself said in a 1988 *NPR* interview that the military had tested some of the ideas presented in the patents. According to Dr. Begich and Jeane Manning:

"*Eastlund said in a 1988 radio interview that the defense department had done a lot of work on his concepts, but he was not at liberty to give details. He later told Manning that after he had worked within ARCO for a year and applied for patents, Defense Advanced Research Project Agency (DARPA) had combed through his theories then gave out a contract for him to study how to generate the relativistic (light speed) electrons in the ionosphere."*

Here's more about Eastlund from *Angels*:

"Eastlund told Chadwick of National Public Radio that the patent should have been kept under government secrecy. He said he had been unhappy that it was issued publicly, but, as he understood it, the patent office does not keep basic 'fundamental information' secret. 'You don't get a patent if you don't describe in enough detail to another person how to use it,' he said. Specifics of military applications of his patent remain proprietary (secret), he added."

Even the technocratic European Parliament found serious concerns about HAARP. A 1999 European Parliament committee report, after hearing Dr. Nick Begich and others, concluded:

"[the Committee on the Environment, Public Health and Consumer Protection] Regards the US military ionospheric manipulation system, HAARP, based in Alaska, which is only a part of the development and deployment of electromagnetic weaponry for both external and internal security use, as an example of the most serious emerging military threat to the global environment and human health, as it seeks to interfere with the highly sensitive and energetic section of the biosphere for military purposes, while all of its consequences are not clear, and calls on the Commission, Council and the Member States to press the US Government, Russia and any other state involved in such activities to cease them, leading to a global convention against such weaponry;"

HAARP may be remotely operated from Lawrence Livermore National Laboratory, Los Alamos National Laboratory, a Stanford University VLF Group facility, and/or Wright-Patterson Air Force Base. These four organizations have produced much of the leading research and development. HAARP is an incredibly high-tech machine. One needs highly skilled scientists and engineers to run it. The best place to run HAARP would be from a laboratory where the technology was developed. Most people (top scientists included) are generally not so hot about relocating to the wilds of Alaska. There may be other command centers capable of controlling HAARP as well.

We know ionospheric heaters such as HAARP can modify the weather. If one could, don't you think they would? Weather control is god-like power. Chemtrails are sprayed to enhance the effectiveness of these operations. Our Government's pattern of lying and obfuscation about HAARP makes perfect sense and is consistent with behavior exhibited by people associated with every aspect of the New Manhattan Project.

Photographic evidence

In an article posted by Dane Wigington on his website *GeoengineeringWatch.org* titled "NASA Satellite Imagery Reveals Shocking Proof Of Climate Engineering," we see proof that electromagnetic energy is being used to influence today's weather. The article features about 15 satellite images showing cloud formations that are signatures of electromagnetic manipulation. A few of these images are reproduced here. While weather systems are the definition of chaos, these photos show large-scale, absolute symmetry. The symmetrical ridging and smooth wave formations are the product of electromagnetic waves generated by antennas such as HAARP. After looking at these images and considering the other evidence presented in this book, it is ridiculous to assert that man cannot modify the weather and that the New Manhattan Project does not exist.

Off the San Francisco Bay Area coast

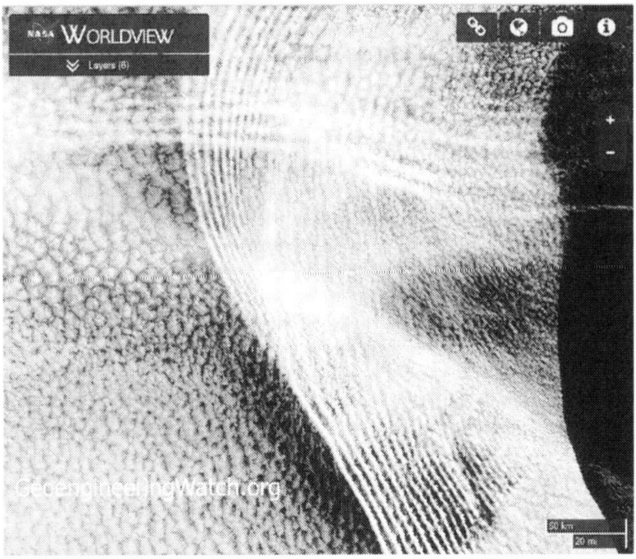

Off Africa's west coast

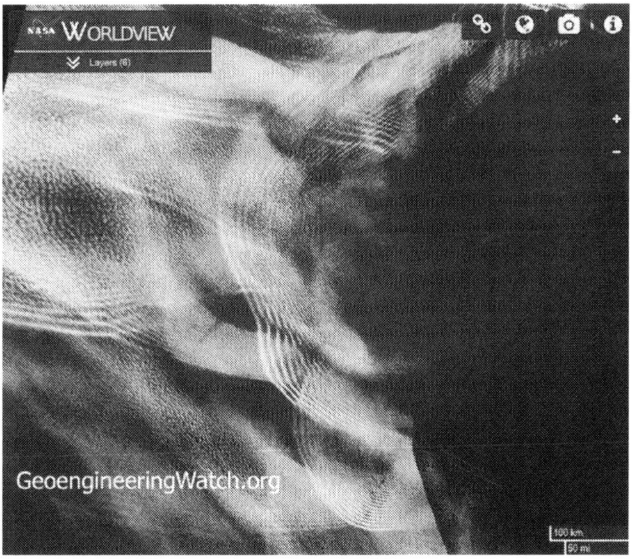

Off Africa's west coast

References

Angels Don't Play this HAARP: advances in Tesla technology a book by Jeane Manning and Dr. Nick Begich, published by Earthpulse Press, 1995

Tesla: Man Out of Time a book by Margaret Cheney, published by Simon & Schuster, 1981

My Inventions: The Autobiography of Nikola Tesla a book by Nikola Tesla, published by SoHo Books

Prodigal Genius: The Life of Nikola Tesla a book by John J. O'Neill, published by Adventures Unlimited Press, 2008

Tesla vs. Edison: The Life-long Feud that Electrified the World a book by Nigel Cawthorne, published by Chartwell Books, 2016)

Tesla: Inventor of the Electrical Age a book by W. Bernard Carlson, published by Princeton University Press, 2013

The Trumps: Three Generations that Built an Empire a book by Gwenda Blair, published by Simon & Schuster, 2001

Surreptitious Entry a book by Willis George, published by D. Appleton-Century Company, 1946

"War and Weather" a report by Dr. Irving P. Krick, published by the Army Air Force Scientific Advisory Group as a volume in the series *Toward New Horizons*, 1946

Prophecy Fulfilled: 'Toward New Horizons' and its legacy a book by the United States Air Force, published by Progressive Management, 1994

"Where We Stand" a report by Dr. Theodore von Kármán, published by the Army Air Force Scientific Advisory Group as a volume in the series *Toward New Horizons*, 1946

Architects of American Air Supremacy: General Hap Arnold and Dr. Theodore von Karman - Conceptualizing the Future Air Force, Covering Rockets, Missiles, Jet Airplanes, Atomic Warfare, Propulsion by Dik A. Daso, published by Progressive Management Publications, 1997

"Theodore von Kármán 1881-1963: A Biographical Memoir" by Hugh L. Dryden, published by the National Academy of Sciences, 1965

Soldiers of Reason: The Rand Corporation and the Rise of the American Empire a book by Alex Abella, published by Harcourt, 2008

Prophecy Fulfilled: 'Toward New Horizons' and its legacy a book by the United States Air Force, published by Progressive Management, 1994

Spy Capitalism: Itek and the CIA a book by Jonathan E. Lewis, published by the Yale University Press, 2002

"The USAF Scientific Advisory Board: Its First Twenty Years 1944-1964" a report by Thomas A. Sturm, published by the USAF Historical Division Liaison Office, 1967

"Early History of Cloud Seeding" a paper by Barrington S. Havens, published by the Langmuir Laboratory at the New Mexico Institute of Mining and Technology, the Atmospheric Sciences Research Center at the State University of New York at Albany and the Research and Development Center of the General Electric Company, 1978

Adventure into the Unknown: The First 50 Years of the General Electric Research Laboratory a book by Laurence A. Hawkins, published by William Morrow and Company, 1950

"Weather Control" by Marc Leepson, published by Congressional Quarterly, Inc., 1980

The Brothers Vonnegut: Science and Fiction in the House of Magic a book by Ginger Strand, published by Farrar, Straus, and Giroux, 2015

Under an Ionized Sky: From Chemtrails to Space Fence Lockdown a book by Elana Freeland, published by Feral House, 2018

“Bernard Vonnegut, 82, Physicist Who Coaxed Rain From the Sky” an article by Wolfgang Saxon, published by the *New York Times*, Apr. 27, 1997

Scientists Against Time, a book by James Phinney Baxter, 3rd, published by the M.I.T. Press, 1968

"Technique for Introducing Low-Density Space Charge into the Atmosphere" a paper by B. Vonnegut, K. Maynard, W.G. Sykes, and C.B. Moore, published by Arthur D. Little and the *Journal of Geophysical Research*, volume 66, number 3, March, 1961

U.S. patent #1,279,823 "Process and Apparatus for Causing Precipitation by Coalescence of Aqueous Particles Contained in the Atmosphere," 1918

"The Future" a paper by Bernard Vonnegut, Vincent Schaefer, J. S. Barrows, and Paul MacCready, published in the *Final Report of the Advisory Committee on Weather Control*, 1958

Storms Above the Desert: Atmospheric Research in New Mexico 1935-1985 a book by Joe Chew with the assistance of Jim Corey, published by the University of New Mexico Press, 1987

“Preliminary Attempts to Influence Convective Electrification in Cumulus Clouds by the Introduction of Space Charge into the Lower Atmosphere” a paper by Bernard Vonnegut, Charles B. Moore and Arthur D. Little, Inc. as it appeared in the *Proceedings of the Second Conference on Atmospheric Electricity*, published by Pergamon Press, 1958

U.S. patent #2,881,335 "Generation of Electrical Fields" by Bernard Vonnegut and Arthur D. Little, 1959

U.S. patent #3,019,989 "Atmospheric Space Charge Modification" by Bernard Vonnegut and Arthur D. Little, 1962

"Electricity in Volcanic Clouds" an article written by C.B. Moore, B. Vonnegut, Robert Anderson, Sveinbjörn Björnsson, Duncan C. Blanchard, Stuart Gathman, James Hughes, Sigurgeir Jónasson, and Henry J. Survilas, published by *Science*, May 28, 1965, vol. 148, issue 3674, p1179-1189

"When Will We Change the Weather?" an article by Bernard Vonnegut, 1967

"Technique for the Introduction into the Atmosphere of High Concentrations of Electrically Charged Aerosol Particles" a paper by B. Vonnegut, C.R. Smallman, C.K. Harris, W.G. Sykes, and Arthur D. Little, Inc., published in the *Journal of Atmospheric and Terrestrial Physics*, 1967, Vol. 29, pp. 781-792, reproduced by the Pergamon Press Ltd.

Interdepartmental Committee for Atmospheric Sciences reports 1960-1978, published by the Federal Council for Science and Technology

"Department of the Interior Program in Precipitation Management for 1970" a paper by Dr. Clement J. Todd as it appeared in the *Proceedings of the Twelfth Interagency Conference on Weather Modification*, 1970

"Weather as a Weapon" an article by Howard T. Orville as it appeared in *Popular Science* magazine, June, 1958

1st National Science Foundation annual weather modification report, 1960

"Present and Future Plans of Federal Agencies in Weather-Climate Modification" a report by the Interdepartmental Committee on Atmospheric Sciences Select Panel on Weather Modification, 1966

"The Interdepartmental Committee on Atmospheric Sciences: A Case History" a paper by Robert E. Morrison

U.S. patent #3,456,880 "Method of Producing Precipitation from the Atmosphere and Apparatus Therefor" 1969

"National Science Foundation Program in Weather Modification for FY 1970" a report by P.H. Wyckoff, as it appeared in the P*roceedings of the Twelfth Interagency Conference on Weather Modification* a report by the U.S. Department of Commerce, 1970

"Use of Electrostatic Precipitation Technology for Cloud Seeding" a paper by Paul L. Smith, Jr. as published in the *proceedings of the Third American Meteorological Society Conference on Weather Modification*, 1972

U.S. patent #4,686,605 "Method and Apparatus for Altering a Region in the Earth's Atmosphere, Ionosphere and/or Magnetosphere," 1987

"Weather as a Force Multiplier: Owning the Weather in 2025" a report by Col. Tamzy J. House, Lt. Col. James B. Near, Jr., LTC William B. Shields (USA), Maj. Ronald J. Celentano, Maj. David M. Husband, Maj. Ann E. Mercer and Maj. James E. Pugh, published by the United States Air Force, 1996

Air Force 2025 a series of reports by the U.S. Air Force, published by the U.S. Air Force, 1996

"An Operational Analysis for Air Force 2025" a report by the U.S. Air Force, published by the U.S. Air Force as part of *Air Force 2025* by the U.S. Air Force, published by the U.S. Air Force, 1996

"Ronald R. Fogleman, Retired United States Air Force General, Joins Galen Capital Group as a Senior Advisor" an article published by BusinessWire, September 19, 2005

"AAR Names General (ret.) Ronald R. Fogleman Lead Director of Its Board" an article published by PRNewswire, May 21, 2013

MITRE Corporation 2012 annual report

"Workshop Report on Managing Solar Radiation" a report by the National Aeronautics and Space Administration, 2007

"Electric Rainmaking Technology Gets Mexico's Blessing" an article by Samuel K. Moore, published by *IEEE Spectrum*, April 1, 2004

Poker Flat Research Range Range User's Handbook a handbook by the Geophysical Institute of the University of Alaska, published by the University of Alaska, March 1989

"The Chatanika and Sondrestrom Radars - a brief history" a paper by M.A. McCready and C.J. Heinselman of The Center for Geospace Studies, SRI International, 2013

The World as a Mathematical Game: John von Neumann and 20th Century Science a book by Giorgio Israel and Ana Millán Gasca, published by Birkhäuser, 2009

"Science & Space" an article by Leonard David of *space.com*, published by *USA Today*, October 31, 2005

"Systems Considerations of Weather Modification Experiments Using High Power Electromagnetic Radiation" a paper by Dr. Bernard Eastlund as it appeared in the *proceedings of the Workshop on Space Exploration and Resources Exploitation - Explospace* 20-22 October, 1998, Cagliari, Sardinia, Italy

"BAE Systems Space Weather Research Facility Wins Top Award" an article published by *Business Wire*, September 25, 2007

U.S. patent #4,712,155 "Method and Apparatus for Creating an Artificial Electron Cyclotron Heating Region of Plasma," assigned to ARCO Power Technologies Incorporated, 1987

"Atmospheric Heating as a Research Tool" a paper by Bernard Eastlund and Lyle Jenkins

U.S. patent #4,686,605 "Method and Apparatus for Altering a Region in the Earth's Atmosphere, Ionosphere and/or Magnetosphere" assigned to ARCO Power Technologies Incorporated, 1987

United States patent #5,068,669 "Power Beaming System" assigned to ARCO Power Technologies Incorporated, 1991

"HAARP Research and Applications" a report by the Air Force Research Laboratory and the Office of Naval Research, 1998

The Creative Ordeal: The Story of Raytheon a book by Otto J. Scott, published by Atheneum, 1974

"The First Practical Uses of Underwater Acoustics: The Early 1900s" an article published in DOSITS.org

The Jasons: The Secret History of Science's Postwar Elite a book by Ann Finkbeiner, published by Penguin Books, 2007

"Extremely Low Frequency Transmitter Site Clam Lake, Wisconsin" a report by the United States Navy, 2001

"Sending Signals to Submarines" an article by David Llanwyn Jones, published in *New Scientist*, July 4, 1985

"Signaling Subs" an article by T.A. Heppenheimer, published in *Popular Science*, April, 1987

Planet Earth: The Latest Weapon of War a book by Rosalie Bertell, published by Black Rose Books, 2001

"Proposal for HIPAS Facility Upgrade for Global ELF Communications and Ozone Conservation" a report by the HIPAS Observatory and the University of California at Los Angeles Plasma Physics Laboratory, 1989

"Pulsed Energy Storage Antennas for Ionospheric Modification" a report by the European Geosciences Union, 2005

Interdepartmental Committee for Atmospheric Sciences report number 15, published by the Federal Council for Science and Technology, March, 1971, p46

"Project Sanguine: Michigan site" a report by the United States Navy, 1973

U.S. patent #4,839,661 "Guided Wave Antenna System and Method," 1989

European parliament report on the environment, security and foreign policy: Committee on Foreign Affairs, Security and Defense Policy, January 14, 1999

"HAARP HF Active Auroral Research Program: Joint Services Program Plans and Activities" a report by the Air Force Geophysics Laboratory and the Navy Office of Naval Research, 1990

“NASA Satellite Imagery Reveals Shocking Proof Of Climate Engineering” an article by Dane Wigington, published by *Geoengineeringwatch.org*, Jan. 16, 2019

Chapter 4
VICE ADMIRAL RABORN

The New Manhattan Project had a prophet. His name was United States Navy Vice Admiral William Francis Raborn, Jr. (1905-1990). His story gives us insight and provides us with compelling investigatory leads into the biggest scientific effort ever.

In the January 1963 edition of the *U.S. Naval Institute Proceedings*, Vice Admiral Raborn outlined a program using electromagnetic energy to modify the weather. His article was entitled "New Horizons of Naval Research and Development." In this paper, underneath the heading of 'Environmental Warfare' he wrote:

"The possibilities for the military employment of the 'weather weapon' may be as diverse as they are numerous. An ability to control the weather could introduce greater changes in warfare than those which occurred in 1945 with the explosion of the first nuclear weapons.

"A severe storm or hurricane striking a naval force may well inflict greater damage than could an enemy. The capability to change the direction of destructive storms and guide them toward enemy concentrations may exist in the future arsenal of the naval tactical commander.

"Ground, sea, air and amphibious operations might be supported by the dissipation of fog or clouds, or by the production of rain or drought. Conversely, the creation of solid, low overcasts might be used to conceal troop concentrations, movements, and task force deployments. Large-scale weather control techniques might be used to cause extensive flooding in strategic areas or even to bring a new 'ice age' upon the enemy. By influencing the ionosphere and atmosphere simultaneously, magnetic, acoustic, and pressure effects might be generated in such a way that ocean-wide sweeping of mines would occur.

"Creating or dissipating atmospheric temperature/humidity ducts might modify the refractive index of the atmosphere enough to influence radar or radio transmission. Artificially-induced ionospheric storms might produce a blackout of communications.

"Certain electromagnetic waves are unable to pass through an area of precipitation. A cloud seeding generator could be employed under appropriate meteorological conditions to produce precipitation that would interfere with the operation of radio-guided or remotely-

controlled devices or vehicles. We already have taken our first steps toward developing an environmental warfare capability. We are using satellite weather data from Tiros II for current, tactical operations and more accurate, long-range weather predictions. Some experiments in fog dissipation have shown promise, and some exploratory research has been conducted on ways to change the heading of major storms.

"For these reasons - and because our advances in science make it reasonable - we are now engaged in planning a ten-year, comprehensive study of the atmosphere, a study which we will designate ATMOS. This plan will be co-ordinated with our TENOC oceanographic studies."

NASA's Jet Propulsion Laboratory writes that the ATMOS [Atmospheric Trace Molecule Spectroscopy] program mentioned by Raborn was a, "space-borne investigation designed to obtain fundamental information related to the chemistry and physics of the Earth's upper atmosphere (20 to 120 km altitude)." My science advisor says that the ATMOS program was conducted at Two Rivers, near Fairbanks, Alaska at UCLA's HIPAS observatory noted in the previous chapter. By the early 1990s, the ATMOS program was about measuring atmospheric chemistry from remote sensing satellites. Although a 1961 report pertaining to the Navy TENOC (Ten Year Program in Oceanography) program also mentioned by Raborn did not contain any specific information pertinent to the New Manhattan Project. It did, however, make mention of another, classified TENOC report.

Vice Admiral W. F. Raborn, Jr.

The United States Navy, of which Mr. Raborn was a vice admiral, was one of the managers of the HAARP facility in Alaska. We learned in the previous chapter that HAARP is capable of controlling the weather. Further, in 1965, Vice Admiral Raborn became the director of the Central Intelligence Agency (CIA). During Raborn's CIA directorship, the Lyndon Johnson white house produced the first top-level document to assert both the theory of man-made climate change and the SRM geoengineering theses. Further still, in 1970 Mr. Raborn became a member of the board of directors of a company called LTV Electrosystems to which he also served as a consultant. The 'L' in LTV Electrosystems stood for Ling - James Ling (1922-2004). James Ling, in the days before he ran big businesses, served as a Navy electrician. In 1972 LTV Electrosystems' parent corporation Ling-Temco-Vought sold the company and it became E-Systems. Admiral Raborn stayed on the board of directors throughout. In the 1990s, E-Systems built the first version of HAARP. It appears that Vice Admiral Raborn's 'horizons' have been met and this is why we devote a whole chapter here to William Raborn.

William Raborn appears to have been central to the development of the New Manhattan Project. As we have seen, he was deeply involved in the development of the Project's defining element, electromagnetic energy, and specifically the HAARP antenna. As we will see, Raborn may have also came up with the plan used for the New Manhattan Project's continuing organization and development. Raborn appears to have been a most important administrator to the biggest scientific effort in Human history.

Polaris

In the mid-1950s, under their Special Projects Office, the United States Navy began developing a nuclear powered submarine capable of launching a ballistic (i.e. thermonuclear) missile while submerged. These efforts went by many different names and stages of development early on, but the effort that was ultimately successful was called Polaris. Early in this development, then Rear Admiral William Raborn took command. Prior to this assignment, Raborn had distinguished himself in the development of the Navy's Sparrow air-to-air missile.

The eponymous Polaris missile was developed by former Nazi scientists at the Army's Redstone Arsenal outside of Huntsville, Alabama, at Lockheed in Sunnyvale, CA, and at Aerojet-General facilities near Sacramento, CA. These Nazi scientists were brought to America as part of Operation Paperclip. In 1942 it was Nazi scientists, including the most famous Wernher von Braun (1912-1977), who first successfully launched a missile from a submerged submarine. The Polaris missile's development can be traced back to the Nazi V-1 rocket which terrorized Britain and other areas of Europe during

WWII. Author Harvey Saplosky provides us with details of the Polaris project's Nazi roots:

"The Germans, appropriately, first thought of combining a missile and a submarine. In the summer of 1942, technicians from the German Army Weapons Department experiment station at Peenemündee fitted a U-boat with short-range bombardment rockets and launched several salvos from depths of thirty to fifty feet. Refinements on the concept continued until several schemes for a submerged launching of the longer-range and heavier V-2 missile had been prepared. Jurisdictional disputes that arose between the German Army Weapons Department and the German Navy Weapons Department prevented the concept from receiving serious attention, and no operational equipment was prepared."

Saplosky continues:

"German weapon research ideas and even German scientists and engineers were absorbed into the American defense effort. Within months of the end of the war, parallel long-range missile projects, some based on V-2 technology, were established in the Navy and the other services. Soon included was a project to develop a cruise missile capable of being launched from a submarine."

During the production of the Polaris, Raborn carried out extensive public relations campaigns extolling the virtues of the project. The 1957 Russian launch of the Sputnik satellite served as a catalyst for the development of Polaris. Twice in the late 1950s, after Sputnik, the congress authorized several hundred million dollars more than what was requested for the development of the project.

Polaris submarine

Polaris missile launching from submerged Polaris submarine

Vice Admiral Hyman Rickover (1900-1986) was in charge of developing the propulsion system while Vice Admiral Raborn was in charge of the overall development. Raborn had limited technical experience and relied heavily upon his scientific advisors. A man by the name of Harold Brown was the designer of Polaris.

General Electric not only worked extensively with Ernest Lawrence and other former Manhattan Project scientists on isotope separation for U.S. Navy nuclear reactors as they developed the first nuclear powered submarines, they also worked with Hughes on the Polaris fire control systems. Edward Teller and Lawrence Livermore Laboratories (LLNL) developed the Polaris thermonuclear warhead. The development of the Polaris warhead was instrumental to the early growth of LLNL. There ended up being three iterations of the Polaris missile produced by LLNL: an A-1, an A-2, and an A-3. Boeing, Lockheed, Aerojet-General, and the Aviation Corporation of America (Avco) contracted for the development of the Polaris missile body with Aerojet-General developing the rocket motor. The guidance system of the Polaris missile was developed by Raytheon, General Electric, and the Instrumentation Laboratory of the Massachusetts Institute of Technology. As we can see, Raborn was running with the people that brought us the New Manhattan Project.

Between 1956 and 1967, $10,778,202,000 was spent on Polaris. In adjusted dollars, that's roughly $94B. That's about *three times* the cost of the original Manhattan Project and that's not even accounting for construction of facilities, military personnel, or tangential warhead development costs. From 1959 to 1964, yearly expenditures on Polaris accounted for about 10% of the Navy's total annual budget.

The first successful test firing of a submerged Polaris submarine occurred on July 20, 1960. On November 15, 1960 the first submarine-launched ballistic missile (SLBM) nuclear submarine produced by the Polaris project, the U.S.S. George Washington made its maiden voyage. By the end of 1960, two Polaris submarines were on patrol and twelve more were in production. Production accelerated throughout the early 1960s. A total of 41 Polaris subs were eventually produced, each carrying 16 missiles (656 total missiles). The Polaris program was considered an outstanding success as the system was deployed several years ahead of schedule without any cost overruns. Since 1964 there has been a joint US/UK Polaris project and Polaris submarines have been committed to NATO.

In March, 1962 William Raborn became Deputy Chief of Naval Operations (Development) - the head of naval research. Raborn then retired from the navy to become vice president for program-management at the aforementioned Aerojet-General.

Project B/S/S/S/ELF and Clam Lake

As noted in the previous chapter, project B/S/S/S/ELF involved the construction of a massive antenna for communications with deeply submerged submarines thousands of miles away. Being that our Vice Admiral Raborn developed a submarine which required these types of communications, it is reasonable to assume that Raborn knew about, or was involved in the B/S/S/S/ELF project. For these reasons, we will now take a closer look at the B/S/S/S/ELF project, Raborn's potential involvement, and the implications for the New Manhattan Project.

Before the advent of nuclear powered submarines such as those produced by the Polaris project, submarines powered by hydrocarbon combustion needed to stay partially submerged or relatively near the surface because internal oxygen supplies needed for propulsion and Human respiration became depleted quickly. Submarines powered by nuclear reactors were suddenly able to dive to depths previously unheard of and stay submerged for extended periods of time because nuclear reactions do not require oxygen. Our boys could dive deep, quickly becoming invulnerable, then spirit away unseen as nuclear subs are also much quieter than their diesel-burning predecessors.

Clam Lake station

But at the time of the first nuclear subs (Polaris had a couple of predecessors), our military had no way of communicating with a deeply submerged submarine. Submarines at or near the surface used similar communications systems as the ships of their day, but standard radio systems did not penetrate significant aquatic depths. Our military needed to devise a way to communicate with far-away, deeply submerged submarines. Enter the facilities at Clam Lake and project B/S/S/S/ELF.

It was at these facilities at Clam Lake, straddling the state line with Wisconsin and Michigan, where a way to communicate with deeply submerged submarines thousands of miles away was developed. Beginning in 1958, a series of projects were planned to establish that a deeply submerged submarine thousands of miles away may be sent messages with electromagnetic pulses. These electromagnetic pulses were eventually generated by Clam Lake's enormous grid of cables 28 miles long strung across the tops of telephone pole-like poles and buried 3 to 6 feet beneath the ground covering 4,000 square miles. The area sits on top of the Laurentian Shield - a large, flat rock formation which greatly increases the efficiency of the antennas. The pulses generated by these wires were bounced off the ionosphere and received by distant, deeply submerged nuclear submarines as ELF waves effectively penetrate water. As the submerged submarines had no way to respond, this was a one-way communications system. The submariners could respond when they eventually surfaced using standard communications systems.

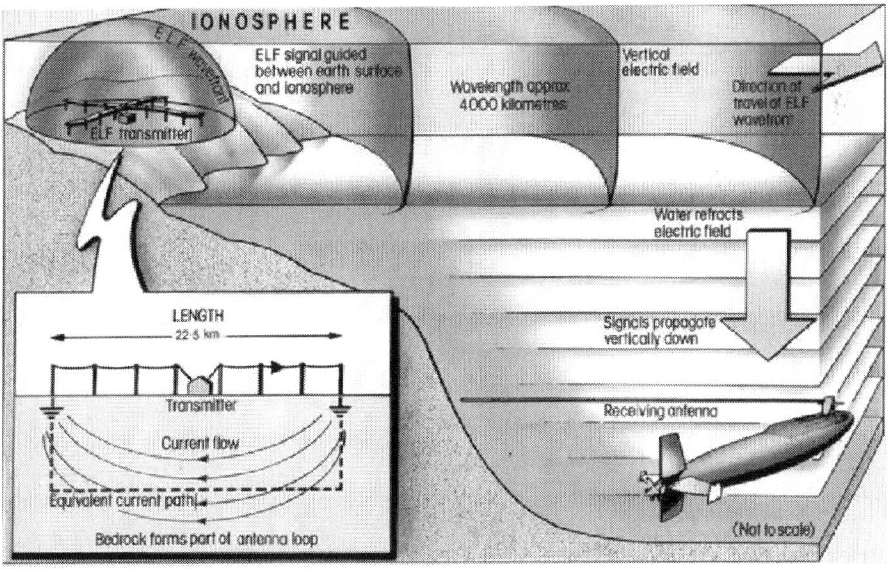

ELF submarine communications

The facilities at Clam Lake, as they are known to be able to bounce powerful signals off of the ionosphere, have played a part in the history of the development of today's ionospheric heaters which are known to be able to modify the weather. Therefore, the facilities at Clam Lake might also have been used in classified weather modification experiments. Although classified projects are known to have existed at Clam Lake, the only specific information that has been declassified concerns submarine communications. But weather modifications, such as making fog or rain for the purpose of troop obscuration, can be very strategic. Over-the-horizon (OTH) radar, as mentioned in chapter 2, is a part of the development of today's ionospheric heaters as well. So it is interesting to note that it was a Polaris ballistic missile that was remotely sensed as part of a breakthrough OTH radar experiment.

Raborn has a significant connection to Clam Lake. In late 1959 a company called Continental Electronics Manufacturing Company won the $50 million Navy contract to build the facilities at Clam Lake, beating out the Radio Corporation of America (RCA). Earlier in that year, the founder of Continental Electronics Manufacturing Company, a man named James O. Weldon, his wife, and another shareholder had sold Continental Electronics Manufacturing Company to a division of Ling-Temco-Vought called Ling-Altec for around $3.5M. James O. Weldon came along to run it. Weldon would later become chairman of the board of the aforementioned LTV [Ling-Temco-Vought] Electrosystems. Ling-Temco-Vought was a giant corporation, one division of which was LTV Electrosystems. As noted earlier, Raborn was on the LTV Electrosystems Board of Directors. Clam Lake contract winner Continental Electronics Manufacturing Company was a pioneer in the design and installation of extremely powerful radio transmitters for such customers as the CIA's Voice of America.

It is also interesting to note that in the same year that plans were being drawn up for Clam Lake, an early ionospheric heater was also getting off the ground. In 1958 Professor William Gordon of Cornell University proposed to the Advanced Research Projects Agency (ARPA) that they support the construction of the aforementioned Arecibo ionospheric heater in Puerto Rico.

The CIA and "Restoring the Quality of Our Environment"

On April 22, 1965 retired Vice Admiral Raborn became the Director of the Central Intelligence Agency (CIA). This was to be a short-lived directorship. By all accounts, Raborn was oblivious in the role. He was in over his head. He was the president's buddy, but stories of his inadequacy at the post are legion. 14 months later he resigned.

It is interesting to note that in November of 1965, during Raborn's CIA directorship, LBJ's Environmental Pollution Panel of the

President's Science Advisory Committee issued the landmark document titled "Restoring the Quality of Our Environment." "Restoring the Quality of Our Environment" was the first top-level U.S. Government document to state the theory of man-made climate change. The authors also theorized about all the bad stuff that could happen if the Earth's average temperature were to fluctuate by two degrees Fahrenheit. Not only did the authors of "Restoring the Quality of Our Environment" provide us with the problem, they also provided us with the solution. They wrote that small particles in the atmosphere could reflect sunlight back into space, thus cooling the Earth and saving us from the problem of climate change. Way back in 1965, the authors of "Restoring the Quality of Our Environment" advocated for the SRM geoengineering thesis which is code for today's New Manhattan Project. Although the CIA does not take credit for "Restoring the Quality of Our Environment," they have produced lots of similar material. In light of the totality of this information, doesn't it make sense that the CIA was involved in the production of "Restoring the Quality of Our Environment?" One of the document's co-authors, Gordon J.F. 'How to Wreck the Environment' MacDonald, was CIA.

After leaving the CIA, Raborn returned to the Aerojet-General Washington D.C. office for three years, then served as president of his own D.C. consulting firm called W.F. Raborn Co., Inc. until retiring in 1986. William Francis Raborn died in 1990.

PERT

During the development of the Polaris submarine, in early 1958, at the urging of his administrative assistant Gordon Pehrson, Raborn initiated a special study of potentially powerful new administrative techniques. The study group included representatives of the Special Projects Office, Lockheed, and the consulting firm Booz, Allen, and Hamilton. Ideas put forth by Pehrson formed the basis of what was worked out over the next six months. Raborn's study group came up with something called the Program Evaluation Review Technique (PERT).

The book *Polaris!* by Baar and Howard describes how, as part of PERT, computers were used to aid the administrative process. It reads:

"The steps needed to develop each part of the Polaris system were put on huge blueprints 15 to 20 feet long. The blueprints were

translated into numbers and fed into computers. The computers swiftly told Raborn when each job would be finished, and he could act on this information.

"Change a step. Lose a day. Lose an hour. Gain a week. Invent a new kind of anything. The new information was fed into the computers. The computers gave Raborn a revised estimate."

The book *The Secret War for the Ocean Depths* by Thomas Burns describes how complex charting was also part of the PERT. It reads:

"The top layer of the chart covered the entire system. The next layer was subdivided into 5 charts, one for each major component of the system. The next layer had 50 charts, and the next layer 100 - each in the same format, each detailing a single activity or project and each tracing a direct relationship to the achievement cycle and timing of the overall program. Each of the charts showed all of the steps required to track a particular job and a timetable based on them. The charts showed who was responsible for seeing that each step was completed. Vertical black lines on the graph showed how the timing coincided with the flow of project activity. Colored markers were placed beside the steps to indicate where the project was with respect to schedule: a blue mark ahead, a red triangle behind, a green circle on schedule, an orange square an uncertainty - and so on."

PERT caught on quickly and to this day is still used extensively. American industry moved rapidly to adopt the PERT, including small businesses. Saplosky writes:

"Within the government, eight different PERT systems were reported in use by 1962 in the Army, Air Force, Navy, and National Aeronautics and Space Administration. Outside the government, consulting firms were established solely on their knowledge of PERT techniques and soon thirty or so versions of PERT were being sold. The peak year for the introduction of private and public PERT substitutes, 1962, became known in the aerospace industry as the 'Year of the Management Systems.'"

The PERT is duly noted here because it may have been subsequently applied to the New Manhattan Project. Since Raborn and his people were so successful with PERT during Polaris, then it stands to reason that the same successful administrative techniques would be applied, especially if Raborn became the project manager for the NMP. Further, PERT systems became required for programs conducted by government agencies known to have been involved in the development of the NMP such as the Department of Defense, the National Aeronautics and Space Administration (NASA), and the Atomic Energy Commission.

Many consider the development of the PERT system as more important than the development of Polaris. PERT has been widely credited with cutting two years off the development of the Polaris project. At the time of the first missile launched from a Polaris sub, the press coverage of the PERT rivaled that of the Polaris project.

References

"New Horizons of Naval Research and Development" an article by William Francis Raborn, published in the *U.S. Naval Institute Proceedings*, January, 1963

"Atmospheric Trace Molecule Spectroscopy (ATMOS)" a paper by C.B. Farmer and the Jet Propulsion Laboratory
"ATMOS: Long-term Atmospheric Measurements for Mission to Planet Earth" a report by NASA, technical memorandum 108720, 1992

"Ten Year Program in Oceanography: TENOC" a report by the U.S. Navy, March 13, 1961

Air America: The Explosive True Story of the CIA's Secret Airline a book by Christopher Robbins, published by Cassell, 2012

"One Vast Conglomerate" an article by Don A. Schance, published in the *Saturday Evening Post*

U.S. patent #4,686,605 "Method and Apparatus for Altering a Region in the Earth's Atmosphere, Ionosphere, and/or Magnetosphere" by Bernard J. Eastlund, assigned to APTI Inc., 1987

The Secret War for the Ocean Depths: Soviet-American Rivalry for Mastery of the Seas a book by Thomas S Burns, published by Rawson Associates, 1978

Polaris! a book by James Baar and William E. Howard, published by Harcourt, Brace, and Company, 1960

History of the Office of the Secretary of Defense: Vol 4: Into the Missile Age 1956-1960 a book by Robert J. Watson and the Department of Defense, published by the Department of Defense, 1984

Von Braun: Dreamer of Space Engineer of War a book by Michael J. Neufeld, published by Random House, 2007

"The Polaris System Development: Bureaucratic and Programmatic Success in Government" a report by Harvey M. Sapolsky, published by Harvard University Press, 1972

"30 Years of Technical Excellence: 1952-1982" a report by Lawrence Livermore National Laboratory, published by Lawrence Livermore National Laboratory, 1982

Hearing Before the Committee on Armed Services, United States Senate, Eighty-ninth congress, first session on the nomination of Vice Admiral William F. Raborn, Jr., USN (retired) to be director of Central Intelligence, April 22, 1965

Call Me Pat: The Autobiography of the Man Howard Hughes Chose to Lead Hughes Aircraft a book by L.A. 'Pat' Hyland, published by The Donning Company Publishers, 1993

Hughes After Howard: The Story of Hughes Aircraft Company a book by D. Kenneth Richardson, published by Sea Hill Press, 2012

The Creative Ordeal: The Story of Raytheon a book by Otto J. Scott, published by Atheneum, 1974

The Education of a College President: A Memoir a book by James R. Killian, Jr., published by the MIT Press, 1985

Teller's War: The Top Secret Story Behind the Star Wars Deception a book by William J. Broad, published by Simon and Schuster, 1992

The General Electric Story: A Heritage of Innovation 1876-1999 a book by the Hall of Electrical History, Schenectady Museum, published by Hall of Electrical History Publications, 1999

"NATO Facts and Figures" a report by the North Atlantic Treaty Organization Information Service, published by the North Atlantic Treaty Organization, 1969

"Extremely Low Frequency Transmitter Site Clam Lake, Wisconsin" a report by the United States Navy, published by the United States Navy, 2001

Ling: The Rise, Fall, and Return of a Texas Titan a book by Stanley H. Brown, published by Beard Books, 1999

The Rise and Fall of Air Force Cambridge Research Laboratories a book by Edward E. Altshuler, self-published, 2013

LTV Electrosystems Inc. v. National Labor Relations Board U.S. Supreme Court Transcript of Record with Supporting Pleadings

"Restoring the Quality of Our Environment" a report by the Environmental Pollution Panel of the President's Science Advisory Committee, published by the U.S. Government Printing Office, 1965

"Raborn Resigns as Director of CIA" a newspaper article by Laurence Burd, published in the *Los Angeles Times*, June 19, 1966

"Former CIA Director William Raborn Jr. Dies" an article by Bart Barnes, published in the *Washington Post* March 13, 1990

Chapter 5
The CHEMTRAIL FLEETS

Today's chemtrail spraying operations necessarily require a large number of jumbo jet airliners operating covertly. The volume and frequency of chemtrail reports from all over the world, the author's own observations and the massive task of controlling Earth's weather in the fashion of the New Manhattan Project suggest that these planes number in the thousands.

Evidence presented in this chapter indicates that there is a proprietary, specialized fleet of aircraft designed for the constant spraying of chemtrails *and* the evidence indicates that common commercial jet aircraft have been outfitted with equipment which allows them to participate in the spraying program. These two chemtrail fleets are referred to here as the 'smaller chemtrail fleet' and the 'larger chemtrail fleet,' respectively. Being that the history of the smaller, specialized chemtrail fleet goes back further, we will examine that history here first. The second section of this chapter covers the larger chemtrail fleet. Not only all that, but evidence indicates that jet fuel is being adulterated as well. We have a lot of information to go over here, so let's begin.

The smaller, dedicated chemtrail fleet

The aircraft of the smaller New Manhattan Project chemtrail fleet have been specifically designed to constantly and exclusively participate in the New Manhattan Project, and therefore they are more effective than those of the larger chemtrail fleet. The aircraft of the smaller, dedicated chemtrail fleet are the workhorses of the Project and probably number in the hundreds.

These planes must necessarily be of a certain breed. Any old plane rigged up with some spraying equipment or even with spiked jet fuel simply will not do. A commercial passenger airliner following a predetermined route is not nearly as effective as a dedicated chemtrail spraying plane. An abundance of chemtrail spray needs to be emitted at specific locations at a moment's notice. The super high-tech nature and payload requirements of the New Manhattan Project demand specialization. The fuselage of these most effective type of chemtrail spraying aircraft needs to be loaded up not with passengers and luggage, but with chemtrail spray, spraying equipment, communications gear, computers and atmospheric monitoring equipment.

As we are interested in exposing chemtrails, we are interested in the origins, development, and current state of this covert, high-tech, jumbo jet air force routinely contaminating our environment and damaging our health. Not surprisingly, when one looks back in history for fleets of covertly operating American aircraft, one finds plenty. It all began in 1940s China.

Many of us have heard of or seen the 1942 war epic starring John Wayne called *Flying Tigers*. It is a finely crafted propaganda piece based on the real Flying Tigers. The real Flying Tigers were an officially sanctioned mercenary air force operating out of Burma and consisting of former U.S. military personnel fighting for Chiang Kai-shek's Nationalist Chinese military against imperial Japan in 1941 and 1942. They were officially known as the American Volunteer Group (AVG). Although they did not operate beyond the purview of your average American (in fact, the media coverage was quite impressive), the Flying Tigers were the beginnings of what eventually became today's top-secret, proprietary New Manhattan Project chemtrail fleet.

Claire Chennault

The Flying Tigers was co-founded by a man named Claire Chennault (1893-1958) and another by the name of William D. Pawley (1896-1977). Chennault was the military aviation commander and Pawley was the adventurous, entrepreneurial executive. In *Flying*

Tigers the film, Commander Chennault is played by John Wayne. Pawley later worked closely with the CIA.

Mr. Pawley set up a company called the Central Aircraft Manufacturing Company (CAMCO) which operated in Burma and supplied Curtiss-Wright aircraft to the Flying Tigers. CAMCO was owned by Intercontinent Corporation - a jointly owned Chinese and American company. A vice president of the Intercontinent Corporation, a former naval officer by the name of Bruce G. Leighton (1892-1965), had the ear of some high Navy brass in Washington. The Flying Tigers were also supported by Chiang Kai-shek's Chinese government and an *ad hoc* political lobby in Washington D.C. doing business through a firm called the Universal Trading Corporation. This *ad hoc* lobby consisted of men such as: Whiting Willauer (1906-1962), brothers David and Thomas 'Tommy The Cork' Corcoran (1900-1981), Bill Youngman, Joseph Alsop (1910-1989), Quinn Shaughnessy, and Lauchlin Currie. Many of the members of this so-called Washington Squadron were graduates of Princeton University. The mascot of Princeton University is a tiger and this is how the Flying Tigers got their name.

William Pawley

Lauchlin Currie (1902-1993) was one of the leading Keynesian economists of the New Deal era. During the formation of the Flying Tigers, he was an administrative assistant to President Franklin Roosevelt (FDR) and was directing lend-lease operations for China. Currie's earliest involvement with the Flying Tigers dates back to April of 1941 when he sent Owen Lattimore (1900-1989) to China in order to meet with Chiang Kai-shek to promote civilian and military programs including the Flying Tigers. Currie's biographer Roger Sandilands writes, "The supply program to Claire Chennault's American Volunteer Group, or 'Flying Tigers,' was one of Currie's major preoccupations. Chennault enjoyed the confidence of the Chiang family, who favored the air force, and in Washington, Currie was asked to support a supply program for Chennault and also to organize a training program in Arizona for thousands of Chinese pilots." Sandilands continues, "Proposals that stressed the need to expand the air force in China under Chennault's command, and its possible use for offensive as well as defensive missions, had been endorsed by President Roosevelt in December 1940." Regarding the training program for thousands of Chinese pilots in Arizona, this was carried out at the then Marana Army Air Field. As this section unfolds, we will have much more about Marana.

Shortly after the end of WWII, Currie launched a firm engaged in consultancy and the export-import business called Lauchlin Currie and Co.. Currie maintained his Washington connections through the Council on Foreign Relations (where he was a member) and something called the National Planning Association, a non-government organization established in the early 1940s to promote Keynesian fiscal policies. Currie was also a board member of something called the Council for Italian American Affairs. The Council for Italian American Affairs supported Italian anti-communist efforts in 1947 & 1948.

Lauchlin Currie

In his biography of Curtis LeMay, Warren Kozak describes the pro-China lobby in FDR's White House as being 'led' by Time/Life publisher Henry Luce (1898-1967). Luce was born in China to missionary parents.

In his 2012 biography of William Pawley, author Anthony Carrozza gives us a detailed look at the American corporations that supported the Flying Tigers: the aforementioned Universal Trading Corporation and an associated business known as China Defense Supplies, Inc.. Carrozza writes:

"The Universal Trading Corporation was indeed a good setup. Headquartered at Rockefeller Center in New York City (the location of Intercontinent's office), Universal was manned by former U.S. government employees, and its sole purpose appeared to be China's go-between purchasing agent for war materials from the United States and other countries. The corporation had handled the transactions of repaying the $25 million U.S. loan in 1939, with exports of metals and minerals, and using the funds to purchase war supplies even though Universal was forbidden to do so by the China Trade Act.

"When the United States began a more aggressive program of military aid to China, China Defense Supplies, Inc. (CDS) - a new private corporation chartered in Delaware - was entrusted with the funds being advanced. Organized by [T.V.] Soong at the request of Madame Chiang, CDS set up its offices at 1601 V Street, Northwest, in Washington, D.C.. As China's appointed purchasing agent and its sole client, CDS attracted a roster of officers that included the cream of the Washington elite. Tommy 'the Cork' Corcoran served as legal counsel and his brother David was installed as president. David served as the manager for General Motors Corporation in Tokyo but resigned after Japan invaded Manchuria in 1931. In 1941, David, along with Nelson Rockefeller, formed the Sydney Ross Company, a subsidiary of Sterling Drugs that supplied CDS with medical drugs throughout the war, and later, quinine, caffeine, and drugs to South America. Frederic A. Delano, FDR's uncle, was named a director of CDS; William Brennan was its congressional liaison; and Whiting Willauer, a reserve officer in naval intelligence (and Chennault's future partner in the postwar Civil Air Transport, which evolved into Air America) was the corporate secretary. Others serving CDS included Gordon Tweedy,

William Youngman from the Federal Power Commission who acted as general counsel and a go-between for Chaing Kai-shek, and Quinn Shaughnessy, a Harvard lawyer and Marine Corps intelligence officer who had worked for the Reconstruction Finance Corporation's legal division."

The Rockefellers' Sterling Drugs was connected to I.G. Farben.

Of course, the Flying Tigers operation, an American air force waging war for China against a foreign power, was totally illegal. It violated the Neutrality Act. But on April 15, 1941, in support of the Flying Tigers, President Roosevelt signed a secret, unlisted executive order which mollified the situation by authorizing a private corporation holding a contract with a foreign government to hire U.S. military officers. The fact that many other ethical and legal issues remained with these U.S. military officers engaging an enemy for a foreign power was hence made irrelevant. Roosevelt supported China and the Flying Tigers and Roosevelt was the President. Henry Luce made it look good in the media, and that was that.

In 1942, the Flying Tigers were disbanded and integrated into the U.S. Army Air Force when the United States armed forces officially entered the ongoing China/Japan conflict in support of Chiang Kai-shek's Nationalist Chinese. After the dropping of the atomic bombs and the subsequent Japanese surrender in 1945, U.S. troops were pulled out of the theatre. The aforementioned *ad hoc* lobby in Washington D.C., along with some new players from both America and China, then formed something called Civil Air Transport (CAT).

Let us reference a passage from *Flying Tigers: Claire Chennault and his American Volunteers* by Daniel Ford. Mr. Ford writes:

"*After the war, Chennault continued in the service of the Chiangs, organizing Civil Air Transport with Whitey Willauer and others of the AVG [American Volunteers Group] 'Washington Squadron.' CAT started with mercy flights and evolved into a paramilitary force during the civil war that ended with Chiang Kai-shek's ouster by the Communists in 1949.*"

Christopher Robbins, author of *Air America*, describes the founding of Civil Air Transport this way:

"*Chennault returned to China to found an airline in which he undertook to move the desperately needed relief supplies that had been sent out by the United Nations Relief and Rehabilitation Administration - UNRRA - but which were accumulating on docksides and in warehouses along the coast instead of being delivered to the interior. In return for flying supplies into the country, the airline was given permission to fly commercial goods out.*"

The ‘commercial goods’ flown out of China were most notably opium.

During the ouster of Chiang Kai-shek's Nationalist Chinese to the island then known as Formosa in late 1949, CAT airlifted thousands to safety. Once on the island we today call Taiwan, in early 1950, CAT established their headquarters in Taipei and began operating out of Songshan Airport; more commonly known today as Taipei International Airport. A large maintenance base was also established at Tainan in southwestern Taiwan.

It was also in early 1950 when the Central Intelligence Agency (CIA) became more extensively involved in CAT’s operations. In early 1950 CAT was broke. With some help from Major General Lyman Lemnitzer (1899-1988), the CIA bailed out CAT in March with an option to purchase. The CIA then began inserting lots of their agents into the operation. They were assigned to the traffic and sales, public relations, and financial divisions of the company. The author of *The Flying Tiger*, Jack Samson writes, “The infusion of CIA personnel into the airline was done so slowly and so subtly that many regular employees never suspected the company was becoming a front for covert activities. To look legitimate, CAT continued its regular commercial operations and many employees had no idea it was anything but a cargo and passenger airline.” This is all not surprising, as William ‘Wild Bill’ Donovan (1883-1959), the head of the CIA’s predecessor, the Office of Strategic Services (OSS), had been involved since the days of the Flying Tigers.

By June of 1950, the CIA had purchased CAT outright. It was reorganized as CAT Inc., under the CIA-controlled Airdale Corporation. Chennault was made chairman of the new CAT Inc. Whiting Willauer became president. A man by the name of James J. Brennan became a financial vice president. CIA man Al Cox became vice president for management and CIA man Bob Terhaar became treasurer. Operational management of CAT fell to a man by the name of Joseph C. Rosbert who later went on to become one of the original financial backers of Robert Prescott's Flying Tiger Line.

While the CIA owned the Airdale Corporation, a group of investors owned stock in CAT Incorporated itself. CAT Incorporated was partly owned by American and Chinese interests. The American ownership consisted of: a company called Rio Cathay (represented by Thomas Corcoran, David Cocoran, and William Youngman) with 28.88%, Whiting Willauer with 17.64%, Claire Chennault with 14.46%, and James J. Brennan with 8.46%. The remaining 30.6% of the company was owned by Chinese interests consisting of: a man named Wang Wen-san, another man named L.K. Taylor, Yunnan People's Development Corporation, and the Shensi Provincial Government. Four years later, the Chinese-owned portions of CAT split off, becoming another company, called Asiatic Aeronautical Company Limited which continued to do contract work for CAT.

Throughout the 1950s CAT continued its combination of normal commercial business as well as clandestine ops. During the 1960s the CIA relegated CAT to a 'cover' role, flying passenger routes extending from Tokyo to Bangkok. An accident and the resulting controversy over responsibility finally put CAT out of business in 1968.

Believe it or not, CAT performed weather modification operations. This is the author's thesis. Christopher Robbins writes:

"*...during a severe drought in Japan, CAT was chartered to carry a group of rainmakers over Osaka in an effort to sow the clouds with silver iodide. A fine sprinkle of rain fell over the countryside as a result, but the skeptical farmers were more inclined to credit nature than CAT for the phenomenon. On the next trip, the pilot took the precaution of mixing a little emerald-colored dye with the other chemicals. Startled Japanese were treated to a chlorophyll cloudburst and CAT collected wherever it rained green.*"

The reorganization of CAT in 1950 was actually more complicated than what has been detailed above. When the CIA took full control of CAT in 1950, the next generation of CIA proprietary airlines collectively known as Air America was also created. Instrumental in the creation of Air America was a man by the name of George A. Doole, Jr. (1909-1985), an experienced airline executive and longtime CIA employee.

George A. Doole

Persuaded by his longtime friend, Air Force General Charles P. Cabell (1903-1971), Doole accepted the assignment. Under something called the Pacific Corporation, Doole reorganized the former CAT's operations into three segments: CAT, Air America, and Air Asia. As noted previously, CAT was mostly responsible for standard commercial business such as passenger and freight operations, Air America handled covert operations, and Air Asia ran the aforementioned maintenance facility in southwestern Taiwan. Doole's efforts were a smashing success and he brought these CIA air proprieties to self-sufficiency, if not profitability - apparently even without the profits derived from the shipping of contraband.

Before running the Pacific Corporation, in the early 1940s, Doole had been the chief pilot for a company called Avianca which was previously known as SCADTA; a German-owned airline in Columbia. In 1946 Doole gave up flying and became Pan Am's regional director for the Middle East and Asia. The author of *Perilous Missions*, William Leary writes, "A colonel in the Air Force Reserve, Doole was called to active duty in 1951 and served under Cabell as chief of estimates for the Middle East, working closely with the CIA." Doole was a good company man. He denied any involvement with the CIA until at least 1970.

Although it was, in itself, an airline, Air America is also a generic term for many airlines owned and operated by the CIA during the time that Air America the airline was in operation. CIA airlines such as Intermountain Aviation, Air Asia, Southern Air Transport, the aforementioned Civil Air Transport and others were all organized under the aforementioned Pacific Corporation and were all often referred to as 'Air America.' The Pacific Corporation ran airlines that collectively were larger than any other single airline in the world at the time. In the 1960s and 70s, the operation was worth hundreds of millions of dollars. By 1973, near the end of their reign, Air America had contracts with the defense department worth 41.4 million dollars.

Pilots of the airlines organized under the Pacific Corporation smuggled everything from currency to drugs to precious metals to human beings. Air America's slogan was, "Anything, anywhere, anytime." Of these different contrabands, opium was the most prevalent.

At least one person has experienced an Air America that leaves any reasonable individual with a lot of questions. The following passage is from an Asian language magazine quoting a former Air America stewardess as recounted by Christopher Robbins, "Before this, I flew with Air America, but that was different, very different. Once, I went to work and had to fly on a plane full of dead bodies . . . better not say anything about that."

The aforementioned Southern Air Transport was involved in the Iran-Contra scandal. Investigative journalist Pete Brewton writes in his book *The Mafia, CIA, & George Bush*:

"Southern Air Transport was a CIA proprietary until 1973, when it was sold to the man who had been fronting the ownership for the CIA. The airline company was later sold to James Bastian, who had been an officer in Air America's parent company [the Pacific Corporation] and Southern Air Transport's in-house lawyer.

"When Southern Air's role in the Iran-Contra scandal was exposed - it had been the airline of choice to haul arms to Iran and to the Contras - Bastian denied any connections to the CIA. What he couldn't deny were connections to the CIA's Iran-Contra cutout, Oliver North,

who went to his old buddy Oliver 'Buck' Revell, then Assistant FBI Director (and now special agent in charge of the FBI's Dallas office), to try to stop an FBI investigation into Southern Air Transport's suspicious activities."

In 1972 the Director of Central Intelligence, Richard Helms (1913-2002) had concluded that the CIA no longer needed a large, covert airlift capability in Asia and ordered the agency to divest itself of ownership and control of Air America and related companies - but the planes kept flying. In 1973, the secret war in Laos ended and a major portion of Air America's business operations vanished with it. In 1974, all the airlines of the Pacific Corporation were finally disbanded due to federal legislation passed as a result of congressional hearings conducted by Idaho Senator Frank Church. Shortly before it was broken up, the Agency's Pacific Corporation and all its proprietary airlines employed about 20,000 people, more people than the CIA itself. The Church Committee Hearings also uncovered that the Pacific Corporation was majority-owned and controlled by Chinese interests.

As previously noted, one of the airlines formerly under the Pacific Corporation was Intermountain Aviation. In 1975, after the Pacific Corporation had been broken up, Evergreen International bought Intermountain Aviation. Evergreen's president and founder Delford Smith (1930- 2014) got a sweetheart deal. Intermountain's 12 aircraft were repainted in the colors of Evergreen International Aviation, Incorporated. The aforementioned Intermountain Marana base at Pinal Air Park in Marana, Arizona became the Evergreen Air Center and their new headquarters. Most Intermountain employees and executives stayed on to become Evergreen employees and executives. Most notably, on the board of directors at Evergreen was none other than the former boss of the CIA's Pacific Corporation, George A. Doole, Jr. Mr. Doole stayed on Evergreen's board of directors until his death in 1985.

Evergreen's accounting firm Arthur Andersen reported that assets almost doubled from the Intermountain acquisition. Even though *The Oregonian* reports that the Intermountain acquisition cost Evergreen less than three million dollars, assets went from a reported $25 million in 1975 to a reported $45.5 million in 1976. *The Oregonian* newspaper did a series of investigative reports on Evergreen and all their shady connections back in 1988.

It appears that, as a result of the Intermountain/Evergreen deal, much of the former Air America became Evergreen. Christopher Robbins writes that shortly after the Intermountain acquisition, "The company [Evergreen] expanded rapidly to own almost 100 aircraft operating on four continents." In 1979, the company headquarters were moved to Newberg, Oregon and then in 1981, moved again to McMinnville, Oregon.

Delford Smith was a master airplane trader and this is how he made the majority of his money. In the book *The Evergreen Story* by Bill Yenne, Evergreen founder and president Del Smith recalls, "'It turned out that 66 percent of the profit made over the 25 years from 1960 through 1985 was in asset sales, and 33 percent was operating profits.'" Evergreen Aircraft Sales and Leasing was established in 1983 for the purpose of buying and selling aircraft and aircraft parts. In 2008, Evergreen Aircraft Sales and Leasing was renamed Evergreen Trade.

Evergreen Supertanker

Delford Smith

Evergreen's business activities, aside from trading airplanes, included charter passenger and cargo service. Most notably, though, they also did a lot of spraying or dumping substances from aircraft. *The Evergreen Story* by Bill Yenne states, "...75 percent of its [Evergreen's] revenue flying hours in the mid-sixties were devoted to application [spraying] work..." Evergreen's mechanics even designed and patented aircraft spraying equipment.

Evergreen did a lot of work for the United Nations (UN). The operations the UN was bankrolling often involved the spraying of insecticides in locations all over the world. Just as today's geoengineers say they will save us from the dreaded global warming, one can rest assured that Evergreen's global spraying operations were all humanitarian efforts as well. Starting in 1974, Evergreen sprayed insecticides in Africa for the UN's River Blindness Control Program.

Evergreen also provided support to the UN's River Blindness Control Program in Nepal. Evergreen has ferried cargo into Africa for the UN's High Commission for Refugees and the UN's World Food Program. Evergreen has also supported the UN's 'peacekeeping' efforts in: Angola, Liberia, Mozambique, Somalia and the Western Sahara.

In 2014 Evergreen Aviation International filed for bankruptcy and the company has since been sold. At the age of 84, Evergreen founder Del Smith passed away on November 7 of that same year. We'll have more about Evergreen Aviation and their Evergreen Maintenance Center a little later.

In 1975, the year Evergreen acquired Intermountain, two CIA agents arranged a deal that allowed Evergreen to participate in an operation known as Condor. Operation Condor was a U.S. State Department program involving seventy-six planes and aid worth at least $115 million. The airplanes of Operation Condor were supposed to eradicate coca plant production by the aerial spraying of herbicides. Evergreen's part in the program involved supplying airplanes and pilots as well as the training of Mexican pilots. But, as with much of what the CIA does, all was not as it seemed. According to Peter Dale Scott and Johnathan Marshall, "DEA [Drug Enforcement Agency] learned that pilots were either spraying fields with water or unloading their herbicides in the desert. Informants reported that some Mexican officials used the planes for joy rides and pleasure trips while other officials shook down drug cultivators in exchange for protection from spraying."

It is for the combination of these reasons: their successful trading of aircraft, their spraying programs, their company lineage, their continued association with the CIA, and more, that we connect Evergreen Aviation with the New Manhattan Project. Evergreen's aforementioned maintenance center in Marana, Arizona is a prime suspect for having produced the aircraft of both the smaller *and the larger* New Manhattan Project chemtrail fleets. We are not nearly done with Evergreen.

~ ~ ~

Evidence indicates that today's smaller chemtrail fleet consists (at least partially) of overhauled, older-model large jet aircraft. In congressional testimony, geoengineer Alan Robock suggested retrofitting existing U.S. Air Force planes. Also, Aurora Flight Sciences published a 2010 cost analysis study titled "Geoengineering Cost Analysis" which states, "The goal of this study is to use engineering design and cost analysis to determine the feasibility and cost of a delivering [sic] material to the stratosphere for solar radiation management (SRM)." In this report, the authors write about utilizing different kinds of aircraft, including blimps, to spray stuff into the atmosphere. Of readily available solutions, it was found that retrofitting old jumbo jet airplanes would be the most cost effective choice. Not only all that, but some chemtrail fleet planes have been expertly identified as retrofitted jumbo jets. Allan Buckmann is a former United States Air Force weather observer who worked on the Tiros III weather satellite system with NASA, RCA and the U.S. Navy. In a 2012 *ActivistPost.com* article titled "Chemtrail Whistleblower Allan Buckmann: Some Thoughts on Weather Modification," he wrote that he has repeatedly witnessed Boeing KC-135s, Boeing 707s and Lockheed C-141s spraying chemtrails. In the first chapter, the KC-135 was also identified by an air traffic controller as being a plane that regularly sprays chemtrails.

In total, the preponderance of evidence indicates that Boeing KC-135s and 707s as well as Lockheed C-141s have been refurbished and retrofitted to become airplanes of the smaller, dedicated chemtrail fleet. These three models of aircraft are known as 'tanker' jets or 'supertankers.' They are referred to this way due to their sizable fuselages which provide a large cargo carrying capacity. Being that the planes of the New Manhattan Project necessarily need to be loaded up with electronics and lots of chemtrail spray, it makes sense that planes with lots of interior space and hauling capacity are used.

If one is to assemble a fleet of refurbished supertanker aircraft, one needs a source of suitable planes. The largest repository of aircraft in the world is located at the Davis-Monthan Air Force Base in Tuscon, Arizona. It is called the 309th Aerospace Maintenance and Regeneration Group. It is commonly referred to as the Davis-Monthan boneyard; or simply 'The Boneyard.' This is where old KC-135s, 707s and C-141s (among many others) go to die. That is, unless they are re-animated. This facility probably provided at least a partial source of the planes used in today's smaller New Manhattan Project chemtrail fleet.

The Davis-Monthan Boneyard

The developmental timeline of the New Manhattan Project suggests that these offending airplanes were retrofitted beginning in about 1980. As evidenced by reductions in government spending, the mid-1970s was around the time that the initial flurry of basic scientific research in support of the New Manhattan Project began to wind down. This suggests that by the mid-70s, the majority of the basic atmospheric research necessary was completed and it was time to transition more into the New Manhattan Project's implementation phases such as assembling a fleet of jumbo jets.

If we give development of these airplanes 5 years, then the bulk of these planes were probably regenerated starting in about 1980. We will have more about development shortly. Being that this is a continuing and probably expanding program, these aircraft regeneration activities may continue to the present day.

More evidence for the assertion of this timeline has been provided by the former head of Lockheed's Skunk Works, Ben Rich. The Skunk Works produced airplanes. In his book *Skunk Works*, Rich notes a sudden dearth of airplane production materials and personnel in the early 1980s. He writes, "...I suddenly found myself on the short end of materials, subcontracting work, machine shop help, and skilled labor. Without warning, there was a dire shortage of everything used in an airplane. Lead times for basic materials stretched from weeks to literally years." Production of the New Manhattan Project chemtrail fleet probably contributed to these deficiencies.

Further evidence for the assertion that large-scale regeneration of aircraft to be used as the New Manhattan Project's original, dedicated chemtrail fleet began around 1980, is found in a book by Tim Weiner titled *Blank Check*. Weiner writes of an explosion in the U.S. military's black budgets at the beginning of the 1980s. Our military's black budgets have probably provided significant funding for the NMP overall and production of its fleet of proprietary aircraft would probably be the single most expensive aspect. With the production of the proprietary chemtrail fleet, production of many other aspects of the NMP would probably need more funding as well. Weiner writes:

"Something new transformed the black budget as the 1980's began. A government obsessed with secrecy began to conceal the costs of many of its most expensive weapons. It enshrouded them in the deep cover once reserved for espionage missions. Under that cover, the black budget exploded. By 1989, at the end of the Reagan administration, it had grown to $36 billion a year. It was bigger than the federal budget for transportation or agriculture, twice the cost of the Education Department, eight times more expensive than the Environmental Protection Agency. It was bigger than the military budget of any nation in the world, except the Soviet Union's."

Weiner goes on to note that in the early 1980s, while the Pentagon's overall budget doubled, our military created hundreds of black programs. Specifically, during this time, we saw the biggest increases in spending on secret research and development which multiplied sixteen fold in the 1980s. Weiner describes the CIA as being, "At the center of this realm."

Also of note to our discussion, 1981 was the year that George Herbert Walker Bush (Bush Sr.) assumed the vice presidency. Members of the Bush family such as Bush Sr. are strongly implicated in the development and production of the New Manhattan Project.

So, we have a large source of planes, the types of planes, and a time frame: the Boneyard, KC-135s, 707s and C-141s, and 1980-today respectively. Let us apply this information. Some Brits by the names of Barry Fryer, Danny Bonny, and Martin Swann, for a long time now, have been recording and publishing the Boneyard's airplane inventories. Their books are compiled from Davis-Monthan's publicly available records. Let's take a look at them.

For the years 1982-2005, the books by Fryer, Bonny and Swann show: a company called Tuscon Iron and Metal received 24 Boeing 707s and 54 Lockheed C-141s; something called National Aircraft Incorporated received 64 Boeing 707s and 25 Boeing KC-135s; an outfit calling itself HVF West took on 99 Lockheed C-141s; and lastly, Tinker Air Force Base got 13 KC-135s. The Tuscon Iron and Metal website says that they are a metal recycling company. Ostensibly, Tuscon Iron and Metal recycled the 24 Boeing 707s and the 54 Lockheed C-141s they got between 1982-2005 and turned them into sheet metal, pipes and the like. According to what is available online, Tuscon's National Aircraft Incorporated looks like some type of small, private boneyard near the big boneyard at Davis-Monthan. God knows what they did with the 64 Boeing 707s and the 25 Boeing KC-135s they received between 1982 and 2005. Maybe they turned around and sold them to a regeneration center as a way to launder the planes. It's just a thought. HVF West is another outfit headquartered in Tuscon, AZ. They have a decent website. They say their specialty is the 'demilitarization' of old military aircraft. This involves dismantling the plane piece by piece, selling valuable parts and recycling recyclable metals. So, ostensibly this is what they did with the 99 Lockheed C-141s they received between 1982 and 2005. Coming up shortly in the 'production' section, we'll have more about Tinker AFB and the 13 Boeing KC-135s they got between 1982 and 2005. Yes, it appears that

Tinker AFB may have churned out some of the New Manhattan Project's proprietary aircraft. Can you smell what I'm cooking?

~ ~ ~

Evidence and logic suggest that the aircraft of the New Manhattan Project's smaller chemtrail fleet are operated remotely as drones or as the military calls them: unmanned aerial vehicles (UAVs). Remotely piloted aircraft are advantageous in the context of the New Manhattan Project because pilots are full of liabilities. Pilots have ethics. Pilots might blab about the project - and, as we saw in the first chapter, at least one already has. Pilots can only fly for a certain number of hours. Pilots are unionized. Pilots are expensive. Many robot planes could be controlled by a single operator or a centralized computer. Overall, robot planes would make the project run more smoothly.

Why not? The conventional weather modification industry is starting to use them. According to *AccuWeather.com*, in late 2013 the Federal Aviation Administration selected six states for participation in a pilot program using drone aircraft for the purpose of spraying the conventional weather modification industry standard, silver iodide. Although we know that silver iodide is not what the New Manhattan Project aircraft spray, the concept is the same.

Jumbo jet airplanes such as those used in the New Manhattan Project have been capable of remote operation for a long time now and the history of American drone aircraft dates back to World War One. Late in WWI, the aforementioned Henry 'Hap' Arnold teamed with a group of civilian scientists to produce the first guided missile known as the 'Flying Bug.' The Flying Bug was capable of carrying 200-300 pounds of explosives. It was launched from a wagon-like contraption that ran along a section of portable track. Once launched, the Bug would fly up to a preset altitude controlled by a barometer hooked up to a system of cranks and bellows designed to maintain the preset altitude. Once the engine achieved the predetermined number of rotations required to bring the craft as far as it needed to go, the wings folded and the Bug dived down upon its prey. Although the Bug was not exactly accurate (Arnold said it flew like 'a thing possessed of the devil'), nor was it ever used in battle, it was the first American drone aircraft.

Replica of The Bug

In his book *Call Me Pat*, the former head of Hughes Aircraft writes that as early as 1926 Carlos Mirick of the Naval Research Laboratory was working on a remote controlled aircraft project. Automatic pilots have been in use since about 1933. For target practice, the Air Force produced radio controlled aircraft dating back to at least 1947. In his book *The Business of Science*, former Hughes executive Simon Ramo makes note of a remotely piloted bomber in use around 1950.

As part of Project Cloudbuster, our Air Force was using drones for weather modification by 1964. During Congressional testimony conducted in 2009 and 2010, geoengineer Alan Robock suggested that chemtrail aircraft be drones. The aforementioned 1996 document "Owning the Weather in 2025" describes drone aircraft spraying substances for the purpose of weather modification.

Today's dedicated New Manhattan Project chemtrail fleet probably also utilizes other automated systems. U.S. Patent #6,131,854 "Ground Handling Apparatus for Unmanned Tactical Aircraft" describes a system whereby a drone can land on a runway and then be remotely hooked onto a track system that pulls the airplane through a series of automated stations that diagnose the aircraft, arm or disarm the plane, and lastly refuel the aircraft.

Another document in the series containing the infamous "Owning the Weather 2025" is titled "Aerospace Sanctuary in 2025 - Shrinking the Bull's Eye." This Air Force document speaks to extensive automation of all aspects of aircraft handling. It reads:

"Use of robotics can synergistically increase effectiveness by reducing manpower, infrastructure, and other support systems typically required for 24-hour operations on the flight line, in the bomb dump, and around the base. Today's robot is a mere infant compared to what may be available in 2025. The present advantages of implemented parallel processing in robots allow rudimentary cognitive skills. When combined with the limited motor sensory skills available today, robotic structures are able to perform limited inspections on aircraft. By 2025, robots could be designed as advanced 'tools' with the capability to fully inspect, diagnose, and maintain aircraft, as well as most other base systems for day-to-day operations. Robots could be expected to perform refueling operations, buildup, transport and loading of weapons, security functions, and even explosive ordnance disposal. The large computing capacity expected to be available in 2025 suggests that a single robot may be capable of alternating among the aforementioned tasks for every aircraft on the flight line, including those of our allies. Each robot could be engineered to be resistant to varying types of environmental extremes to retain its functionality; that is, ultraviolet rays, precipitation, cold and heat, EMP, and chemical/ biological attacks."

The authors continue:

"The remaining airfield operations would be automated and fine tuned specifically to accommodate a combat turn type of ramp function. On approach, aircraft would automatically report their status to the base system through an unmanned control center. The control center would distribute the data throughout the base. The distributed information would include tasking specified support functions. For instance, fuels would know how much fuel is required. Munitions storage would know which armaments to select and transport to the

aircraft for loading. Aircraft maintenance would know whether they need to respond to the aircraft to conduct system checkouts, repair, or replace components, and exactly what they need to bring with them. The base would be totally coordinated in supporting the aircraft's next mission. The aircraft would be directed to a specific location on the airfield where all support functions would automatically converge. The entire operation would be handled predominantly by robots. Delivery and loading of munitions, refueling, and final system checkout would all be automated. The only humans involved in the operation would be those necessary to perform high dexterity operations and to visually supervise the activities."

These types of systems would be advantageous to a covert project like the New Manhattan Project in that, as with drone aircraft, these systems reduce the number of Humans needed to conduct operations and therefore reduce the probability that the project will be exposed.

~ ~ ~

The proprietary aircraft of this new Manhattan Project may be powered wirelessly and have no need for conventional jet fuel. For this assertion, there is plenty of evidence. If ground-based antennas like HAARP are powering the proprietary chemtrail airplanes, this would be a great logistical advantage as the planes would not need to be grounded for refueling. The airplanes would probably only need to be grounded for payload and maintenance and therefore could remain in the sky, performing their functions without interruption for much longer periods of time. This would also be an advantage because the more time these planes spend in the air, the less chance there is of the program being exposed. It's incredibly difficult (as this author has learned) to expose something going on at 40,000 ft. in the sky. Also, the space usually required for fuel can be used to store chemtrail spray instead. The power beaming need not be constant as the airplanes could utilize rechargeable batteries.

Wireless power is not a new thing. Today, on the Apple website, one can purchase a wireless charger for their iPhone for crying out loud. Although the aircraft of today's proprietary chemtrail fleet would necessarily need their power transmitted through the air, over vast distances, the concept is the same. A leading expert in the field, William C. Brown (1916-1999) notes, "Little power is lost in the transmission of energy through the atmosphere and the vacuum of space provides a lossless, superconducting medium."

The wireless transmission of power goes back to Heinrich Hertz, James Maxwell (1831-1879), and Nikola Tesla. Heinrich Hertz (1857-1894) first demonstrated it, Maxwell contextualized it, and Nikola Tesla later popularized it. Tesla demonstrated wireless lamps in 1892 and often suggested remotely powered aircraft. With the exception of the earliest works by Hertz and Maxwell, all of the significant early work in the field happened in America. At a dinner at the Waldorf Astoria early in 1908, Tesla predicted the pre-eminence of air forces. He also envisioned the prominence of drone warfare. Later in 1908 Tesla revealed that he was working on an aircraft that would be remotely powered by electricity transmitted from ground stations. All this while he was going about powering cars and planes remotely from his Wardenclyffe facility.

William C. Brown

William Brown writes that, "The basic technology for wireless power transmission was introduced during World War II." In the late 1950s, George Goubau, F. Schwering and others demonstrated that microwave power could be transmitted with efficiencies approaching 100 percent. This was later confirmed by Raytheon. In 1964, at their Spencer Laboratory, Raytheon demonstrated a 5 lb., wirelessly-powered helicopter flying at an altitude of 50 ft. for 10 hours. About this, William Brown, the Director of Raytheon's Spencer Laboratory wrote:

"*In 1963, the Rome Air Development Center of the United States Air Force became interested in supporting a feasibility study to construct and fly a microwave-powered helicopter and awarded a contract to the Raytheon Company in June 1964. In October of the same year, the Rome Air Development Center and the Raytheon Company jointly sponsored a public demonstration of a microwave-powered helicopter which received all power needed for its flight from a beam of microwave energy. This flight which established the technical feasibility of the microwave-powered helicopter is the most recent completed development in this area. There are now plans to go ahead with the next phase of development.*"

This October, 1964 demonstration was covered by Walter Cronkite and the CBS evening news. Later experiments kept the drone aloft for months at a time.

Beginning in 1977, the field of wireless power transmission got a big boost from a joint Department of Energy (DOE) and National Aeronautics and Space Administration (NASA) assessment study of a solar-powered satellite concept. The concept of the solar-powered satellite had a great impact on the field of wireless power. Arthur D. Little and Textron worked in these areas as well. A solar panel is somewhat like a wireless power-receiving antenna known as a 'rectenna.' Along with Raytheon and the Air Force again, corporations such as Boeing and Rockwell International became involved. Conveniently, the study concluded in 1980 - which would fit right in with the author's suggested NMP proprietary chemtrail fleet production schedule. Following the conclusion of the joint DOE/NASA study, NASA furthered work in the area of wireless power transmission by sponsoring studies on microwave-powered, high-altitude atmospheric platforms.

These studies conducted by DOE and NASA in the late 1970s produced a greatly improved rectenna which was demonstrated in a modified form, beginning in 1983, as part of a Canadian development program known as SHARP (Stationary High Altitude Relay Platform). SHARP involved the remote powering of an airplane. Brown writes:

"I witnessed an impressive demonstration of the free-flying, microwave-powered, scaled down SHARP airplane ... in Ottawa, Canada in October 1987. The electric motor powered airplane had a 14 ft wing span and a rectenna consisting of a portion under the wings and a larger portion contained in a large disk mounted on the bottom of the fuselage. The airplane was hand launched and with the aid of a battery climbed to an altitude of about 100 m. There it encountered the microwave beam and the battery was switched off. The flight of the airplane was then radio controlled from the ground and the microwave beam automatically tracked the airplane."

In 1984, Raytheon's William Brown wrote pertaining to the wireless powering of satellites:

"It is interesting to note that the technology has now matured to the point where it is seriously being proposed as an alternate approach to the use of solar photovoltaic arrays in applications where hundreds of thousands of kilowatts of continuous power may be desired for propulsion or for payload."

In this same piece, William Brown also depicts and writes of a rectenna attached to the wing of a small airplane said to be achieving 85% efficiency. This new technology is described as, "...applicable to vehicles in space and to vehicles in the Earth's atmosphere."

SHARP airplane

The 1991 United States patent #5,068,669 titled "Power Beaming System" is one of the original HAARP patents (ch 3). This patent outlines the technical details of how to remotely power airplanes. Let us refer once again to a passage from the seminal book *Angels Don't Play this HAARP*. Authors Dr. Nick Begich and Jeane Manning reference an *Aviation Week* article:

"This 'Star Wars' technology developed by ARCO Power Technologies, Incorporated [patent #5,068,669's assignee] was used in a microwave-powered aircraft. The aircraft was reported to be able to stay aloft for up to 10,000 hours at 80,000 foot altitudes in a single mission. This craft was envisioned as a surveillance platform. The craft had no need for refueling because the energy was beamed to it and then converted to electrical energy for use by the aircraft. Flight tests were undertaken at Tyendinga Airport near Kingston, Ontario, Canada in the early 1990's. This test by APTI most likely involved this patent..."

FIG. 1

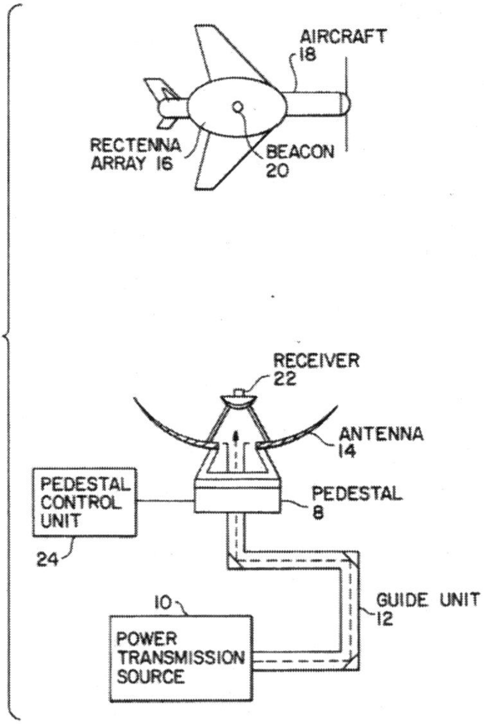

Figure 1 of "Power Beaming System with Printer Circuit Radiating Elements Having Resonating Cavities"

In 1992 Peter Koert of Arco Power Technologies (APTI) reported a successful breakthrough in rectenna design. It led to a great deal of interest in keeping aircraft aloft using wireless power. Koert's work in this area resulted in a 1993 patent titled "Power Beaming System with Printer Circuit Radiating Elements Having Resonating Cavities." The patent was assigned to APTI and it features a system diagram of how an aircraft can be remotely powered by a ground-based power transmission source. In 1996 Brown wrote that the evaluation of the practical application of Koert's new rectenna was, "now in progress." Let us recall that APTI was the assignee of all of the HAARP patents and that the large-scale, domestic execution of the New Manhattan Project began in 1996.

In 1996, Brown also noted that NASA's Jet Propulsion Laboratory was then currently completing a study of 'a microwave powered aircraft system.' At the time, this was being performed for a company called Skysat Communications Network Corporation. Brown went on to continue his work at the University of Alaska Fairbanks (you know, the outfit that runs HAARP), where he developed his SABER (Semi-Autonomous Beam Rider) program designed to remotely power helicopters.

There are scores of other patents pertaining to the wireless powering of airborne vehicles. From the early U.S. patent #3,464,207 "Quasi-Corona-Aerodynamic Vehicle" to the Queen of England's U.S. patent #4,955,562 "Microwave Powered Aircraft" to another patent of the same name (U.S. patent #5,503,350) to the U.S. Navy's patent #6,364,253 "Remote Piloted Vehicle Powered by Beamed Radiation" to Power Beaming Corporation's patent #6,534,705 "Methods and Apparatus for Beaming Power" to U.S. patent #7,711,441 "Aiming Feedback Control for Multiple Energy Beams" to U.S. patent #7,929,908 "Method and System for Controlling a Network for Power Beam Transmission" to the U.S. patent "Photovoltaic Receiver for Beamed Power" to the U.S. patent "Unmanned Vehicle and System" to U.S. patent #8,876,061 "Methods and Systems for Beam Powered Propulsion," the reality of this technology, applicable to the New Manhattan Project, is apparent.

Recent developments continue to provide support for the assertion that the dedicated aircraft of today's NMP are remotely powered, unmanned e-jets. In 2019, *Breitbart* and the *BBC* separately reported on the reality of electrically-powered aircraft. In 2018 *The Daily Mail* reported on Airbus' Zephyr S solar-powered drone. In 2017 *The Telegraph* reported that Rolls-Royce and Airbus were teaming up to produce the next generation of 'electric jets.' Lastly, a company named Powerlight Technologies today produces technology for the remote powering of drone aircraft.

It should be noted that my science advisor tells the author that the current state of power-beaming technology is insufficient for the propulsion of large jet aircraft. But due to the amount and quality of evidence for the assertions made in this section, coupled with the fact that it would be of tremendous logistical advantage to the New Manhattan Project, the author is sticking to his speculation. Also, much more advanced technologies may exist than what is publicly known.

~ ~ ~

The highly specialized aircraft described here would necessarily need to be developed. These are obviously not off-the-shelf aircraft. Extensive research, development and testing would be required. This research, development, and testing would best be suited to a single base of operations. Multiple bases of development separated by significant distance would not be desired because people work better together in close proximity and multiple development centers would increase security risks. Anyway, as far as the development phase goes, all you need is one or two planes, so all you need is one development location. Although we have seen evidence of the Rome Air Development Center developing pertinent technology, and Lockheed's famous Skunk Works as well as the Arnold Engineering and Development Center may also be seen as possibly responsible, an even more logical choice exists.

One of the types of aircraft expertly identified earlier by Mr. Buckmann is the Boeing KC-135. Wright-Patterson Air Force Base has an extensive history of retrofitting KC-135s for activities pertaining to weather modification and the atmospheric sciences. The Boeing KC-135 is a close relative of the Boeing 707 which is also noted by

Mr. Buckmann as one of today's offending aircraft. Scientific reports going back to 1960 are rife with instances of our Air Force retrofitting KC-135s for use in programs involving weather modification and the atmospheric sciences.

Boeing KC-135

Let us reference Robert S. Hopkins' book *Boeing KC-135 Stratotanker: More than Just a Tanker*. Mr. Hopkins expounds upon the easily modifiable KC-135 and how Wright-Patterson AFB was the best place to modify them:

"*...the KC-135 was the logical choice as a test bed platform [experimentally modified airplane] due to its size, capacity, speed, endurance, high altitude capability, and suitability for extensive modification. With a dozen early production KC-135s serving in a temporary test and evaluation role at Wright-Patterson AFB, Ohio, the nexus of Air Force research, their jump from basic flight test airplanes to research platforms was simple and immediate.*

Wright-Patterson Air Force Base

"From their arrival in 1957 at Wright-Patterson AFB for operational test and evaluation, until their transfer in 1994 to Edwards AFB as part of the sweeping reorganization of the Air Force, test-bed KC-135s have long been synonymous with 'Wright-Pat.' These airplanes were assigned to the Flight and All-Weather Test Division at WADC [Wright Air Development Center] at Wright-Patterson AFB."

In the 1960s, automated take-offs and landings of the KC-135 were evaluated. This is consistent with the author's previous assertion that the proprietary New Manhattan Project chemtrail fleet consists of drone aircraft.

In the development of the dedicated airplanes of the New Manhattan Project, Wright-Patterson AFB is a prime suspect. Development of a super-cool, cutting edge, next-generation jet plane in the land of the Wright brothers might go over well. Wright-Patterson AFB is located just outside of Dayton, Ohio where the aviation-pioneering Wright brothers lived. This also may have been the location to develop and produce the first of today's dedicated New Manhattan Project chemtrail planes and the Boeing KC-135 Stratotanker appears to be the likely airplane of choice.

As a result of the Base Closure and Realignment Act of 1988, the Davis-Monthan boneyard came under the command of Wright-Patterson. Later, the aforementioned Rome Air Development Center was renamed the Rome Laboratory and placed under Wright-Patterson's command as well. If you will recall, Rome Air Development Center is a location where power beaming technology was developed. Lastly, as noted in chapter 3, Wright-Patterson AFB is also one of the main developers of HAARP.

Once Wright-Patterson established a standardized process for turning old bones into awesome new meta-jets, many other maintenance and regeneration facilities around the country and around the world may have then replicated these same activities. During the chemtrail fleet production phase, multiple modification and regeneration centers would be called for because many of these facilities don't have the capacity to work on more than 2 or 3 planes of this size and level of secrecy at a time. The production of chemtrail fleet aircraft would require extraordinary levels of manpower, technology, planning, infrastructure, and security for every production location involved.

Wright-Patterson itself may have produced many of these regenerated aircraft. Other facilities involved may have included: the Aerospace Maintenance and Regeneration Group at Davis-Monthan AFB, Tinker AFB, a Taiwanese maintenance facility, and the Evergreen Air Center.

The aforementioned Davis-Monthan boneyard has world class aircraft maintenance, modification, and regeneration facilities. Maybe New Manhattan Project (NMP) airplanes have been conveniently overhauled and retrofitted on site.

Tinker Air Force Base in Oklahoma may have been another location to produce chemtrail fleet aircraft. Tinker AFB has an extensive history of maintenance and modification of the Boeing KC-135 Stratotanker. As we have seen, Tinker AFB was the recipient of 13 mothballed KC-135s between 1982 and 2005. These mothballed planes may have served as bones for the proprietary New Manhattan

Project chemtrail fleet. From 1959-2001, Tinker worked on 3,336 KC-135s. In the 1980s, just in time for NMP dedicated aircraft production, Tinker AFB got greatly expanded and updated aircraft repair and modification facilities.

Tinker Air Force Base

Another facility which may have been involved in the production of dedicated chemtrail fleet aircraft is the aforementioned maintenance base at Tainan in southwestern Taiwan - today known as Tainan Airport. These were the maintenance facilities of the CIA proprietary, Air Asia. These gigantic facilities have been described by the authors of *The CIA and the Cult of Intelligence* as, "The largest air repair and maintenance facility in the Pacific region." These land-based facilities were established in 1955 and have regularly employed about 8,000 people. These facilities continued operations long after the 1975 break-up of Air America and are apparently still in operation today. Marchetti and Marks write, "It not only services the CIA's own planes, it also repairs private and military aircraft. The U.S. air force makes heavy use of Air Asia and consequently has not had to build a major maintenance facility of its own in East Asia, as would have been necessary if the CIA proprietary had not been available. Like Air America, Air Asia is a self-sustaining, profit-making enterprise."

Air Asia started in about 1950 as a floating maintenance facility of the aforementioned Civil Air Transport and, interestingly, Air Asia has a history of providing the CIA with completely untraceable aircraft. The former military liaison to the CIA, Fletcher Prouty writes:

"As the fortunes of war drove them from one base to another, someone decided to put the maintenance facility on board a big war surplus ship. Finally, with the defeat of the forces under Chiang Kai-shek, this shop with its facilities and stockpile of equipment sailed to Taiwan and anchored beside a dock in Tainan. There this most unusual aircraft maintenance facility performed maintenance for a fast-growing and very busy fleet of planes for many years.

"One could walk through that ship absolutely amazed at the beehive-like activity on board. Hundreds, perhaps thousands, of Chinese worked in that ship on stages, rather than floors or decks, joined by narrow catwalks. Many of those workers worked in small basket-like spaces, barely large enough for a small Chinese. Parts and materials were brought to them and poured into each work space as through a funnel. The worker would finish his special task and then drop the part through a short chute, where it would end up for the next worker to do his part. The whole operation worked on a sort of force-of-gravity basis, with the finished item falling out at the bottom, ready for an alert runner to carry it to the packaging room."

Prouty continues:

"Even instruments were rebuilt, and as they were, the faces and decals were changed to have Chinese or English markings, as required. There were propeller shops and wheel shops. Planes could be completely rebuilt from this one facility. As a matter of fact, the CIA had obtained master transparent film slide sets of the aircraft manufacturers parts and supplies kits, and for such planes as the

DC-6, Air America could make every part just about as well as Douglas Aircraft. The CIA justified this irregular and perhaps illegal operation on the basis that it was working with sanitized engines and aircraft and that it could not put such items back in the supply line of the services. As a result, instead of buying from Douglas, through the services, it simply made the parts in its Tianan facility. It is entirely possible that complete small aircraft were made in this manner and that Air America or its subsidiaries ended up with more aircraft in operation than it had had in the first place.

"This technique is 'justified' by the nature of air registry, which precludes the availability and even the existence of 'extra' aircraft. Every aircraft built and flown must be registered. Once it has been registered, that serial number stays with it for the rest of its existence. Therefore, if the Agency wishes to remove all traces of identity and ownership from an airplane in order to make it plausibly deniable, it must also arrange to cover that plane in the registry. This is done in many ways, one of which is to assemble an extra plane from the parts available. To begin with, the CIA may be able to salvage a destroyed aircraft and have it declared discarded. Then from the frame or some other essential part it will rebuild the plane from parts not having any serial numbers at all. This method must be used with larger aircraft; but the Tainan facility had the capability to build smaller aircraft from scratch, just by assembling spare parts, many of which it would have made itself right at the plant."

Tainan Airport

In 1975, these vast, world class maintenance facilities in Taiwan were purchased from Air Asia by the aforementioned E-Systems. If you will recall, E-Systems built HAARP. E-systems has an extensive history with the CIA and has had at least three CIA agents among its senior executives. In 1987 E-Systems sold these Taiwan aircraft maintenance facilities to Precision Airmotive Corp., of Everett, Washington which describes itself as, "a world leader in the manufacturing of fuel controls for general aviation." We will have much more about E-systems in the next section.

It was the establishment of these Air Asia facilities on the island of Taiwan that created the opportunity for Flying Tigers Airlines to come into existence. Flying Tigers Airlines got its start as a CIA front company organized and operated by former members of the original Flying Tigers and other military veterans packing and shipping gear needed at the Air Asia maintenance facilities in Taiwan. A man named Erwin Rautenberg and his boss Paul Williams at a company called Air-Sea Forwarders worked with Chennault to establish the company.

As previously noted, Evergreen Aviation operated a maintenance and repair facility at their Evergreen Air Center in Marana, Arizona. This location is a prime suspect for having produced airplanes of the proprietary New Manhattan Project chemtrail fleet. This location had and has world class maintenance facilities, ample storage and a history of accommodating the U.S. military and the CIA. Also as previously noted, this air center was acquired from the CIA's Intermountain Aviation. The CIA offered it only to Evergreen; nobody else. Before that, the base originated as Marana Army Air Field; a United States Army Air Force base. The United States Army Air Force is the predecessor of today's United States Air Force. The Intermountain Marana base at Pinal Air Park in Marana, Arizona is the aforementioned base where Chinese pilots were trained during WWII and immediately after. This air base, known today as Marana Aerospace Solutions, has quite a history.

Evergreen Marana Air Maintenance Center

Investigative journalist Wayne Madsen wrote a 2015 piece that extensively reveals Marana's history. Madsen writes:

"Ever since 1948, when the U.S. Government turned the Marana field back to Pinal County, the CIA has maintained a presence there. Originally built by Sun City developer; Poston, Arizona Japanese-American relocation center builder; and former New York Yankees owner Del Webb in 1942, the airfield trained 10,000 pilots during World War II. Dinah Shore entertained the trainee pilots during a USO tour.

"After the war, Pinal County leased Marana back to Sonoran Flight Services, which was contracted to the U.S. Forest Service to fight forest fires. However, the airfield soon gained the attention of the newly-inaugurated CIA, which piggybacked on the Forest Service contract to begin operating CIA proprietary airline companies from the location much prized for its remoteness. A fairly old massive pile of dirt located outside the main gate of Pinal Air Park is said to be the soil excavated for a massive tunnel system installed under the base by the CIA.

"George Doole, the CIA's guru on establishing proprietary aviation companies, decided that Marana was the best place to center the operations of a number of famous and infamous spook airline

outfits, including Civil Air Transport, Atlantic General Enterprise, Air America, Southern Air Transport, Intermountain Aviation (for which Phil Marshall flew in the early 1980s), Continental Air Services (part of Continental Airlines that supported CIA and U.S. military operations in Southeast Asia), and, finally, Evergreen Trade. The latter is a diversified company started by Del Smith that is involved in everything from chemical spraying for weather modification to growing Christmas trees and operating a vineyard. Its aviation arm, Evergreen Aviation, for years operated Pinal Air Park for the CIA as a major center for maintaining and refurbishing aircraft for special operations.

"Those 'special operations,' as Phil Marshall was undoubtedly aware from his days as a pilot for DEA-informant Barry Seal, included narcotics smuggling, for which Air America and Southern Air Transport were the most infamous."

Of the Intermountain Marana base, Marchetti and Marks write:

"A particularly mysterious air proprietary is known within the agency as Intermountain Aviation. Its public dealings are through firms called Aero Associates and Hamilton Aircraft. Intermountain specializes in charter flights, airplane repair, reconditioning of old military planes, and the shipment of these planes overseas. It is located on a large private airfield near Tuscon, Arizona, which looks much like an air force base: housing is provided for senior personnel; there is an impressive officers' club, a swimming pool, and other sports facilities-all purchased and maintained at the CIA's expense. (One senior agency official often speculated that the two most pleasant assignments he could think of to finish his career in luxury were to be chief of station in Johannesburg, South Africa, and director of Intermountain Aviation.)

"Intermountain was founded by the agency in the 1950s primarily for the maintenance of CIA aircraft, but it soon became a parking and storage facility for planes from other agency proprietaries. Additionally, the agency used it for the training of both American and foreign mercenaries. When the CIA brought Tibetan tribesmen to the United States in the late 1950s to prepare them for guerrilla forays into China, the agency's Intermountain Aviation assisted in the training program.

"Then, in the early 1960s CIA air operations grew by leaps and bounds with the expansion of the wars in Southeast Asia and the constant fighting in the Congo.

"Intermountain rapidly expanded its operations to the point where its cover as a commercial air charter and repair company became difficult to maintain. If nothing else, its parachute towers looked suspicious to the casual viewer. The problem of cover was partially solved, however, when Intermountain landed a Department of the Interior contract to train smoke jumpers for forest fire control. But a reporter visiting Tucson in 1966 still wrote, 'Anyone driving by could see more than a hundred B-26s with their armor plate, bomb bays, and gun ports.' Not long after this disclosure appeared in the press, CIA funds were made available to Intermountain to build hangers for the parked aircraft. Prying reporters and the curious public soon saw less."

It wasn't just Intermountain providing military training to foreign nationals either. Beginning in 1957, American efforts to strengthen the militaries of overseas allies resulted in a large number of foreign nationals coming to Arizona for military training. At various Arizona locations, foreign nationals from: West Germany, the United Kingdom, Canada, Iran, Norway, the Philippines, South Korea, South Vietnam, Thailand, Saudi Arabia, and Turkey were trained in various aerospace-related disciplines.

Jason Howard Gart, in his 2006 PhD dissertation, writes that Intermountain Aviation, Inc. was founded in 1961 by Orme Lewis and Robert C. Kelso of the Lewis Roca Scoville Beachamp & Linton law firm. Gart also notes that a CIA man by the name of Garfield 'Gar' M. Thorsrud (1928-2014) managed Intermountain on a daily basis. Gart writes:

"With upward of $2 million in assets and a cadre of CIA employees, including aircraft conversion experts, aerial delivery technicians, paramilitary specialists, master parachutists, and pilots, Intermountain began a host of clandestine assignments." ... "Intermountain also provided operational support for the 'secret war' in northern Laos. During the summer of 1962, Marana Air Park served as a training facility for CIA case officers assisting Hmong tribesmen in the insurgency campaign against the North Vietnamese.

Intermountain offered instruction on low-level parachute drops and pilot certification in short takeoff and landing (STOL) aircraft."

Reminiscing about the former Intermountain Marana base, the author of *The Evergreen Story* Bill Yenne quotes Evergreen founder and president Delford Smith:

"'*It was a natural fit,' Del Smith recalls. 'Intermountain approached us and explained that they had a repair station with an unlimited, Class I through IV certification, which gives Federal Aviation Administration authority to work on any type of aircraft. We really bought Marana for our own planned needs, but as we grew, it would change into more of a third-party operation.*'"

Let us reference another passage from *The Evergreen Story*:

"*The Evergreen Maintenance Center (EMC), located in Marana, Arizona, is one of the most comprehensive and diversified maintenance facilities in the Western Hemisphere, including the world's largest storage and preservation facility for private and commercial aircraft. The Maintenance Center also has an on-site engineering department to support routine maintenance, modification and conversion work.*"

The Evergreen Story continues:

"*Evergreen Maintenance Center is one of very few companies in the United States with several Federal Aviation Administration Certificated Repair Station ratings, as well as European Joint Aviation Authorities (European Equivalent of FAA), JAR-145 (Joint Aviation Requirements 145) ratings and those of other international agencies, such as the International Organization for Standardization (ISO), European Aviation Safety Agency (EASA) and the Chinese Civil Aviation Authority (CAAC).*"

The Evergreen Story also recounts, "The early years of majority in-house work [working on Evergreen aircraft] shifted to majority third-party work in the eighties and early nineties." This is consistent with the author's proposed proprietary chemtrail fleet development and production timeline. Also, a former manager at the Evergreen Maintenance Center notes that at the time of acquisition, Evergreen started with about 40 center employees. Eight years later in 1982, there were 350 working out of the EMC. As we have seen, 1980 is probably about the time when the New Manhattan Project chemtrail fleet went into production. In 1985, Evergreen's George A. Doole Aviation Center at the Evergreen Maintenance Center came online which expanded and updated the EMC's capabilities. The aforementioned George A. Doole, if you will recall, was the guy who founded and ran the CIA's Pacific Corporation. Also as previously mentioned, George A. Doole was on the Evergreen Aviation board of directors. *The Evergreen Story* states that, "...the entire Pan American World Airways fleet of Boeing 707's was stored at Evergreen Maintenance Center." As we have seen, the Boeing 707 is one of the airplanes expertly identified as a chemtrail spraying aircraft.

As the crow flies, the former Pinal Air Park (and thus the Evergreen Maintenance Center) is only 17 miles from the Davis-Monthan Boneyard. If aircraft stored at the Boneyard were airworthy or even towed out to and along the highway, transferring these old dogs from the Boneyard to Pinal would be relatively easy. Let us again reference a passage from *The Evergreen Story*. Bill Yenne writes:

"*Evergreen Maintenance Center also was doing military and government work for the United States and other countries. These projects included a contract for overhauling the transmission and rotor-heads of Peruvian helicopters and reconditioning aircraft that came out of long term storage at the US Air Force's Military Aircraft Storage & Disposition Center at Davis-Monthan AFB south of Tuscon. The proximity of Evergreen Maintenance Center to Davis-Monthan has always put Evergreen in a good position for military contracts.*"

The Oregonian was more explicit: "Under both Intermountain and Evergreen, planes from the boneyard have been short-hopped to Marana's giant shops and reconditioned for assorted customers, including, under Evergreen, the Colombian Air Force."

As this book continues to unfold, we will continue to see that today's chemtrail spraying operations are part of a military program, so it makes sense that Evergreen International would have an extensive history of working for what passes for our U.S. military. As noted, Evergreen has conducted lots of military aircraft maintenance and regeneration work. Also as previously noted, Evergreen's headquarters in Marana, Arizona was originally a U.S. military base. Evergreen has also worked with the United States Air Force spraying rice crops in Pakistan. In 1978, Evergreen had a passenger service contract with the U.S. Navy. In that same year of 1978, Evergreen started flying lots of personnel and materiel for the former U.S. Air Force Military Airlift Command. It helps when the former chief of the Military Airlift Command, General William G. Moore (1920-2012) is on your board of directors. Evergreen operated small planes in Panama under contract with the Department of Defense. Evergreen participated in something called the Civil Reserve Air Fleet which augmented U.S. military operations such as those in the Middle East. Just as the Western-instigated wars in the Far East were good for Air America's business, the Western-instigated Middle East wars of late were very good for Evergreen's bottom line too.

Evergreen has also done work for the National Aeronautics and Space Administration (NASA). Evergreen maintained the jumbo jet that carried the Space Shuttle piggyback. This is significant because NASA has a long history of contributing to weather modification and the atmospheric sciences and we continue to note their contributions to the New Manhattan Project.

Evergreen had a division which supplied customers with the services of drone aircraft called Evergreen Unmanned Systems. As we have seen, there is plenty of evidence for the assertion that the airplanes of today's New Manhattan Project chemtrail fleet are operated remotely as unmanned aerial vehicles.

The security at the Evergreen Maintenance Center was extraordinary. Being that the regeneration of airplanes to be used in the New Manhattan Project is highly secret work, this level of security would be advantageous. *The Oregonian* writes:

"The presence of so many expensive aircraft is the reason given for the visibly high level of security at Evergreen Air Center. Visitors don't wander in casually. Guards in military-green uniforms with Evergreen patches on the shoulders control traffic through the entry gate. A candy-striped barrier like a railroad-crossing gate swings down to stop vehicles the guards don't recognize.

"Visitors' identities are checked, their appointments verified with phone calls and passes are issued to dangle on rear-view mirrors and clip to lapels. Behind the neatly kept guard building is a dog run housing German shepherds that help patrol the area at night."

As we have seen, Evergreen had extensive ties to the CIA. The CIA can provide cover for covert operations such as those described here. We have already seen the capabilities of the CIA's maintenance facility at Tainan. The CIA may have covered the tracks of today's chemtrail spraying aircraft. Let us refer again to the book *Air America*. Christopher Robbins writes:

"*All planes have tail numbers, and their engines and instruments are numbered as well, which makes disguise difficult. The CIA would keep a list of the aircraft that had crashed and then create two or three airplanes with the same tail number, and two or three with no tail numbers at all. Then, in an exceedingly complex operation using very careful manipulation and scheduling, they would cause aircraft to show up in places two at a time, making them impossible to follow. On top of this, Air America had the capacity on Taiwan and at Udorn to manufacture their own planes. The idea was to create a plane that did not exist, one that even the manufacturer back in the States would swear had never been made. Instruments and engines were produced with no serial numbers and no decals, which was a problem in itself because an engine is dye-stamped, and even if the numbers are erased, the stamp can be seen on the metal.*"

Investigative journalist Wayne Madsen writes that the Evergreen Maintenance Center had the appropriate capabilities for untraceable aircraft production. Madsen writes, "Pinal not only has the capability to maintain aircraft but, according to a source familiar with the operations at the facility, it can manufacture aircraft from the cannibalized parts from other planes. In addition, the facility can fabricate new parts for special purposes."

In short, Evergreen was the right operation with the right pedigree and the right connections at the right time to be involved in assembling the proprietary airplanes of the New Manhattan Project. Just as the New Manhattan Project operates on a global scale, so did Evergreen Aviation. Maybe George Doole, Delford Smith, and Evergreen Aviation were instrumental in not only regenerating the airplanes, but also in brokering the deals for the airplanes. As noted earlier, Evergreen Aviation made most of its money buying and selling aircraft and had a special division (the aforementioned Evergreen Aircraft Sales & Leasing) devoted to the task.

In 2011, Evergreen sold the maintenance and repair facilities at Pinal Airpark, Marana, Arizona to a New York based private equity firm called Relativity Capital. The facilities now go by the name Marana Aerospace Solutions. In 2017 Marana Aerospace Solutions merged with Ascent Aviation Services.

Many other aircraft maintenance facilities may have also been involved in the production of the New Manhattan Project's dedicated chemtrail fleet, but those noted here are the most probable.

~ ~ ~

Aircraft maintenance and modification centers dedicated to work in the area of the atmospheric sciences have also existed. Most notably, the National Center for Atmospheric Research (NCAR) Research Aviation Facility and the National Oceanic and Atmospheric Administration (NOAA) Research Flight Facility have historically modified many airplanes. The modifications involved outfitting the airplanes with atmospheric monitoring and modification equipment and in some cases reinforcing an aircraft to be able to withstand certain inclement weather conditions such as large hail. These facilities may have played a role in the development and/or production of the New Manhattan Project chemtrail fleets.

Specifically, being that today's offending airplanes are probably used to collect atmospheric data as they saturate our atmosphere, the electronic equipment used to accomplish these objectives may very well have come from these atmospheric flight research facilities. For example, Dr. Robert H. Simpson of ESSA [the Environmental Science Services Administration], the predecessor of NOAA, was the leading pioneer in the field of airborne atmospheric observation, and ESSA's planes were capable of measuring: pressure, temperature, humidity, cloud particle size and distribution, wind direction and speed, and other atmospheric phenomena.

ESSA's Research Flight Facility provided aircraft for ESSA as well as other groups such as the Atomic Energy Commission, the University of Chicago, and the University of Stockholm. ESSA's Research Flight Facility supported National Science Foundation efforts as well as those of the World Meteorological Organization.

ESSA writes:

"The Weather Bureau first used instrumented airplanes for hurricane research in 1956. Three aircraft (two WB-50 and one WB-47 aircraft) were furnished and operated by the U.S. Air Force, Air Weather Service. Meteorological equipment was installed and operated under the direction of the Weather Bureau. These aircraft were used during the 1956, 1957, and 1958 hurricane seasons; sensors employed were mostly of the type having some previous service history in other activities. During the 3-year program, considerable strides were made in melding together a variety of sensors and supporting instrumentation to yield meteorological data in digital form. This program, combined with the cumulative experience in the processing of aircraft data, provided a valuable foundation upon which the present RFF [Research Flight Facility] program has been built."

In 1965 the RFF was transferred to ESSA.

A Department of Commerce flying laboratory

The ESSA Research Flight Facility's support of activities pertaining to the tracing of radioactive atmospheric particles is particularly relevant. In 1965 and 1966, the RFF conducted research supported by the Atomic Energy Commission pertaining to the tracing of both naturally occurring and artificial radioactive particles. Also of note to our discussion, in April and May of 1962, RFF aircraft participated in the air-to-air tracking and control of meteorological drones.

The interior of a flying laboratory

The larger, dual-purpose chemtrail fleet

The 'larger chemtrail fleet' as defined here is a bigger army of seemingly normal and innocuous large, commercial passenger and freight jet aircraft which has been secretly retrofitted to spray chemtrails. While the smaller, dedicated chemtrail fleet probably consists of a few hundred aircraft, the larger, dual-purpose chemtrail fleet probably consists of thousands. While the aircraft of the smaller, dedicated chemtrail fleet have commonly been observed as being all white and otherwise unmarked, the aircraft of the larger chemtrail fleet have been observed to be flying the colors of seemingly every variety of airline.

The video doesn't lie. As far as video evidence of commercial jets spraying chemtrails is concerned, there's really only one web page one needs to know: Mike Decker's *YouTube* channel. Mike Decker has lots and lots of close-up videos of commercial aircraft spraying chemtrails. His expert marksmanship has produced hard evidence of: Hawaiian Airlines, Austrian Airlines, Southwest Airlines, British Airways, Air France, Delta, United, and many others all spraying chemtrails.

If you've ever tried filming an airplane in flight at altitude, much less tried to have it large enough in the frame to be able to distinguish its markings, then you know just how difficult this type of photography is. Kudos to you, Mike. Our noble anti-geoengineering movement thanks you. Mike says that his original *YouTube* channel was terminated along with 1,000 videos, so please watch his videos while you can! And please do a brother a favor by hitting both his 'subscribe' button and notifications bell.

It should also be noted here that a man claiming to be a German aerospace engineer came forward on May 12 of 2014 to talk publicly about how he and his co-workers had installed chemtrail spraying equipment on many aircraft at a German military base. What he apparently describes (it's translated from German) is consistent with rigging planes for payload spraying. The video should be available on *YouTube* as well.

Now that we have seen evidence supporting the assertion that today's commercial aircraft are spraying chemtrails, let us go back in time in order to see from whence all this came. We now go back to a simpler time when we were only being sprayed a little bit because today's chemtrail fleets, both large and small, were only in the research and development phase. Temco Aircraft Corporation was apparently a key player in these early years.

Immediately following WWII, two aeronautical engineers by the names of Robert McCulloch and H.L. Howard founded the Texas Engineering and Manufacturing Company (Temco) on November 17, 1945. After attracting some investors, McCulloch and Howard leased the same gigantic, 5-year-old, government-built, $30M production facilities in Dallas, TX where the two engineers had run military aircraft production operations during the war. Part of their deal with the government was that they would quickly re-convert the facilities for wartime production, if needed. The founders of Temco also leased the vast majority of their tools (some $850K worth) from the federal government. From the beginning, the U.S. Government tacitly owned Temco. Of the two Temco founders, the more notable is McCulloch. Robert McCulloch was widely known in the industry as an astute executive as well as a sharp aeronautical engineer.

ROBERT McCULLOCH
President & General Manager

H. L. HOWARD
Executive Vice-Pres. & Treasurer

David Harold Byrd (1900-1986) was one of the original investors in McCulloch and Howard's Temco. David Harold Byrd, cousin of the famous Admiral Richard E. Byrd, was an aircraft enthusiast who, over the course of his lifetime, among many other aircraft-related activities, owned 58 aircraft and organized and operated three commercial aircraft services. He helped found the Texas Civil Air Patrol and his picture hangs on the wall of the Civil Air Patrol 'Hall of Honor' at Wright-Patterson AFB in Ohio. Byrd was famously good friends with the driving political force behind the creation of the National Aeronautics and Space Administration (NASA), the then Majority Leader of the Senate Lyndon Baines Johnson (LBJ). In fact, in July of 1957, one year before the bill that created NASA was signed into law,

Byrd had fortuitously organized and become chairman of the board of something called Space Corporation. The Space Corporation specialized in aerospace ground support equipment. NASA and their many former Nazis have apparently played a central role in the production of the New Manhattan Project. Equally as curious, Byrd was the owner of the book depository building in downtown Dallas where Lee Harvey Oswald supposedly fired the magic bullet that killed President John F. Kennedy.

David Harold Byrd

Temco's combination of McCulloch and Byrd garnered the critical support of the local Texas congressional delegation which helped bring lots of business to the fledgling Temco. In 1945 the new Temco began producing the Fairchild F-24 and the Globe Swift aircraft as well as engaging in the conversion of surplus military aircraft. In late 1946, the firm was incorporated as TEMCO, Inc. After the Globe Aircraft Corporation went bankrupt on June 23, 1947, TEMCO gobbled up its assets and inventory as TEMCO continued production of the Globe Swift, a classic light aircraft, until 1950.

Fairchild F-24

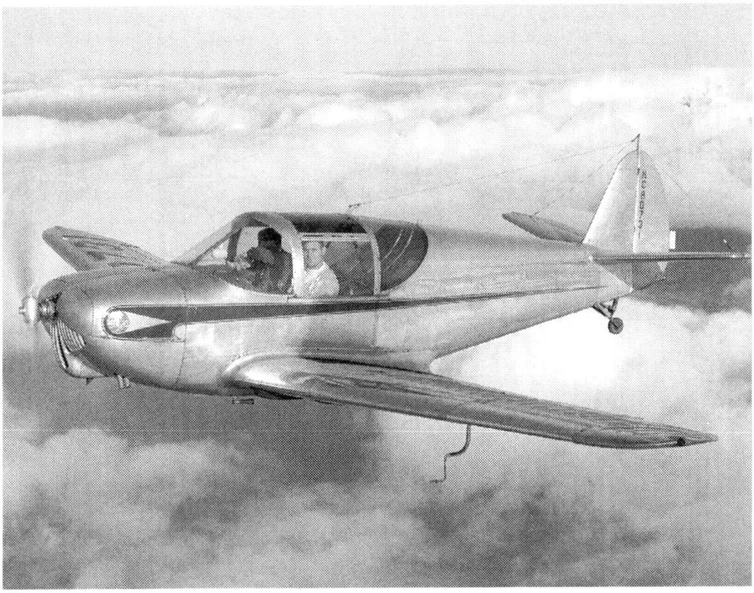

Temco Globe Swift

It is worth mentioning that Fairchild Industries, which developed the Fairchild F-24 and sub-contracted the production of said airplane to Temco, is the company where the famous Operation Paperclip Nazi Wernher von Braun went. In the early 1970s, not long before his death, von Braun was the Fairchild vice president for engineering and development.

Furthermore, it is interesting to note in the provided image of the Temco Globe Swift, the presence of what appears to be some sort of spray nozzle almost directly below the pilot, on the underbelly of the plane. What is this appendage for? Might the Temco Globe Swift have been an aircraft used for spraying chemtrails as part of early New Manhattan Project experiments? As we are about to see, Temco is the logical company to have produced such an airplane. In fact, all of the Temco-produced aircraft known to the author are pictured here due to the fact that all of them may have been used as early New Manhattan Project chemtrail spray planes.

In 1949 TEMCO acquired the assets of the bankrupt Luscombe Airplane Corporation. A year later, TEMCO resumed production of Luscombe's Silvaire airplane. TEMCO was only able to produce 50 Silvaires before the Korean War started in 1950 and all production was shifted back to military needs.

Luscombe Silvaire

The Korean War lasted three years. As production shifted back to the civilian market, TEMCO began developing their own, original designs. TEMCO designed the T-35 Buckaroo trainer airplane right around the time that they reorganized themselves as the Temco Aircraft Corporation in 1952. When the Korean War ended in 1953, Temco Aircraft acquired additional factory space at their Greenville, Texas location. Due to the fact that Temco Aircraft failed to win a certain contract, only 20 T-35 Buckaroos were produced. Another company run by a man named Jack Riley subsequently picked up the T-35 designs and converted them for production of an airplane called the Twin-Navion. Temco Aircraft bought Riley's Twin-Navion operations in 1953 and further developed the conversion as the Riley Twin. The Riley Twin was then produced by Temco Aircraft in significant numbers.

T-35 Buckaroo

Riley Twin

Here's where the New Manhattan Project comes in. We are most concerned with Temco Aircraft due to their connections to the aforementioned LTV Electrosystems. LTV Electrosystems was partly a product of the Temco Aircraft Corporation. Because of this, Temco Aircraft Corporation is the logical company to have produced the early New Manhattan Project chemtrail fleet aircraft.

In 1960 the Temco Aircraft Corporation merged with an electronics company called Ling Altec. The resulting company was known as Ling-Temco. The following year, Ling-Temco absorbed the Chance Vought Corporation which produced Corsair aircraft for the Navy. After this absorption, the resulting company was known as Ling-Temco-Vought. As previously mentioned, LTV Electrosystems was a publicly traded spin-off of Ling-Temco-Vought, hence the 'LTV.' In fact, although it was cobbled together from a few different internal sources, the biggest part of the new LTV Electrosystems came from the old Temco Aircraft.

Vought Corsair

From the mid-1960s to the early 1970s LTV Electrosystems contracted for the United States Air Force in the repair, overhaul, and modification of aircraft. Might these modifications have had anything to do with aerial spraying of substances and the collection of atmospheric data? The Air Force and the Navy have been ubiquitous throughout the development of the New Manhattan Project. Furthermore, beginning in 1970, on the LTV Electrosystems board of directors sat the aforementioned Navy Vice Admiral William F. Raborn who, for all the world, appears to have been a very central player in the historical development of today's New Manhattan Project. In 1972 LTV Electrosystems was bought and taken private by a group of investors led by a man by the name of John W. Dixon. The company subsequently changed its name to the aforementioned E-Systems. Also as previously noted, throughout this transition, Raborn stayed on the board.

John W. Dixon was a Kentucky native and a longtime employee of LTV Electrosystems before he became the first E-Systems CEO in 1972. Gregg Jones of *The Dallas Morning News* writes, "From the beginning, E-Systems sought top-notch engineers and experts with connections to the intelligence community. Drawing on contacts made as deputy Pentagon comptroller and at LTV, Mr. Dixon hired men such as a former CIA deputy chief of science and technology and a National Security Agency signals-intelligence expert." As mentioned earlier, in 1975, after congress ordered the CIA to divest itself of their airlines, E-Systems bought Air Asia for $1.9M. With this purchase, E-Systems got the largest aircraft repair and maintenance facility in Southeast Asia and loads of CIA affiliations. Peter A. Marino, a former director of the CIA's technical services division and a defense industry executive became an E-Systems senior vice president in 1991.

Consistent with the author's proposed timeline of NMP development, in Reagan's first year in office (1980), E-Systems' order backlog went up 34 percent, topping $1B for the first time. It was only upside from there. Jones writes, "In an 18-month period in 1985-86, the company hired 2,100 people in the Dallas area alone. Revenue broke the $1 billion mark for the first time in 1986, the year that earnings topped the $100 million plateau." Because of all their insider connections, Dan Peterson, a senior vice president at Martin Marietta said that E-Systems is, "Able to win contracts no one else even knows about."

Because of all the history detailed here, it most probably has been the Temco Aircraft Corporation and all of its mutations over the years which have been responsible for the production of many airplanes used in the development of the New Manhattan Project. At least one of those mutations is still operating today. That mutation is known as L3 ISR Systems. L3 ISR Systems claims their lineage as such: Temco Aircraft Corporation 1951-1960, Ling-Temco-Vought 1961-1971, E-Systems 1972-1994, Raytheon E-Systems 1995-2001, L3 Communications, Integrated Systems 2002-2011.

Today, the aforementioned, Greenville, Texas location formerly inhabited by the Temco Aircraft Corporation is known as Majors Field and it is inhabited by L3 ISR Systems. On their website, L3 ISR Systems writes about how E-Systems took over their Greenville location in 1972. Moreover, L3 ISR Greenville's history is filled with other connections to the New Manhattan Project. They write that starting in 1961, their Majors Field location, under Ling-Temco-Vought, worked on Boeing KC-135s. They write that after E-Systems took over in 1972, they worked on Boeing 707s - another aircraft that has been expertly identified as a common chemtrail sprayer. They proceed to write about how they do work for the Department of Defense and 'select U.S. government intelligence agencies.' L3 ISR Systems Greenville describes itself as, "The world's premier SIGINT [signals intelligence] systems developer and integrator of airborne reconnaissance platforms." They also write that their core capabilities consist of 'specialized aircraft, maintenance and modification' along with many other specialized and highly technical capabilities which would be necessary to retrofit and otherwise produce aircraft for use in the New Manhattan Project. From their website they write:

"L3 ISR Systems Greenville is an internationally recognized systems integration organization specializing in the modernization and maintenance of aircraft of all sizes, and the study, design, development and integration of special mission systems for military and commercial applications. L3 Greenville's expertise spans the design, development and integration of advanced avionics, special purpose airborne systems and aircraft maintenance and modernization for government and military customers worldwide. L3 Greenville specializes in the

modification, modernization and maintenance of a diverse group of mission aircraft. The division is a leader in advanced technologies for complete systems integration, signal processing, electronic countermeasures, sensor development and aircraft self-protection. With more than 6,500 employees, it has the capability to support programs from initial development through total life-cycle management."

L3 recently merged with a company named Harris to form L3 Harris. Harris also comes from a background with relevance to the New Manhattan Project as they have produced components for weather satellites and missile systems. Today they describe themselves as, "a proven leader in tactical communications, electronic warfare, avionics, air traffic management, space and intelligence, and weather systems." One can read all about them on their website *Harris.com*.

Although this section focuses on L3's potential involvement in the production of the larger, multi-purpose chemtrail fleet, they may have been involved in the production of the smaller, proprietary chemtrail fleet as well. It appears that their facilities have been and continue to be large enough to do both.

Conversely, it is logical to speculate that the facilities which were previously mentioned as being potential locations for the production of the smaller chemtrail fleet have also been involved in the production of the larger chemtrail fleet because retrofitting planes for the larger chemtrail fleet requires less resources than producing planes for the smaller chemtrail fleet. There may very well be other American, New Manhattan Project aircraft retrofitting sites yet beyond all of those implicated here. But because of the history pertaining to Majors Field, E-Systems, and Temco Aircraft, and until what has been hidden in darkness is exposed to light, whoever operates the aircraft modification facilities at Greenville, Texas is a prime suspect.

Spiking the jet fuel

It appears that there is a third way that geoengineers are increasing the atmospheric saturations of particulate matter. It appears that common jet fuel has been spiked with additives that do just that.

The former director of the Center for Energy and Combustion Research at the University of California at San Diego, a man named

Stanford Solomon Penner (1921-2016), was a proponent of the catastrophic theory of man-made global warming who worked in the area of commercial jet fuel exhausts producing more and better particles. These particles, so he wrote, would create artificial cloud-cover which would bounce sunlight back into space and thus cool the planet - the SRM geoengineering thesis. In a 1993 paper, Penner suggested a 'jet-fuel-coal mixture' for such purposes. In the early 1960s, Penner had previously served as director of the Research and Engineering Division of the Institute for Defense Analyses. Penner had also worked closely with the aforementioned Theodore von Kármán. Penner's Center for Energy and Combustion Research was funded by: NASA, the Department of Defense, the Department of Energy, and the National Science Foundation - the usual suspects.

Let's look at more examples. Wright-Patterson AFB conducted a series of studies involving aluminum nanoparticle exposure. One of these studies noted that we are prone to inhale aluminum nanoparticles because they are used in jet fuels. Raytheon's infamous "Stratospheric Welsbach Seeding for Reduction of Global Warming" patent advocates for aluminum-spiked jet fuels. In their paper "Modification of Cirrus Clouds to Reduce Global Warming," David Mitchell and William Finnegan of the Desert Research Institute, writing of jet planes, suggest that, "the seeding material could either be dissolved or suspended in their jet fuel and later burned with the fuel to create seeding aerosol." In geoengineer Alan Robock's 2009 paper "Benefits, Risks, and Costs of Stratospheric Geoengineering," he and his co-authors write of adding sulfur to jet fuels as a means of dispersing particles suitable for geoengineering.

There are many other examples. Jim Lee over at *ClimateViewer.com* has been doing exceptional work in the area of jet fuel spiking and his website is a tremendous resource for this type of information as well as many other topics related to the New Manhattan Project. Specifically, Lee even produces U.S. patents which indicate that additives can be added to a jet's fuel mixture *in flight*. This would account for some variability in the length and presence of chemtrails due to fuel spiking.

References

Flying Tigers: Claire Chennault and his American volunteers, 1941-1942 a book by Daniel Ford, published by Smithsonian Books, 2007

William D. Pawley: The Extraordinary Life of the Adventurer, Entrepreneur, and Diplomat Who Cofounded the Flying Tigers a book by Anthony R. Carrozza, published by Potomac Books, 2012

Thomas G. Corcoran: The Public Service of Franklin Roosevelt's "Tommy the Cork" a PhD thesis by Monica Lynne Niznik, 1981

The Devil's Chessboard: Allen Dulles, the CIA, and the Rise of America's Secret Government a book by David Talbot, published by Harper Perennial, 2015

The Flying Tiger: The True Story of General Claire Chennault and the U.S. 14th Air Force in China a book by Jack Samson, published by Lyons Press, 2012

The Life and Political Economy of Lauchlin Currie: New Dealer, Presidential Adviser, and Development Economist a book by Roger J. Sandilands, published by the Duke University Press, 1990

Lemay: The Life and Wars of General Curtis LeMay a book by Warren Kozak, published by Regnery Publishing, 2009

Perilous Missions: Civil Air Transport and CIA Covert Operations in Asia a book by William M. Leary, published by Smithsonian Institution Press, 2002

Air America a book by Christopher Robbins, published by Cassell, 2012

Prelude to Terror: the Rogue CIA and the Legacy of America's Private Intelligence Network a book by Joseph J. Trento, published by Carroll & Graf, 2005

The Mafia, CIA, & George Bush a book by Pete Brewton, published by Shapolsky Publishers, 1992

The Politics of Heroin: CIA Complicity in the Global Drug Trade a book by Alfred W. McCoy, published by Lawrence Hill Books, 2003

Cocaine Politics: Drugs, Armies, and the CIA in Central America a book by Peter Dale Scott and Jonathan Marshall, published by The University of California Press, 1998

"Final Report of the Select Committee to Study Governmental Operations with Respect to Intelligence Activities" a report by the U.S. Senate Foreign and Military Intelligence Committee, published by the U.S. Government Printing Office, 1976

The Evergreen Story a series of reports appearing in the Orogonian Newspaper from August 4, 1988 to August 22, 1988

The Evergreen Story a book by Bill Yenne & Delford Smith, published by Evergreen International Aviation, 2008

"Geoengineering Cost Analysis" a report by Aurora Flight Sciences, 2010

"Chemtrail Whistleblower Allan Buckmann: Some Thoughts on Weather Modification" an article by Allan Buckmann, published by ActivistPost.com, August 16, 2012

"Geoengineering: Parts I, II, and III" hearings before the Committee on Science and Technology House of Representatives, 2009-2010

Interdepartmental Committee for Atmospheric Sciences reports 1960-1978, published by the Federal Council for Science and Technology

Skunk Works a book by Ben R. Rich & Leo Janos, published by Little, Brown and Company, 1994

MASDC II AMARC: Aerospace Maintenance & Regeneration Center, Davis-Monthan AFB, Arizona 1982-1997 a book by Barry Fryer and Martin Swann, published by Aviation Press Ltd and British Aviation Research Group, 1998

AMARC (MASDC III): Aerospace Maintenance & Regeneration Center, Davis-Monthan AFB, Arizona a book by Danny Bonny, Barry Fryer and Martin Swann, published by British Aviation Research Group, 2006

Prodigal Genius: The Life of Nikola Tesla a book by John J. O'Neill, published by Adventures Unlimited Press, 2008

Call Me Pat: The autobiography of the man Howard Hughes chose to lead Hughes Aircraft a book by Lawrence A. 'Pat' Hyland, published by The Donning Company Publishers, 1993

Tinker Air Force Base: A Pictorial History a book by the Office of History, Oklahoma City Air Logistics Center, Tinker Air Force Base, Oklahoma, Air Force Logistics Command, published by the US Government Printing Office, 1983

The CIA and the Cult of Intelligence a book by Victor Marchetti and John D. Marks, published by Dell Publishing Co., 1980

The Business of Science: Winning and losing in the high- tech age a book by Simon Ramo, published by Hill and Wang, 1988

Blank Check: The Pentagon's Black Budget a book by Tim Weiner, published by Warner Books, 1990

"Weather Modification: Fifth annual report" by the National Science Foundation, 1964, p23

"Weather as a Force Multiplier: Owning the Weather in 2025" a report by Col. Tamzy J. House, Lt. Col. James B. Near, Jr., LTC William B. Shields (USA), Maj. Ronald J. Celentano, Maj. David M. Husband,

Maj. Ann E. Mercer and Maj. James E. Pugh, published by the United States Air Force, 1996

"Drones Offer New Horizon, Solutions for Weather Modification" an article by Katy Galimberti, published by AccuWeather.com, June 10, 2014

Architects of American Air Supremacy: General Hap Arnold and Dr. Theodore von Karman - Conceptualizing the Future Air Force, Covering Rockets, Missiles, Jet Airplanes, Atomic Warfare, Propulsion by Dik A. Daso, published by Progressive Management Publications, 1997

Prophecy Fulfilled: 'Toward New Horizons' and its legacy a book by the United States Air Force, published by Progressive Management, 1994

"Forty Years of Research and Development at Griffiss Air Force Base" a report by John Q. Smith and David A. Byrd USAF, published by the United States Air Force, 1991

U.S. Patent #6,131,854 "Ground Handling Apparatus for Unmanned Tactical Aircraft," 2000

"Aerospace Sanctuary in 2025 - Shrinking the Bull's Eye" a report by Col. M. Scott Mayes, Maj. Felix A. Zambetti III, Maj. Stephen G. Harris, Maj. Linda K. Fronczak, and Maj. Samuel J. McCraw, produced by the U.S. Air Force, 1996

"The History of Power Transmission by Radio Waves" a paper by William C. Brown as it appeared in *IEEE Transactions on Microwave Theory and Techniques*, Vol. MTT-32, No. 9, September, 1984

Tesla: Man Out of Time a book by Margaret Cheney, published by Simon & Schuster, 1981

Prodigal Genius: The Life of Nikola Tesla a book by John J. O'Neill, published by Adventures Unlimited Press, 2008

Tesla: Inventor of the Electrical Age a book by W. Bernard Carlson, published by Princeton University Press, 2013

U.S. patent #787,412 "Art of Transmitting Electrical Energy Through the Natural Mediums," assigned to Nikola Tesla, 1905

U.S. patent #1,119,732 "Apparatus for Transmitting Electrical Energy," assigned to Nikola Tesla, 1914

Tesla vs. Edison: The Life-long Feud that Electrified the World a book by Nigel Cawthorne, published by Chartwell Books, 2016

"The Microwave Powered Helicopter" a paper by William C. Brown, published by Raytheon Company, Burlington, Massachusetts, as it appeared in the premier edition of *The Journal of Microwave Power*, 1964

"The History of Wireless Power Transmission" a paper by William C. Brown, published by *Solar Energy*, vol. 56, no. 1, p 3-21, 1996

U.S. patent #5,068,669 "Power Beaming System," 1991

Angels Don't Play this HAARP: advances in Tesla technology a book by Jeane Manning and Dr. Nick Begich, published by Earthpulse Press, 1995

U.S. patent #5,218,374 "Power Beaming System with Printer Circuit Radiating Elements Having Resonating Cavities," 1993

U.S. patent #3,464,207 "Quasi-Corona-Aerodynamic Vehicle," 1969

U.S. patent #4,955,562 "Microwave Powered Aircraft," 1990

U.S. patent #5,503,350 "Microwave-Powered Aircraft," 1996 139

U.S. patent #6,364,253 "Remote Piloted Vehicle Powered by Beamed Radiation," 2002

U.S. patent #6,534,705 "Methods and Apparatus for Beaming Power," 2003

U.S. patent #20080017239A1 "Photovoltaic Receiver for Beamed Power," 2008

U.S. patent #7,711,441 "Aiming Feedback Control for Multiple Energy Beams," 2010

U.S. patent #7,929,908 "Method and System for Controlling a Network for Power Beam Transmission," 2011

U.S. patent #20120150364A1 "Unmanned Vehicle and System," 2012

U.S. patent #8,876,061 "Methods and Systems for Beam Powered Propulsion," 2014

“Green New Deal Omen: 4-Hour Flying Electric Plane Wants FAA Okay” an article by Penny Starr, published by *Breitbart.com*, July 27, 2019

“‘World’s First’ Fully-electric Commercial Flight Takes Off” published by the BBC, December 11, 2019

“Airbus unveils its massive solar-powered drone that can stay in the air for 45 DAYS, after successful test flight in Arizona” an article by Annie Palmer, published by the Daily Mail, July 18, 2018

“Electric aircraft near take-off as Rolls-Royce and Airbus team up to build ‘e-jets’" an article by Alan Tovey, published by The Telegraph, November 28, 2017

Boeing KC-135 Stratotanker: More than Just a Tanker a book by Robert S. Hopkins III, published by Midland Publishing Limited as part of the Aerofax series, 1997

Tinker Air Force Base: Sixty Years of History 1942-2002 a book by the Oklahoma City Air Logistics Center, 2002

50 Years of the Desert Boneyard a book by Philip D. Chinnery, published by Motorbooks International, 1995

Fixing the Sky: The Checkered History of Weather and Climate Control a book by James Roger Fleming, published by Columbia University Press, 2010

"Nanosized Aluminum Altered Immune Function" a paper by Laura K. Braydich-Stolle, Janice L. Speshock, Alicia Castle, Marcus Smith, Richard C. Murdock, and Saber M. Hussain, published by the American Chemical Society, 2010

U.S. patent #5,003,186 "Stratospheric Welsbach Seeding for Reduction of Global Warming" by David B. Chang, 1991

"New Horizons of Naval Research and Development" a paper by William Francis Raborn, published in *U.S. Naval Institute Proceedings*, January, 1963

"Aviation and the Global Atmosphere" a report by the Intergovernmental Panel on Climate Change (IPCC), 1999

"Engineering the Climate: Research Questions and Policy Implications" a report by The United Nations Educational Scientific and Cultural Organization (UNESCO), November, 2011

Cocaine Politics: Drugs, Armies and the CIA in Central America a book by Peter Dale Scott and Jonathan Marshall, published by the University of California Press, 1998

The Secret Team: The CIA and its Allies in Control of the United States and the World a book by L. Fletcher Prouty, published by Skyhorse Publishing, 2008

"Cloak and Dagger in the Back: Businessman Erwin Rautenberg secretly worked for the CIA, until it dumped him" an article by John Mintz, published in the *Washington Post*, Feb. 26, 1995

The Lady and the Tigers a book by Olga Greenlaw, published by Warbird Books, 2012

Prelude to Terror: The Rogue CIA and the Legacy of America's Private Intelligence Network a book by Joseph J. Trento, published by Carroll & Graf, 2005

"E-Systems Sells Unit" an article published by Dow Jones and Co., June 15, 1987

"9/11: CIA Likely Built Remote-Controlled Commercial Jets in Aircraft Boneyard" an article by Wayne Madsen, published by *Infowars*, October 29, 2015

"Marana and Ascent to Merge" an article published by *PRNewswire*, December 15, 2016

"The ESSA Research Flight Facility: Facilities for Airborne Atmospheric Research" a report by Howard A. Friedman, Frank S. Cicirelli, and William J. Freedman, published by the Environmental Science Services Administration, August 1969

A History in the Making: 80 Turbulent Years in the American General Aviation Industry a book by Donald M. Pattillo, published by McGraw-Hill, 1998

"Temco Tidings" a newsletter published by the Texas Engineering and Manufacturing Company, Oct. 6,1946, Vol. 1, No. 16

I'm an Endangered Species: The Autobiography of a Free Enterpriser a book by David Harold 'Dry Hole' Byrd, published by Pacesetter Press, 1978

The National Aeronautics and Space Administration a book by Richard Hirsch and Joseph John Trento, published by Praeger Publishers, 1973

The Rise and Fall of the Conglomerate Kings a book by Robert Sobel, published by Stein and Day, 1984

"E-Systems Coming Out of the Shadows" an article by Gregg Jones, published by the Dallas Morning News, Oct. 18, 1992

"Nanosized Aluminum Altered Immune Function" a paper by Laura K. Braydich-Stolle, Janice L. Speshock, Alicia Castle, Marcus Smith, Richard C. Murdock, and Saber M. Hussain, published by the American Chemical Society, 2010

"Remembering Stanford 'Sol' Penner" an article by Daniel Kane, published by UC San Diego, July 20, 2016

"Active Measures for Reducing the Global Climatic Impacts of Escalating CO2 Concentrations" a paper by S.S. Penner, A.M. Schneider, and E.M. Kennedy, published in Acta Astronautica, Vol. 11 No. 6, pp. 345-348, 1984

"A Low-Cost Technology for Increasing the Earth's Albedo to Mitigate Temperature Rises" a paper by S.S. Penner and J. Haraden, published by *Energy* Vol. 18 No. 10, pp. 1087-1090. 1993

"University of California, San Diego Center for Energy and Combustion Research" a report by the UC San Diego Center for Energy and Combustion Research

"Modification of Cirrus Clouds to Reduce global Warming" a paper by David L. Mitchell and William Finnegan, published by *Environmental Research Letters*, October 30, 2009

"Benefits, Risks, and Costs of Stratospheric Geoengineering" by Alan Robock, Allison Marquardt, Ben Kravitz, and Georgiy Stenchikov, published in *Geophysical Research Letters*, Oct. 2, 2009

Chapter 6
COAL FLY ASH

Hard and compelling scientific data indicates that the common particulate spray of the New Manhattan Project is coal fly ash. Many times this data has been collected, compiled, peer-reviewed, and published in reputable scientific journals. A world famous scientist is behind the discovery and dissemination of these findings: Dr. James Marvin Herndon, PhD.

What is coal fly ash? Coal fly ash is the smoke from burning coal. It is produced by the megaton as a toxic waste byproduct of the electrical power industry. In order to generate electricity, coal is often burned in electrical power plants. The smoke from the burning coal is often collected by massive machines called 'electrostatic precipitators' which whisk the coal fly ash away using principles of atmospheric electricity which are, in themselves, germane to the New Manhattan Project. If they didn't spray it all over us, it would cost the power companies a pretty penny to dispose of it properly, so it most probably makes dollars and cents to ruin our health and wreck our environment. Do you see how that works?

This chapter examines Dr. Herndon's work and takes a long, hard look at coal fly ash as well as the history of the field of coal fly ash sequestration and utilization. As we will continue to see, the substance itself and the history of the field of coal fly ash utilization fit neatly into weather modification and the New Manhattan Project.

Catching up with Dr. J. Marvin Herndon

Dr. J. Marvin Herndon is an American interdisciplinary scientist. He earned his BA in physics in 1970 from the University of California at San Diego (UCSD) and his PhD in nuclear chemistry in 1974 from Texas A&M University. He has worked extensively with famous scientists such as geochemist Hans Suess (1909-1993) and Manhattan Project physical chemist Harold Urey (1893-1981), who themselves were trained by master scientists, including Nobel Laureate Niels Bohr (1885-1962). Dr. Herndon's scientific papers have been published in world-class journals such as the *Proceedings of the Royal Society of London* and the *Proceedings of the National Academy of Sciences*. He has also been profiled in the popular press worldwide including a *Discover Magazine* cover story. Dr. Herndon is a well-known and respected, politically active citizen.

Although he has written a great many scientific journal articles over his multi-decade career, his first published anti-geoengineering piece appeared in the June 2015 edition of *Current Science*. It is the first published, anti-SRM geoengineering journal article. It is titled "Aluminum Poisoning of Humanity and Earth's Biota by Clandestine Geoengineering Activity: Implications for India." Later that year he published another anti-geoengineering piece in the *International Journal of Environmental Research and Public Health*. Come early 2016, he gave us another; this time published in the *Indian Journal of Scientific Research and Technology*. In the middle of 2016, he published another anti-geoengineering paper in *Frontiers in Public Health*; followed by yet another in the *Journal of Agricultural Science*. A complete bibliography of Herndon's anti-geoengineering journal articles can be found in the 'References' section of this chapter.

Dr. J. Marvin Herndon, PhD

With the start of 2017, Herndon only upped his game. Beginning in 2017, Herndon's papers identifying coal fly ash as common chemtrail spray were published after being *peer-reviewed*. All four 2017 articles were published in the *Journal of Geography, Environment and Earth Science International*. Herndon also co-authored three 2018 peer-reviewed papers which further substantiated the evidence for common chemtrail spray being coal fly ash.

When a paper is peer reviewed, that means it is peer endorsed. This means that one might as well put the peer reviewers' names in place of the author's. The peer reviewers are verifying the work of the author(s). This way, the shills can no longer claim that it's just old J.

Marvin shooting off again. As time has gone on, more have been joining him. They know the scientific establishment is up against them, but they're putting their careers on the line anyway. They are Sparticus! This is huge.

Throughout Herndon's anti-geoengineering articles, one finds a staunch opposition and many bold assertions. We concern ourselves here with his coal fly ash hypothesis because this hypothesis can be effectively used to produce further avenues of investigation. Of particular interest to our movement, though, in multiple articles, Herndon provides us with a definitive word on the chemtrails vs. contrails debate. And the word is that contrails do not persist. Chemtrails do. Please use his information to defeat the trolls when they claim otherwise, as they so often do.

In 2014, Dr. Herndon self-published a book titled *Herndon's Earth and the Dark Side of Science* in which he writes about the ongoing chemtrail spraying. He includes 4 pictures of chemtrails in the sky. He describes his observations such as how the trails can abruptly begin and end, spread out over time, and create an artificial cloud cover. He writes of rainwater samples testing anomalously high in levels of aluminum and barium. He writes of probable adverse health impacts. He writes that the full extent of the potential Human health and environmental consequences are wholly unknown. He notes that scientists involved in malevolent agendas such as this have no concern for Humanity. He writes that the chemtrail spraying operations are indicative of a seriously compromised and corrupted scientific establishment that will take a long time to fix. He finishes his section on chemtrails noting that, due to the unforeseen consequences, "It is a naïve and foolish notion that humans can engineer Earth to the benefit of humanity."

Dr. Herndon's detractors (the shills) like to claim that because two of his early anti-geoengineering journal articles were retracted, that this somehow discredits the good doctor. These retractions were political decisions, not scientific ones. Dr. Herndon says the journals which retracted his articles violated codes of conduct up one side and down the other. In writings posted on his website, he says that these journals (the *International Journal of Environmental Research and Public Health* and *Frontiers in Public Health*) have internal problems that should be resolved. He says that he is pursuing the issue.

Coal fly ash

Coal fly ash is the ash that rises when coal burns. It is an industrial waste byproduct of coal-fired electrical power plants. Today in America there is a vast, inexpensive, and readily available supply of coal fly ash because we use devices called 'electrostatic precipitators' to remove the ash from the smokestacks of coal-fired power plants.

It is interesting to note that when today's coal fly ash is sequestered with an electrostatic precipitator, a mini New Manhattan Project occurs. You see, electrostatic precipitators collect fly ash by first electrically charging the airborne ash particles then attracting them to oppositely charged plates where the ash is then taken away. This is analogous to how, in the course of today's New Manhattan Project, atmospheric particles are ionized (charged) and then manipulated.

Once the fly ash has been sequestered, then the question as to what to do with the stuff arises. It's expensive to dispose of it properly. Enter the field of coal fly ash utilization. For thousands of years, coal ash has been used in cements. Today coal ash is used in construction materials such as roadbeds and high-quality cements. Coal ash enables cementitious building materials to dry faster and set stronger. But only about 45% of today's national coal fly ash production is used for these purposes. Most of it is ostensibly buried as a toxic waste, to the great expense of the power industry.

In attempts to turn their lemons into lemonade, since 1967, international fly ash utilization symposia have been held. Organizations such as: the Calgary Fly Ash Research Group, the Western Fly Ash Research Development and Data Center, the Edison Electric Institute, the Electric Power Research Institute, the American Public Power Association, the Department of the Interior, and the Department of Energy have historically worked in the field of coal fly ash utilization. Today the biggest dog in the space appears to be an industry group called The American Coal Ash Association.

Hunting for the source

Chemtrails are probably coal fly ash derived from a *certain type* of coal. Here we hunt for a single source of coal used to produce the coal fly ash of the New Manhattan Project.

It is reasonable to assume that the coal used to produce today's coal fly ash chemtrail spray is derived from a single source because the Project probably demands that the chemtrail spray be standardized. It would not be advantageous to have inhomogeneous chemtrail spray particle clouds. Inhomogeneous particle clouds of chemtrail spray would be problematic to unworkable in the context of the New Manhattan Project because such *ad hoc* atmospheric saturations would probably complicate operations. There may be multiple power plants and multiple power plant operators producing the coal fly ash, but we are probably looking at a single source of coal because it would be easier to standardize the fly ash from a single source.

We may be able to identify the coal source by matching NMP coal fly ash properties to those of a known, standardized coal fly ash. Identifying the coal source can lead us to the identity of the miner(s) and other people and organizations involved. The final physical and chemical properties of coal fly ash are a function of the type of coal, the combustion system of the power plant, and the collection system used.

We already know some things about this single source of coal. From hundreds of rainwater sample test results and the California Air Resources Board's ambient air samplings (ch 1), we know that this coal and its resultant ash contain barium and strontium and have a high aluminum content. Aluminum is very responsive to electromagnetic energy, so that should be helpful within the context of the NMP. It would probably be a high-calcium ash. High-calcium coal fly ashes have a basic pH and environmental scientists such as Francis Mangels and others from the USDA Soil Conservation Department have confirmed that soils are turning basic. The evidence also suggests that the coal deposit for which we are hunting contains high levels of aluminum and calcium as well as barium, and strontium. So where might we find such a coal? It would probably be a large deposit of American coal. The New Manhattan Project has almost entirely been developed in America. It would be easier to maintain this secret project's cover and more convenient to use coal from an abundant, domestic source rather than any importing. It would need to be a large deposit because of the massive quantities of ash required.

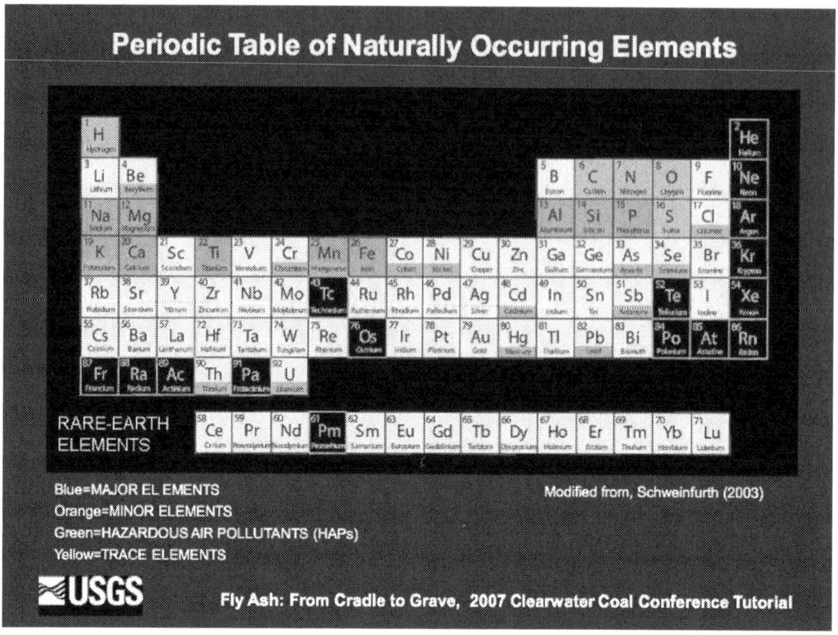

Coal fly ash constituents

NIST SRMs 1633c & 1633a

In order to find the coal in question, we can apply the parameters of our hypothetical coal to standardized coal fly ashes. For many decades now, the National Institute of Standards and Technology (NIST) has been certifying standardized coal fly ashes. Here we compare our hypothetical coal to the NIST standardized fly ashes.

Of these NIST standardized coal fly ashes currently available, the one that most closely matches our hypothetical NMP coal fly ash is NIST Standard Reference Material (SRM) 1633c. This has been determined by comparing the average ratios of aluminum to barium found in the American (mostly) rainwater sample test results posted on *GeoengineeringWatch.org*, to the ratios of same in the standardized NIST coal fly ashes. From the 31 sample test results considered, the average ratio of aluminum to barium is 65 to 1. The ratio of aluminum to barium in SRM 1633c is 121 to 1.

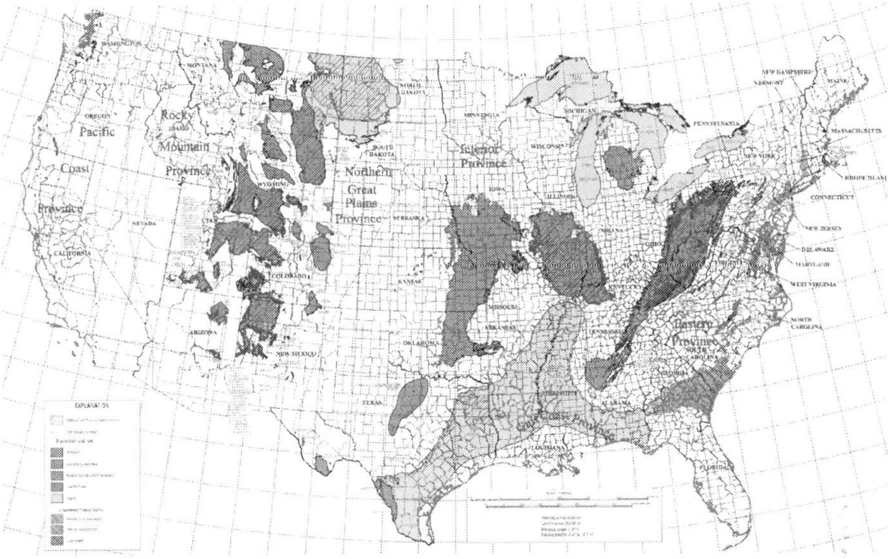

American coal deposits

Only the ratio of aluminum to barium is considered here because that's where the abundance of meaningful data is. Most of the test results included aluminum and barium while test results for the other elements were much more scarce. Average ratios produced from lacking data sets are prone to be less representative and therefore less accurate and are therefore not included here.

NIST describes SRM 1633c only as, "supplied by a coal fired power plant and is the product of Western Pennsylvania bituminous coal." It is interesting to note that NIST notes the exact origins of all the other SRM fly ashes, such as the name and location of the power plant from whence it came and the specific location of the coal deposit, but they do not list such information for 1633c. Is this information omitted in order to thwart our investigation?

A discontinued NIST coal fly ash known as SRM 1633a is an even closer match. Although NIST stopped selling SRM 1633a in 1993 (a few years before large-scale spraying operations began), the certificate for it is still available online and the ratio of aluminum to barium in SRM 1633a is 95 to 1. Again without disclosing the power plant(s) involved or a specific coal source, NIST says only that 1633a is, "a product of Pennsylvania and West Virginia coals." The region around the border of Pennsylvania and West Virginia and up on into western Pennsylvania is the Northern Appalachian region and this is one of the largest deposits of coal in North America.

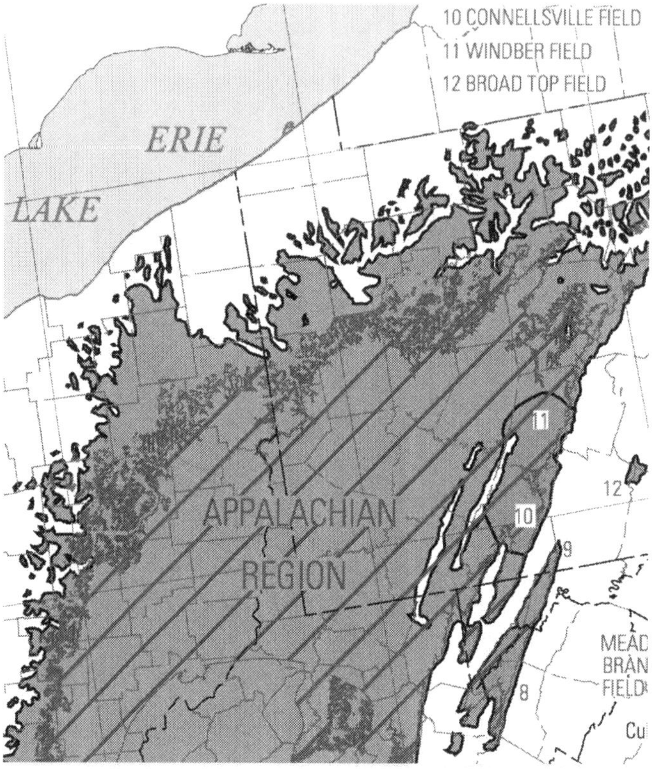

Northern Appalachian coal deposits

Both 1633c and 1633a have other elements noted here which have been consistently showing up in the rainwater sample test results such as calcium and strontium.

Maybe today's chemtrail sprayers are using this discontinued 1633a SRM fly ash. Just because NIST is no longer providing it does not mean that it can't still be in production. Once the production methods have been established, it would be relatively easy to continue to produce the material.

The next logical step is for a reputable scientist such as Marvin Herndon to order a 75g sample of SRM 1633c for $698 and compare its chemical signature to that of collected chemtrail spray. Better yet, maybe someone can get their hands on a sample of 1633a. A comparison might also be made by an analysis of readily available data such as that which has been presented here, but with more detail. However it happens, it would behoove our anti-geoengineering movement to further explore these findings.

Fly ash and weather modification

There is quite a historical precedent for the use of coal fly ash in weather modification and the atmospheric sciences. Many military men have claimed that the smoke from exploded bombs causes precipitation. The early American meteorologist James Pollard 'The Storm King' Espy (1785-1860) claimed that the smoke from forest fires causes rain to fall. The Nobel Prize-winning weather modifier Irving Langmuir and NMP pioneer Bernard Vonnegut (ch 3) were working with fine particle oil smokes way back during WWII. The nucleant discovered by Vonnegut and used through to the conventional weather modification industry of today is the smoke of burning silver iodide. In all of these instances, as with coal fly ash, smoke from some type of fire is either said to cause or is known to cause weather modification. The geoengineers of today often talk about how stratospheric masses of volcanic ash spewing from a fiery volcano can reflect sunlight back into space and therefore cool the planet, saving us from the dreaded global warming - in other words, the SRM geoengineering thesis. In 1963, weather modifier Narayan R. Gokhale of the Department of Earth and Atmospheric Sciences of the State University of New York was experimenting with volcanic ash as a nucleant.

It is possible to produce coal fly ash particles of the proper size for optimum nucleation. Particles optimally sized for nucleation would be the most useful for the New Manhattan Project as those particles attract the most amount of atmospheric water. The proper size is in the neighborhood of one tenth of a micron (.1μ). Therefore it is safe to assume that the coal fly ash particles to be used as part of the New Manhattan Project are in the neighborhood of .1μ in diameter. .1μ is extremely small, but, to attain the proper particle size, before combustion, the coal can be finely pulverized. A coal-fired power plant finely pulverizing their coal before combustion may be an indication that the resultant fly ash is to be used as part of the New Manhattan Project.

The different elements which comprise the coal fly ash of the New Manhattan Project may serve different purposes. Aluminum has been used extensively as a nucleant. Silicon (which presents abundantly in coal fly ash) has too. The barium and strontium probably serve as atmospheric tracers. Geoengineers are on record as using tiny radioactive particles, monitored by satellites and computers, to track atmospheric movements, and both barium and strontium can be naturally radioactive.

The circumstantial evidence for coal fly ash as common chemtrail spray just goes on and on. The First International Coal Fly Ash Utilization Symposium and Dr. Marvin Herndon both mention how the iron oxide in the fly ash makes the fly ash a brownish color. Your author has also observed that chemtrail spray has a brownish tint. Many of the things geoengineers say they want to spray us with are in coal fly ash like silica, sulphates, and aluminum.

The use of coal fly ash as chemtrails has precedent in that it is analogous to water fluoridation. What our local water district calls 'fluoride' is not the fluoride most people think it is. What is put in the water is an industrial waste by-product called hydrofluosilicic acid which, if the population was not forced to ingest it, would be expensive to get rid of. Instead, corrupt governments pay for it to be consumed by the unsuspecting public and a giant disinformation apparatus constantly tells us that anyone opposed to water fluoridation is a crazy conspiracy theorist. Does that sound familiar?

Although it is naturally a powder, the NMP's coal fly ash would probably need to be reconstituted into some type of liquid. A liquid would work well within the context of the New Manhattan Project as a liquid could be more easily and efficiently transported and stored. When dispersed, a liquid would be preferable to a powder because liquids tend to clog-up the necessary tubing much less. A final blast of heat from a jet engine might turn a dense liquid into tiny, finely dispersed particles.

The Welsbach effect

When the atmospheric coal fly ash particles of today's New Manhattan Project are hit with the right electromagnetic energy frequency, they heat up. The most effective heating frequency is known as a particle's 'resonant frequency.' Different materials have different resonant frequencies.

When large masses of atmospheric coal fly ash particles are heated by certain frequencies of applied electromagnetic energy, they exhibit something called the 'Welsbach effect.' This is the effect that happens when electromagnetically heated atmospheric particles make other, similar particles around them heat up (or resonate) as well. It is demonstrated in the mantle of a gas lantern. Applied energy makes the entire mantle light up not because the mantle is soaked with fuel, but because the particles comprising the mantle are resonating together. The New Manhattan Project turns our atmosphere into a gigantic mantle with chemtrail spray comprising the mantle material and electromagnetic energy being the applied energy.

The 1988 U.S. patent #4,755,673 "Selective Thermal Radiators" by Slava A. Pollack and David B. Chang describes how small particles may be energized in this fashion. David B. Chang, one of the inventors listed on "Selective Thermal Radiators," is also the sole inventor noted on the aforementioned "Stratospheric Welsbach Seeding for the Reduction of Global Warming" patent. Hughes Aircraft is listed as the assignee on both.

The former President and CEO of Hughes Aircraft was a man by the name of Lawrence 'Pat' Hyland (1897-1989). He wrote a 1993 autobiography (which appeared posthumously) titled *Call Me Pat*. On this book's cover, Mr. Hyland is pictured lighting a gas lantern, and thus producing the Welsbach effect.

As one can see, the lantern Pat lights has an inscription. Although your author was initially unable to decipher this inscription, a resourceful commenter from *IntelliHub.com* going by the name of Tonya Gray provides us with a readable close-up. The inscription reads, "THE OFFICIAL BICENTENNIAL PAUL REVERE - NORTH CHURCH LANTERN AUTHORIZED BY THE CONCORD ANTIQUARIAN SOCIETY." The inscription means that this lantern is a replica of one of the two lanterns lit on the evening of April 18, 1775 to warn of British soldiers' advancing in boats across the Charles River into Cambridge, Massachusetts. We know it is a replica because in 1775 the Welsbach effect had not yet been discovered.

Call Me Pat book cover

This is interesting because the Boston area has TONS of connections to the New Manhattan Project. Boston is home to the American Meteorological Society which is up to its eyeballs in this thing. Harvard University, the home of geoengineer David Keith, is in the area. Cambridge is home to research firm Arthur D. Little who produced the early New Manhattan Project experiments of Bernard Vonnegut *et al.* Cambridge is also home to the former Air Force Cambridge Research Laboratories which has extensive ties to the New Manhattan Project. The Massachusetts Institute of Technology, which has been heavily involved, is also in Cambridge. This is just scratching the surface. What could be the significance of this inscription, though?

Call Me Pat book cover lantern inscription close-up image

A history of coal fly ash and electrical precipitation

When one looks back in time at the history of coal fly ash sequestration and utilization, appropriately enough, one finds the New Manhattan Project. The history of coal fly ash as it pertains to the New Manhattan Project begins with its sequestration. You see, since the beginnings of the Industrial Revolution, much of the industrialized world (especially Europe and the industrialized centers of America) struggled with poor air quality due to the emissions coming from local factories. This pollution was the byproduct of myriad industrial processes, not the least of which was burning coal.

But necessity is the mother of invention and processes were developed that allowed for the sequestration of the offending particulate matter. Although the measures were costly, these pollution-eliminating practices were eventually enacted mostly due to class-action lawsuits lodged by local residents. These processes of removing particulate matter from the emissions of industrial factories started in Europe with the scientific field known as 'electrical precipitation.'

The principles of electrical precipitation are based on the fact that atmospheric particles are naturally attracted to bare, charged wires. Electrostatic precipitation goes back to Queen Elizabeth's court physician and early geophysicist William Gilbert (1544-1603). In 1600 he noted that atmospheric particles were commonly attracted to 'electricks.' In the early 1800s, M. Hohlfield of Leipzig University demonstrated the effect by clearing smoke from the inside of a glass jar with an electrified wire.

In the late 1800s commercial possibilities for the field arose with the work of Sir Oliver Lodge (1851-1940). Lodge applied electrical precipitation to the recovery of valuable substances hitherto gone 'up in smoke' as one might say. Before Lodge's inventions, significant quantities of valuable materials such as potash, lead, tin, silver, and gold were lost during industrial processes. If the process involved vapors arising from the substance, then Lodge invented ways of positioning electrically charged wires which would attract the vaporized substances and simultaneously purify the air. It was a win win. The industrial producers recovered significant quantities of materials and the factory workers and general public enjoyed cleaner air. In order to commercialize his processes, Lodge formed his own company called the Lodge Fume Deposit Company.

Sir Oliver Lodge

Lodge also created another endeavor called the Agricultural Electric Discharge Company. Here Lodge applied the field of electrical precipitation to agriculture and weather modification. He found that electrical precipitation can be used to eliminate low-lying fogs, mists, and smoke that can interfere with plant growth. This phenomenon is well documented in the historical weather modification literature. The field of electrical precipitation is also directly applicable to the space charge experiments previously detailed in chapter 3.

In America, the main early proponent of these types of technologies was a man by the name of Frederick Gardner Cottrell (1877-1948). In the early 1900s, Cottrell also applied the principles of electrical precipitation to the problems of industrial pollution. Cottrell formed a company called the Western Precipitation Company which built machines for industrial plants that could reduce or eliminate their atmospheric emissions. Cottrell's Western Precipitation Company and its parent company, the International Precipitation Company soon enjoyed tremendous success. Orders for their goods and services poured in from all over the world.

Frederick Gardner Cottrell

Interestingly, Cottrell and his business associates, instead of ruthlessly exploiting the markets they had created, ultimately decided to relinquish most of their interests in these companies for the creation and support of something called the Research Corporation. The Research Corporation, later known as the Research Corporation for Science Advancement, was created as a vehicle for the funding of basic science.

Once the Research Corporation had been established in 1912, Cottrell returned to his native San Francisco Bay Area where he, like Lodge, applied the principles of electrical precipitation to weather modification and the atmospheric sciences by dissipating low-lying fogs with electrified wires.

Immediately following WWI, the manager of Cottrell's International Precipitation Company, a man by the name of Walter Schmidt, went to Europe to rejuvenate International Precipitation's business there. In the course of doing so, he joined forces with Sir Oliver Lodge's Fume Deposit Company to form the British firm Lodge-Cottrell Ltd.. To this day, Lodge Cottrell Ltd. produces and services industrial electrostatic precipitators, mostly outside of America.

Back in the States, Frederick Cottrell was put in touch with members of the Rockefeller and Carnegie families when he served as the chairman of the National Research Council's Division of Chemistry and Chemical Technology. The National Research Council was funded by the Rockefeller and Carnegie Foundations, among others.

Here's where Cottrell's work in coal fly ash utilization comes in. In the early 1930s, in association with the Research Corporation, the Smithsonian had established a laboratory for Cottrell's use. In this laboratory, Cottrell installed a man by the name of Chester Gilbert who was formerly the president of the American Coal Corporation. At their Smithsonian laboratory, after receiving an anonymous donation of $6K, Gilbert initially investigated the production of lime-gypsum plaster based on some Research Corporation patents. Gilbert's lime-gypsum work led him to work pertaining to the use of coal fly ash. This work put Gilbert and Cottrell among the modern pioneers in the field of coal fly ash utilization. Gilbert and Cottrell initially figured on how coal fly ash could be better processed and then used as a filler in cements and plasters. They also went about processing coal fly ash for use as a household cleaning powder.

The author of Cottrell's biography, Frank Cameron describes the Research Corporation's Smithsonian laboratory in a very interesting way. Cameron describes the Smithsonian laboratory as analogous to a phenomenon germane to weather modification and the atmospheric sciences: nucleation. Curiously, when referring to the laboratory where Gilbert and Cottrell performed their pioneering work in the field of modern coal fly ash utilization, Cameron writes, "They [Gilbert and Cottrell] did not foresee it as the mote, the speck of dust around which Cottrell's ideas and those of his protégés, like so many particles of moisture, would begin to coalesce to form the drop of rain."

It may have been during this time that Gilbert, Cottrell, or other pioneers in the modern field of coal fly ash utilization noticed and/or figured that the smoke from coal-fired, electrical power plants causes precipitation. It had been noted for many years previously that explosions and smoke from burning fires cause precipitation. Maybe the Research Corporation and their Smithsonian laboratory attempted to determine the validity of these claims by doing a study about whether or not, under the appropriate atmospheric conditions, the smoke from coal-fired electrical power plants can be scientifically proven to cause precipitation. As previously noted, it is known today that if coal is finely pulverized before combustion, then many of the resultant fly ash particles will be the optimum size for atmospheric nucleation (.1 micron). Maybe they even collected some coal fly ash from one of their electrical precipitators and dumped it out of an airplane to see if it caused precipitation.

It is interesting to note that at around the same time that Gilbert and Cottrell were pioneering the modern field of coal fly ash utilization, the foundations were also being laid for what would later become the New Mexico Tech Research & Development Division (R&DD). This is the organization detailed in chapter 3 that sponsored Vonnegut and Moore's later space charge experiments. In 1935 a physicist by the name of Everly John Workman began his studies of clouds and lightening in the southwest. Workman founded the R&DD. As the R&DD was a division of New Mexico Tech, it is also interesting to note that New Mexico Tech used to be called the New Mexico School of Mines whereas Cottrell had been a physical chemist with the Bureau of Mines and had organized the Bureau's San Francisco office and laboratory.

Somewhere along the line, Cottrell's electrical precipitators were significantly improved with the principles of electrostatics. The resulting precipitators were hitherto known as electrostatic precipitators rather than electrical precipitators. As previously described, rather than simply collecting atmospheric particles upon charged wires, these newer electrostatic precipitators use the principles of static electricity. Electrostatic precipitators are what continue to be in use today.

Today, a company known as Hamon-Research Cottrell exists as the descendant of Cottrell's Western and International Precipitation Companies and they conduct air quality control operations at industrial plants throughout the world.

Dresser Industries and the Bush family

The largest builder and servicer of electrostatic precipitators in America and Europe today is known as Dresser Industries. Their company history goes back to the late 1800s and they have historically worked with organizations heavily associated with the New Manhattan Project. They bought the aforementioned Lodge-Cottrell in 1972. Dresser Industries is also a Bush family business with Nazi connections. Is Dresser Industries providing the coal fly ash needed for today's New Manhattan Project?

Dresser Industries has many connections to the New Manhattan Project. First of all, Dresser's Pacific Pump Works produced equipment which enabled the scientists of the original Manhattan Project to refine the uranium used in the world's first atomic bombs. With the dawning of the Space Age, Dresser positioned itself for work in areas connected to the New Manhattan Project. In 1957, about a month after Sputnik scared the pants off of the Pentagon, Dresser created a division known as Dresser Dynamics and they were positioned to develop new missile navigation technologies. As we have already learned (ch 2), the air traffic control systems of today's New Manhattan Project partially originated from early automated missile battery systems. Another Dresser division produced over-the-horizon radar equipment - technology succinctly applicable to the New Manhattan Project (ch 3). Dresser has also worked extensively with the National Aeronautics and Space Administration (NASA) and the Atomic Energy Commission; two organizations with strong implications for the New Manhattan Project. Dresser's division known as Dresser Electronics did work at New Mexico's White Sands Missile Range; the site of the world's first atomic bomb blast and the original home of the most significant group of former Nazi scientists working with General Electric in areas relevant to the NMP. Many of Dresser's top executives have historically come from the Massachusetts Institute of Technology (MIT) - an institution with only the most significant implications for the New Manhattan Project. Dresser Industries maintained a division in Santa Ana, CA called the Environmental Technology Division which, according to their official company history, "carried out research in air, water, and solid waste pollution abatement, and advised the company on all environmental concerns." The former director of Stanford Research Institute's Southern Laboratories, a man by the name of Dr. Robert D. Englert ran the division. Something called the LTV Tower, which sits next to the giant Dresser Building in downtown Houston, was acquired by Dresser in 1975. LTV stands for the aforementioned Ling-Temco-Vought. As noted in previous chapters, Ling-Temco-Vought created a spin-off called LTV Electrosystems which became E-systems before it built HAARP. Dresser's headquarters have historically been in Houston, TX. The same place from whence Enron came. Enron first developed the weather derivatives market where one can make a wager on almost any type of geophysical condition. The New Manhattan Project can manipulate virtually any type of geophysical condition. What a money-maker! We all know what happened to the Bush family-friendly Enron. Dresser is still around today. They are now known as Dresser-Rand, a Siemens Business. Siemens is a former Nazi company.

Way back in 1928, the Bush family bank bought Dresser. At the time, Dresser was a moderately large and successful business mostly engaged in oilfield production and distribution equipment. Senator Prescott S. Bush (grandaddy Bush) and his cronies from W. A. Harriman and Company turned it into an international behemoth. Grandaddy Bush later passed his regard for Dresser onto his son, George H. W. Bush (Bush Sr.). As we all know, Bush Sr. is now dead.

Perhaps Dresser's mantle has since been passed to George W. Bush (Bush Jr.).

The Bush family in front of a Dresser aircraft

W. A. Harriman and Company is known today as Brown Brothers Harriman; one of the largest private banks in the world and one of the banks closest to the private United States Federal Reserve Bank which makes sure that Brown Brothers Harriman gets only the freshest (and most valuable) capital. The author has seen no evidence indicating that the Bush family or their Brown Brothers Harriman bank has since divested itself from Dresser. Brown Brothers Harriman is a bank known to have done substantial business with Nazi Germany.

Spray production

Is Monsanto producing today's chemtrail spray? This book provides much evidence supporting the assertion that the New Manhattan Project is a Department of Defense (DOD) production, and from Monsanto's production of Agent Orange during the Vietnam War, we see a history of Monsanto producing toxic substances for the military to be sprayed from aircraft. In fact, Monsanto's research director and later President and Chairman of the Board Charles Allen Thomas (1900-1982) was also a key scientist of the original Manhattan Project.

Charles Thomas got his master's degree in chemistry from the Massachusetts Institute of Technology, was a trustee of the Carnegie Corporation, was a director and a trustee of the RAND Corporation, and served as a lifetime member of the Corporation of the Massachusetts Institute of Technology. He was elected to the Monsanto Board of Directors in 1942. He was president from 1951 to 1960. In 1960 he was elected chairman of the Monsanto Board of Directors. Thomas held a total of 86 U.S. and foreign patents. Following WWII, Thomas was a co-author of a report titled "A Report on the International Control of Atomic Energy." The report, known as the Acheson-Lilienthal Report, proposed a master plan for the international control of atomic energy.

During the time of the original Manhattan Project, Monsanto's research laboratory was located in Dayton, Ohio. Dayton is in close proximity to Wright-Patterson Air Force Base. This book has provided many connections between Wright-Patterson and the New Manhattan Project. Specifically, we know of studies performed at Wright-Patterson involving aluminum oxide nanoparticle spray exposure (ch 5).

Although today Monsanto is mostly thought of as a genetic engineering company, it has historically been and still is a chemical producer. Production of genetically modified seeds has largely been done to sell their Roundup weed killer which is designed to be used in combination with their seeds.

As described in the book *The World According to Monsanto*, Monsanto and their employees have historically demonstrated extreme callousness and disregard for our lives, the lives of other living things, and the environment. They would be the perfect company to produce the New Manhattan Project's chemtrail sprays. Monsanto has gone on to lose so many cancer lawsuits that they were rescued by Bayer (another former Nazi company) in a 2018 buyout.

References

Maverick's Earth and Universe: Understanding Science without Establishment Blunders a book by J. Marvin Herndon, Ph.D., published by Trafford Publishing, 2008

"Aluminum Poisoning of Humanity and Earth's Biota by Clandestine Geoengineering Activity: Implications for India" an article by J. Marvin Herndon, published in *Current Science* on June 25, 2015, Vol. 108, no. 12

"Evidence of Coal-Fly-Ash Toxic Chemical Geoengineering in the Troposphere: Consequences for Public Health" an article by J. Marvin Herndon, published in the *International Journal of Environmental Research and Public Health* on August 11, 2015

"Obtaining Evidence of Coal Fly Ash Content in Weather Modification (Geoengineering) Through Analyses of Post-Aerosol-Spraying Rainwater and Solid Substances" an article by J. Marvin Herndon, published in the *Indian Journal of Scientific Research and Technology*, February, 2016

"Human and Environmental Dangers Posed by Ongoing Global Tropospheric Aerosolized Particulates for Weather Modification" an article by J. Marvin Herndon, published in *Frontiers in Public Health*, June 30, 2016

"Adverse Agricultural Consequences of Weather Modification" an article by J. Marvin Herndon, published in the *Journal of Agricultural Science*, 2016, 38(3): p213-221

"An Indication of Intentional Efforts to Cause Global Warming and Glacier Melting" an article by J. Marvin Herndon, published in the *Journal of Geography, Environment and Earth Science International*, January 29, 2017

"Further Evidence of Coal Fly Ash Utilization in Tropospheric Geoengineering: Implications on Human and Environmental Health" an article by J. Marvin Herndon, published in the *Journal of*

Geography, Environment and Earth Science International, February 3, 2017

"Evidence of Variable Earth-heat Production, Global Non-anthropogenic Climate Change, and Geoengineered Global Warming and Polar Melting" an article by J. Marvin Herndon, published in the *Journal of Geography, Environment and Earth Science International*, April 21, 2017

"Contamination of the Biosphere with Mercury: Another Potential Consequence of On-going Climate Manipulation Using Aerosolized Coal Fly Ash" a paper by Dr. J. Marvin Herndon, PhD and Dr. Mark Whiteside, MD, published by the *Journal of Geography, Environment and Earth Science International*, December 2017

"Previously Unrecognized Primary Factors in the Demise of Endangered Torrey Pines: A Microcosm of Global Forest Die-offs" a paper by Dr. J. Marvin Herndon, PhD, Dale D. Williams, and Dr. Mark Whiteside, MD, published by the *Journal of Geography, Environment and Earth Science International*, August 2018

"Aerosolized Coal Fly Ash: A Previously Unrecognized Primary Factor in the Catastrophic Global Demise of Bird Populations and Species" a paper by Dr. J. Marvin Herndon, PhD and Dr. Mark Whiteside, MD, published by the *Asian Journal of Biology*, November 2018

"Previously Unacknowledged Potential Factors in Catastrophic Bee and Insect Die-off Arising from Coal Fly Ash Geoengineering" a paper by Dr. J. Marvin Herndon, PhD and Dr. Mark Whiteside, MD, published by the *Asian Journal of Biology*, August 2018

Herndon's Earth and the Dark Side of Science a book by J. Marvin Herndon, Ph.D., self-published, 2014

"Fly Ash Utilization: Proceedings: Edison Electric Institute-National Coal Association-Bureau of Mines Symposium, Pittsburgh,

Pennsylvania, March 14-16, 1967" a report by the United States Department of the Interior, Bureau of Mines, 1967

Third International Ash Utilization Symposium Proceedings a book by the Department of the Interior, Bureau of Mines, 1973

Denver Fly Ash Symposium Proceedings a book edited by Edwin R. Dustan, published by the Denver Fly Ash Symposium, Inc., 1984

Fly Ash and Coal Conversion By-Products: Characterization, Utilization and Disposal I a book edited by Gregory J. McCarthy and Robert J. Lauf, published by the Materials Research Society, 1984

Fly Ash and Coal Conversion By-Products: Characterization, Utilization and Disposal III a book edited by Gregory J. McCarthy, Fredrik P. Glasser, Della M. Roy, and Sidney Diamond, published by the Materials Research Society, 1986

Fly Ash and Coal Conversion By-Products: Characterization, Utilization and Disposal VI a book edited by Robert L. Day and Fredrik P. Glasser, published by the Materials Research Society, 1989

Fly Ash and Coal Conversion By-Products: Characterization, Utilization and Disposal IV a book edited by Gregory J. McCarthy, Fredrik P. Glasser, Della M. Roy, and Raymond T. Hemmings, published by the Materials Research Society, 1987

"Chemistry and Mineralogy of Coal Fly Ash: Basis for Beneficial Use" a paper by Barry E. Scheetz, as published in the "State Regulation of Coal Combustion By-product Placement at Mine Sites" a report by Southern illinois University and the U.S. Department of the Interior, published by Southern illinois University and the U.S. Department of the Interior, 2004

NIST SRM 1633c certificate

NIST SRM 1633a certificate

Langmuir's World a film by Roger R. Summerhayes, 1998

"Early History of Cloud Seeding" a report by Barrington S. Havens, published by the Langmuir Laboratory, New Mexico Institute of Mining and Technology, the Atmospheric Sciences Research Center, State University of New York at Albany, and the Research and Development Center, General Electric Company, 1978

"Langmuir, the Man and the Scientist" a biography by Albert Rosenfeld, published as part of *The Collected Works of Irving Langmuir* by Pergamon Press, Volume 12

"History of Project Cirrus" a report by Barrington S. Havens, published by the General Electric research laboratory, 1952

U.S. patent #4,755,673 "Selective Thermal Radiators" by Slava A. Pollack and David B. Chang, 1988

Call Me Pat: The Autobiography of the Man Howard Hughes Chose to Lead Hughes Aircraft a book by L.A. 'Pat' Hyland, published by The Donning Company Publishers, 1993

Cottrell: Samaritan of Science a book by Frank Cameron, published by Doubleday, 1952

Initiative in Energy: The Story of Dresser Industries a book by Darwin Payne, published by Simon and Schuster, 1979

Storms Above the Desert: Atmospheric Research in New Mexico 1935-1985 a book by Joe Chew with the assistance of Jim Corey, published by the University of New Mexico Press, 1987

Wall Street and the Rise of Hitler a book by Antony Sutton, published by Buccaneer Books, 1976

Blowback: The First Full Account of America's Recruitment of Nazis, and its Disastrous Effect on Our Domestic and Foreign Policy a book by Christopher Simpson, published by Weidenfeld and Nicholson, 1988

George Bush: The Unauthorized Biography a book by Webster Tarpley and Anton Chaitkin, published by Executive Intelligence Review, 1992

"Science, Engineering and the Quality of Life" a report by the Monsanto Company, published by the Monsanto Company, 1970

"The Rand Corporation: The First Fifteen Years" a report by the Rand Corporation, published by the Rand Corporation, 1963

The World According to Monsanto: Pollution, Corruption, and the Control of the World's Food Supply a book by Marie-Monique Robin, published by The New Press, 2010

The Making of the Atomic Bomb a book by Richard Rhodes, published by Simon and Schuster, 2012

"Charles Allen Thomas Biographical Memoir" by R. Byron Bird, produced by the National Academy of Sciences, 1994

Chapter 7
C4

"An effective technique of weather control will undoubtedly require an elaborate system that tells us minute by minute, and in detail, what the many variables in the atmosphere are doing and how our activities are affecting them."
-Bernard Vonnegut from his article "When Will We Change the Weather?" 1967

The New Manhattan Project requires an incredibly huge command and control apparatus. Hundreds of airplanes need to be commanded. The ionospheric heaters and other highly technical machines need to be operated. Atmospheric conditions need to be monitored and analyzed. Supercomputers are needed to assist throughout. Today's military refers to such an apparatus as C4: command, control, communications, and computers.

The development of these types of technologies used in weather modification and the atmospheric sciences is well documented. This chapter examines the historical development of these technologies, the currently known state of the art, and other, possibly undisclosed operations. This chapter examines the development and current status of C4 technology used as part of today's New Manhattan Project.

Mapping the atmosphere

Before our atmosphere could be commanded and controlled, it was necessary to understand its composition and movements. Pertaining to this quest for understanding, this section recounts some of the most notable efforts.

The weather that we experience is vastly a product of planet Earth, its atmosphere, and the Sun. Our atmosphere consists of many layers. In ascending order, our atmosphere consists of: the troposphere (where we breathe the air and where our weather occurs), the stratosphere, the mesosphere, the thermosphere, the ionosphere, and the magnetosphere. Phenomena in all of these regions, stretching to 32,000 miles above Earth's crust, have a direct effect upon the weather we see every day. Not only that, but the water, ice, volcanoes, calderas, and other features of the Earth's surface and sub-surface have direct relevance here. Because of this, in order to comprehensively modify the weather, all of these regions must be understood and, often, manipulated.

Recording atmospheric conditions was the first step in mapping the atmosphere and it has been largely a military exercise. In the 1830s, the U.S. Navy began recording meteorological data at their yards and from ships. Terrestrial weather data networks began in the 1840s with the development of the telegraph. As *The Weather Changers* by D.S. Halacy notes, "By 1850 there were 150 telegraph stations sending weather data to the Smithsonian Institution in Washington, a kind of central clearinghouse for weather information." Beginning in 1870, the U.S. Army Signal Office began providing daily weather reports and forecasts to mostly commercial and agricultural concerns.

The first scientific explorations of our atmosphere were conducted using balloons, dropsondes and atmospheric sounding rockets. Weather balloons would be floated up to 100,000 ft. into the stratosphere where attached devices would record atmospheric conditions and then fall back to Earth to be collected. Later, balloons used devices producing radio transmissions (radiosondes) to send atmospheric data back to meteorologists on the ground. Dropsondes are devices dropped from aircraft at altitude. A dropsonde will have a parachute that opens up so that the device can more slowly return to Earth as it gathers atmospheric data.

Atmospheric sounding rockets became prevalent starting in the mid-1940s. Rockets can go much higher than balloons or aircraft (200,000 ft.), and therefore gather higher-elevation atmospheric data - thus giving us a more complete picture of our atmosphere. Rockets have also been used to produce smoke trails which are then recorded and analyzed as a means to determine atmospheric wind patterns. A 1976 report by the Air Force titled "Analysis of Smoke Trail Photographs to Determine Stratospheric Winds and Shears" recounts experiments employing rockets which left titanium chlorine trails. Later sounding rocket experiments were designed to coordinate with ionospheric heaters in order to map the auroral electrojet. The auroral electrojet is comprised of the Earth's natural magnetic energy.

The Earth is a giant magnet. Strong magnetic fields enter and exit the Earth at the poles and surround the Earth in a toroidal fashion. Most of the later sounding rocket experiments were conducted near the Arctic Circle because at that high latitude, the auroral electrojet is at lower elevations and therefore it is more easily observed and influenced. These sounding rockets carried payloads consisting of chemicals used to enable observations of the auroral electrojet. When the rocket got to around its apex, the nosecone would explode and the chemical payload would be released. Scientists on the ground and in aircraft used photography and ionospheric heaters to make observations and thus map the auroral electrojet.

Neil Davis' informative and quite entertaining book *Rockets Over Alaska: The Genesis of Poker Flat* details how in the late 1960s and early 1970s the Defense Advanced Research Projects Agency (DARPA), the Defense Atomic Support Agency, and the Atomic Energy Commission jointly conducted a rocket launching program out of Poker Flat, Alaska (ch 3). It was supported by the General Electric Space Sciences Laboratory and the then Stanford Research Institute's Radio Physics Laboratory was heavily involved. Raytheon and many, many others were there. These programs were called SECEDE I & II.

At Poker Flat they had an ionospheric heater on site providing the electromagnetic energy. This was the ionospheric heater mentioned in chapter 3 which started out at Pennsylvania State, was later moved to Platteville, CO, and then came to Poker Flat. Pennsylvania State University continued to operate this heater under the U.S. Navy.

As of 2006 (the year of his book's publication), Mr. Davis says that since 1969 there have been at least 100 chemical payload release rockets launched from Poker Flat. Poker Flat appears to still be in use today. Similar operations by many different experimenters have been carried out all over the world. Worldwide, all time we're talking about dumping at least metric tons of all sorts of highly toxic materials into the atmosphere. There is also, believe it or not, quite a documented history of satellites dispersing chemicals for the purpose of mapping the upper atmosphere. There are also a few patents describing how satellites can be used to directly modify Earth's weather. These patents are listed in the 'References' section for this chapter.

Weather satellites play an important role in today's New Manhattan Project. Over the years, the technology has gotten quite good. Historically, the field has been dominated by the National Aeronautics and Space Administration (NASA). The first meteorological instrument was carried on a satellite in 1959. The first line of dedicated weather satellites, known as TIROS, was launched in 1960. The TIROS line was followed by others such as General Electric's NIMBUS as well as weather satellites from the Defense Department, the Environmental Science Service Administration (ESSA), and the National Oceanic and Atmospheric Administration (NOAA). Globally and domestically there have been many, many other weather satellite programs. The most useful, overt weather satellites today are known as GOES (Geostationary Operational Environmental Satellite).

The following is an illustration from the Interdepartmental Committee for Atmospheric Science (ICAS) report 20b of July, 1977. It shows how the National effort in satellite remote atmospheric sensing between 1963-1979 evolved. As one can see, by 1980 Earth's atmosphere was pretty well covered.

Table 4. Availability of Satellite Data

PARAMETER	PERIOD (63-79)
Atmospheric temperature	(NOAA 2, 3, 4, etc.)
Winds from cloud motions — NESS	(SMS, GOES)
— Univ. of Wisconsin	
Planetary radiation budget	(TIROS 7, Nimbus 2) (Nimbus 3) (Nimbus 6, G)
Cloud Cover	(TIROS) (ESSA, NOAA, U.S. Air Force Satellites)
Snow Cover	(ESSA) (NOAA) (LANDSAT)
Polar sea ice and ice sheets	(ESSA) (NOAA) (Nimbus microwave)
Precipitation	(Nimbus microwave — oceans)
Vegetation and surface albedo	(TIROS) (Nimbus) (LANDSAT)
Sea surface temperature	(TIROS) (Nimbus) (NOAA)
Upper atmosphere (temperatures and winds)	(Meteorological rockets, quarterly intervals, 20-30 sites globally) (Nimbus 4, 5, 6)
Ozone	(Nimbus 4, G, Atmospheric Explorer)
Solar Irradiance — UV only	(Rockets) (Nimbus 3, 4, G)
— total flux	

Legend:

ooooooo Observations exist but probably not useful for data base (poor calibrations, deficient archiving, etc.)

ccccccc Observations exist but retrieval process must be developed to make them applicable to global data base.

▪▪▪▪▪▪▪ Observations and retrieval process exist but operational effort is needed to produce information in format applicable to data base.

●●●●●●● Directly applicable information exists.

♦ Research Satellite

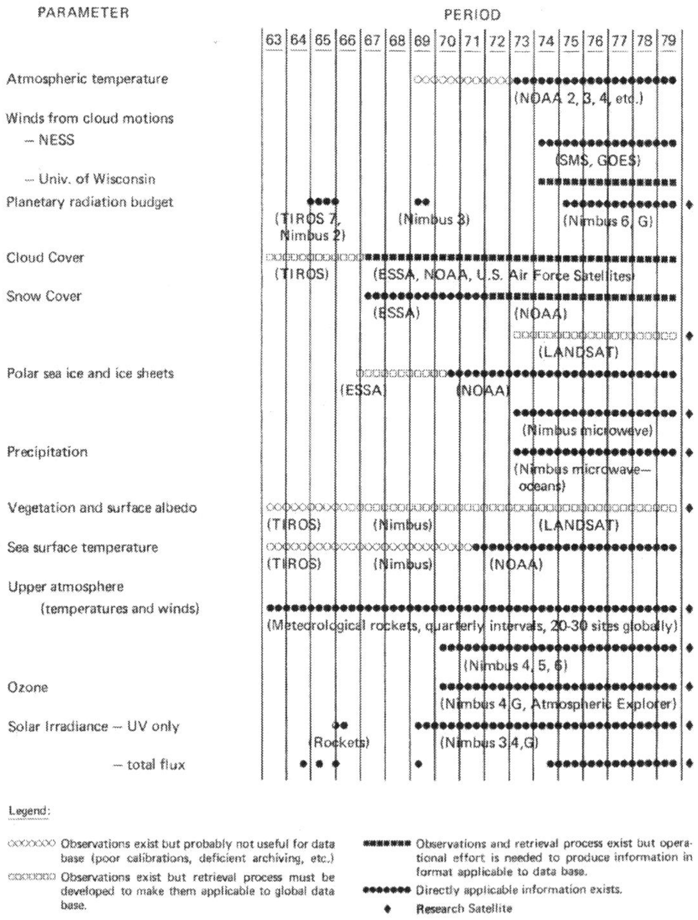

Satellites have traditionally also been used as all-purpose, remote weather data collection sites and pass-throughs. All types of weather data collected from ground stations, buoys, weather balloons, aircraft, etc. is sent wirelessly to satellites which then send the information on to ground-based data centers called 'Earth stations.' Examples of this are pervasive throughout the weather modification literature. The earliest example of this as part of a weather modification effort yet found by the author is in a 1974 Department of the Interior report. In the early 1970s the Bureau of Reclamation's Division of Atmospheric Water Resources used the ERTS-1 satellite in support of cloud seeding operations.

It is interesting to note that one of the most prevalent rockets used to launch satellites and in atmospheric soundings was the V-2; a Nazi rocket originally designed to carry munitions. The V-2 was the first sounding rocket used to help map the auroral electrojet. There was something called the V-2 Panel; later called the Rocket and Satellite Research Panel. This was a group of scientists formed at Princeton University in 1946 who were interested in high-altitude rocket research. Their activities largely revolved around modifying the V-2 for atmospheric sounding use. Out of 9 seats, the panel included the famous scientist J.A. Van Allen (1914-2006) and three board members from General Electric. The board's chairman, E.H. Krause left in 1947 to go work on nuclear bomb tests.

Over the years, many large-scale, international programs such as: the 1969 Barbados Oceanographic and Meteorological Experiment, the 1969 Tropical Meteorological Experiment, the 1974 GARP (Global Atmospheric Research Program) Atlantic Tropical Experiment (GATE), and the 1978 First GARP Global Experiment have been conducted. These coordinated programs were executed in order to gain deeper understandings of our Earth's atmospheric processes and involved all types of atmospheric monitoring activities, methods, and equipment. Participating Federal agencies included the usual suspects: the Department of Defense, the Department of Commerce, the National Aeronautics and Space Administration, the National Oceanic and Atmospheric Administration, the National Science Foundation, etc. In 1980, the Joint Organizing Committee of GARP voted to transform itself into the Joint Scientific Committee of the World Climate Research Program.

A high-level, rhetorical shift from 'weather forecasting' to 'climate modeling' began during the 1960s. In 1977, the National Oceanic and Atmospheric Administration (NOAA) started writing about weather data collection networks in a different way. Under the heading of "Global Monitoring for Climatic Change," NOAA writes:

"The purpose of NOAA's Global Monitoring for Climatic Change (GMCC) Program is to provide quantitative data needed for predicting climatic changes. These consist of (a) dependable measurements of existing amounts of natural and manmade trace constituents in the atmosphere, (b) determination of the rates of increase or decrease in these amounts, and (c) possible effects these changes may have on climate. The present U.S. network of baseline observatories consists of four stations located at Barrow, Alaska; Mauna Loa, Hawaii; American Samoa; and South Pole, Antartica. These stations are designed to supply information on atmospheric trace constituent concentrations using identical instrumentation and established measurement techniques. The GMCC Program is the U.S. portion of a planned world network of stations in the EARTHWATCH program of the United Nations Environment Program."

As evidenced by this passage, by this time the dialog had changed from 'weather modification' and the 'atmospheric sciences' to 'climate change' and 'atmospheric chemistry.' 1977 was also the publication year of the aforementioned paper titled "On Geoengineering and the CO2 Problem" which employs the earliest mention of the word ‘geoengineering.'

So who is in charge of monitoring the atmosphere for today's New Manhattan Project? It may be the United Nations and their World Meteorological Organization (WMO). They would be in a position to share this information with other participating nations. A 2004 document titled "The Changing Atmosphere: An Integrated Global Atmospheric Chemistry Observation Theme" depicts on the cover an unmarked, all white jet airplane emitting visible trails and outlines a program for comprehensive and continuous global atmospheric observation. The authors of this report describe a system of atmospheric observation and analysis which would be sufficient for conducting today's New Manhattan Project. Their proposed system involves ground based radar, aircraft-based measurements, satellite-based applications and a comprehensive data modeling system. This document was produced by the International Global Observing Strategy (IGOS) which is run by the United Nations' WMO and the European Space Agency. Here in the United States, today's dominant atmospheric data collection organization is NOAA's National Centers for Environmental Information.

There is a pre-existing network of satellites which would be especially suited to the New Manhattan Project. Although this network bills itself as a global media telecommunications network, it would be relatively easy to covertly install the necessary equipment. The network is called 'Iridium.' It is particularly suited to the New Manhattan Project because it produces coverage, "...to all parts of the globe, including the polar regions." Global coverage is what this global weather modification project requires.

Let us refer to the textbook *Satellite Technology: Principles and Applications*. It reads:

"One important application of LEO [low Earth orbit] satellites for communication is the project Iridium, which is a global communication system conceived by Motorola. A total of 66 satellites are arranged in a distributed architecture, with each satellite carrying 1/66 of the total system capacity. The system is intended to provide a variety of telecommunication services at the global level. The project is named 'Iridium' as earlier the constellation was proposed to have 77 satellites and the atomic number of iridium is 77. Other applications where LEO satellites can be put to use are surveillance, weather forecasting, remote sensing and scientific studies."

The 2009 and 2010 U.S. patents of the same name "System and Method for Using Iridium Satellite Signals for Meteorological Measurements" describe how one or more satellites of the Iridium network can gather meteorological data and share that data with aircraft.

Not only is this Iridium constellation suitable for the New Manhattan Project's remote sensing needs, a 1996 Air Force document titled "Space Operations: Through the Looking Glass (Global Area Strike System)" describes the Iridium constellation as suitable for *controlling* the weather. Writing specifically of the Iridium constellation, the document reads, "Since water molecules are also known to absorb certain bands of microwave frequencies, it is also possible a properly designed HPMW [high-power microwave] weapon system could be used to modify terrestrial weather."

The New Manhattan Project C4 apparatus probably employs not only satellite-based remote sensors, but ground-based atmospheric remote sensors as well. The previously noted International Global Observing Strategy (IGOS) report of 2004 makes note of an in-use, ground-based network of remote sensors employing, "...passive remote sensing spectrometers operating in various wavelength regions providing total column or low resolution vertical profiles of a number of atmospheric gases and aerosols; and active remote sensing lidar instruments for high-resolution remote sensing of atmospheric components." Also, something called the Network for the Detection of Atmospheric Composition Change (NDACC) is comprised of, "...more than 70 high-quality, remote sensing research sites for: observing and understanding the physical/chemical state of the stratosphere and upper troposphere" as well as, "...assessing the impact of stratospheric changes on the underlying troposphere and on global climate."

Aircraft atmospheric data reconnaissance

As part of its atmospheric observational networks, the New Manhattan Project probably involves direct atmospheric sampling activities performed from aircraft. These types of activities typically include sampling of temperature, humidity, and atmospheric composition. As the airplane flies along through the atmosphere, specialized instruments collect atmospheric samples and take measurements. Direct atmospheric sampling from aircraft can be much more accurate than any remote sensing. The aircraft performing this reconnaissance are probably the New Manhattan Project's proprietary fleet of chemtrail spraying jet airliners as well as commercial passenger airliners.

The 2008 U.S. patent "Airborne Weather Profiler Network" describes how both commercial airliners and drones can gather and report detailed atmospheric data including particulate concentrations. The weather modification literature provides us with information showing that commercial passenger airliners have historically been employed for the routine collection of atmospheric data. According to ICAS report number 20, as early as 1976, a system was being developed to, "...obtain real-time observations from airlines using the

SMS/GOES satellites as a relay..." National Science Foundation scientists also published a 1983 paper detailing this type of operation.

International participation

International participation and cooperation has been extensive. Although America has captained the New Manhattan Project, it wouldn't be a global weather modification project without others doing similar things. We have already noted large-scale efforts such as GARP. Here are some other interesting examples.

The United States, Canada, and Mexico have been coordinating weather modification efforts for a long time. A 1978 report by the U.S. Congressional Research Service notes:

"The North American Interstate Weather Modification Council (NAIWMC) was organized on January 17, 1975, by representatives of the governments of several U.S. States and Canadian provinces and the Mexican Government. Its purpose is to coordinate and serve as a focal point for intrastate, interstate, and international weather modification activities. This would include research into weather modification, legislation and treaties governing weather modification activities, and public information activities as well as its coordination functions. Membership is open to any state or province of the United States, Canada, and Mexico."

This Council was, of course, geared towards the conventional cloud seeding industry.

Because the New Manhattan Project was developed during the Cold War, the weather modification literature shows that proponents of these technologies stressed the importance of achieving such capabilities before the Russians did. Meanwhile, we were *sharing* these technologies with the Russians. If we can be led to believe that a chemtrail is a contrail, then this duplicity makes perfect sense. The U.S. Congressional Committee on Government Operations produced a 1965 report showing cooperation between the U.S.S.R. and the U.S. The report reads:

"A duplex 24-hour circuit directly linking the National Meteorological Center with a similar facility in Moscow, U.S.S.R., is provided in accordance with existing international agreements. It is capable of voice transmission as well as transmission of photo, facsimile, and telegraph signals."

Also, during 1975 Congressional testimony, atmospheric scientist Dr. Currie Downie provided evidence of Russian cooperation in weather modification by stating, "We had a delegation of Russians over here not too long ago and there are some plans for a visit to their sites over there, their hail sites and other weather modification sites."

Most recently, in a 2018 *Sun* article, we have seen evidence of China conducting NMP-like activities. In an article titled "RAIN MAKERS China 'will move clouds and make it rain in different places using satellites' in terrifying weather control plan," the author writes about how China planned to herd clouds from one region of their country to another using electromagnetic energy from satellites.

Atmospheric modeling

Atmospheric models are designed to predict the future. Values assigned to current atmospheric conditions are plugged into a model and the model predicts what will happen over the next hour, day, week, month, year, etc.. Longer term models are known as 'climate models.' When it comes to the use of atmospheric models in the context of the New Manhattan Project, scientists can foretell the effects of man-made atmospheric modifications such as spraying chemicals from airplanes and the use of electromagnetic energy. As we will see, there are many examples.

Today's atmospheric models require ultra-massive raw computing power, hence supercomputers have an extensive history of being used for atmospheric modeling. Supercomputers bring the atmospheric models to life. Just as it is not a coincidence that every aspect of the New Manhattan Project evolved together in a coherent chronological order, it is not a coincidence that supercomputers have evolved simultaneously with the development of atmospheric models. All these things are interdependent. But, before the supercomputers came, there were atmospheric models. In this development, atmospheric models are a horse that has pulled the supercomputer wagon, so we will delve into supercomputing in the next section.

Atmospheric modeling began in the late 1800s with the work of the Swedish scientist Vilhelm Bjerknes (1862-1951), founder of the Bergen School of Meteorology. But just as with the New Manhattan Project itself, things didn't really get going until the 1940s. As noted in chapter 2, the work of John von Neumann and others at the Princeton

Institute for Advanced Studies pioneered today's modern atmospheric computer modeling efforts. In 1954, something called the Joint Numerical Weather Prediction Unit was established to serve the Air Force Weather Service, the civilian Weather Bureau, and the Naval Weather Service. At this Joint Numerical Weather Prediction Unit, the aforementioned Norman Phillips demonstrated the first general circulation model (GCM) to accurately replicate, "...the seemingly permanent, large-scale structures of the global atmosphere, such as the jet stream and the prevailing winds. Researchers in several places began to conduct general circulation experiments after Phillips's demonstration." When these general circulation models began to be developed, man began to have a real understanding of Earth's atmospheric circulations. GCMs were mainly developed at: NOAA's Geophysical Fluid Dynamics Laboratory, NASA's Goddard Institute of Space Science, the National Center for Atmospheric Research (NCAR), the University of California at Los Angeles, and the RAND Corporation.

In 1962, five authors including Yale Mintz published a RAND Corporation document titled "A New Rational Approach to Weather-Control Research." In it they write of, "...altering individual weather systems through the introduction of artificial energy sources directly into the atmosphere." As we have seen in chapter 3, injecting energy directly into the atmosphere for the purpose of weather control is the distinguishing feature of the New Manhattan Project. The document continues a little later:

"The experiment would examine the numerical prediction and pick out a feature in space and time that he wished to modify - for example, a Pacific storm that produced heavy precipitation over Oregon - and that he wanted, instead, to redirect onto a more southerly course so that it would drop its precipitation on the California mountains. Having picked this case 'after the fact,' he would carefully examine the fields, time-step by time-step, going backwards, until he found one or more grid points in space and time where the 'trigger mechanism' hypothesis indicated the artificial energy source should be introduced."

These 'trigger mechanisms' are something written of many times in this book. The term refers to a point in space and time at which one may make a relatively small intervention in order to cause a chain reaction which in turn creates a significant impact upon the weather.

Here, Mintz and the RAND Corporation have provided the most compelling example yet.

In 1963, the National Science Foundation (NSF) also wrote of atmospheric models capable of demonstrating the result of injecting energy into the atmosphere. In the NSF's 1963 annual weather modification report, the authors write, "Finally, although details are not available, mention should be made of the imminent numerical experiment of Smagorinsky and his associates for the STRETCH computer of the U.S. Weather Bureau." The NSF continues:

"It is clear that with numerical models of the kind just described, experiments could be performed that would demonstrate the result of either modifying the earth's surface in some specified way or of injecting energy directly into the atmosphere."

In 1974, a paper appeared in the proceedings of the Fourth Conference on Weather Modification titled "A Numerical Simulation of Warm Fog Dissipation by Electrically Enhanced Coalescence" which outlined an atmospheric computer model which factors in artificial electrical influences. This research is especially significant because it formed the basis for future weather modification by atmospheric heating, including the New Manhattan Project's use of weather modifying electromagnetic energy.

Beginning in the the mid-1970s, the center of the National effort in atmospheric modeling shifted to Lawrence Livermore National Laboratory (LLNL). For many years now, their Program for Climate Model Diagnosis and Intercomparison (PCMDI) has been producing cutting-edge atmospheric models. In 1989, the PCMDI was established at Livermore and has been supported by the Environmental Sciences Division of the U.S. Department of Energy. The PCMDI has collaborated with the National Center for Atmospheric Research (NCAR) in Boulder, CO and the Los Alamos National Laboratory.

A 1999 LLNL report by Edward Teller, Ken Caldeira, Lowell Wood, *et al.* speaks to big developments in global atmospheric models. The authors of this document write about the advent of the CCM3 (Community Climate System Model) developed at NCAR, as well as the Naval Research Laboratory COAMPS (Coupled Ocean/ Atmosphere Mesoscale Prediction System) model. Further, a May, 2005 document by Ken Caldeira *et al.* makes note of a, "...full physical climate system (atmosphere-ocean-ice-land) model." They are writing about the Community Climate System Model 3.0 (CCSM3). As one of the very top geoengineers, Caldeira knows what he's writing about. Today's geoengineers are probably using this Community Climate System Model 3.0 or something like it. A 2006 document by Forrest Hoffman *et al*. details the CCSM3. It reads:

"*The Community Climate System Model Version 3 (CCSM3) is a coupled modeling system consisting of four components representing the atmosphere, ocean, land surface, and sea ice linked through a coupler that exchanges mass and energy fluxes and state [sic] information among these components. CCSM3 is designed to produce realistic simulations of Earth's mean climate over a wide range of spatial resolutions. The modeling system was developed through international collaboration and received funding from the National Science Foundation (NSF) and the Department of Energy (DOE) as well as support from the National Aeronautics and Space Administration (NASA) and the National Oceanic and Atmospheric Administration (NOAA). A portion of DOE's support for CCSM has been through SciDAC Projects, including the multi-laboratory Climate Consortium Project headed by Phil Jones and John Drake*."

If you are wondering... yes, that's Phil Jones of Climategate infamy. If you will recall, Climategate was the hacked email scandal that caught the people responsible for the Intergovernmental Panel on Climate Change's climate data with their pants down. Yeah, that was where they were caught explaining how they cook the data, keep opposing viewpoints out of the official discussion, block Freedom of Information requests, manipulate the peer-review process, and practice other decidedly unscientific activities. It appears that the spirit of the pre-scientific era of weather modification lives on.

Supercomputers

Supercomputers have historically been manufactured by corporations like: IBM, Cray, Control Data Corporation, Texas Instruments, Burroughs, and the Univac Division of Sperry Rand. Today's big dogs in the space are IBM and Cray. Supercomputers by nature take up a lot of space. A single machine is usually about the size of a smallish refrigerator. Supercomputing facilities often employ multiple machines. Large supercomputing facilities consist of a sizable room filled with dozens of big, hot boxes. Over the years supercomputers have, of course, become much faster, but their size has mostly remained a constant.

The earliest examples of supercomputer atmospheric simulations were a product of nuclear bomb testing. Atomic bomb developers knew that radioactive particles were dangerous to the environment and Human health. They also knew that testing nuclear bombs (which was what they did before and after dropping two on Japan) creates lots of radioactive atmospheric particles. Therefore, they wanted a way to track the dispersion of said particles - hence von Neumann's early work in atmospheric modeling.

Observed atmospheric conditions on the day of the detonation such as temperature, pressure, humidity, and wind speed and direction would be entered into von Neumann's model. In this way, the project directors would get a good idea of which adjacent areas would be contaminated and (due to knowledge of the isotope half-lives) for how long. Von Neumann's pioneering work in this area led to the more advanced atmospheric models running on supercomputers such as those in use today.

Founded at the site of an old WWII Naval air station, Lawrence Livermore Laboratories has been involved in supercomputing since their 1952 inception. In that year they purchased their first supercomputer; a Sperry Rand UNIVAC 1 delivered in April of 1953. In the early to mid 1950s, Livermore Labs was the premiere buyer of supercomputers. Let us refer to a passage from *The Supermen: The Story of Seymour Cray and the Technical Wizards behind the Supercomputer*. It reads:

"*For computer manufacturers the needs of the bomb builders created an incredible opportunity. Throughout the early and mid-1950s, Livermore and Los Alamos stepped up their computing efforts until a friendly competition formed between the two labs. They vied for prestige; they vied for funding; they vied for access to the first of every kind of computer. For both labs computers emerged as status symbols, much as they had for giant corporations such as General Electric*."

Although others such as the National Center for Atmospheric Research (NCAR) and the Department of Defense were prolific supercomputer buyers, elsewhere in *The Supermen*, Lawrence Livermore Labs is described as, "...the leader among industry users.

When Livermore purchased a new machine, the other government labs took notice." The Rand Corporation, and NASA have historically been supercomputer buyers and industry supporters as well.

By the mid-1960s, supercomputers were standard meteorological equipment. The Committee on Government Operations tells the story in 1965:

"*High speed digital computers have been used in meteorological data analysis and weather prediction for nearly 10 years. Until 1955 weather data, gathered daily from all over the world, were sorted and processed almost entirely by hand and then entered manually on charts. Furthermore, until recently the job of forecasting was largely a combination of the forecaster's experience, certain statistical relationships, and qualitative or semi qualitative physical reasoning. Today, a large portion of the routine data handling and processing is performed by computers, and certain types of weather forecasts are now produced automatically by these machines. Computers are also used extensively by certain of the Nation's atmospheric science research laboratories such as the Weather Bureau's Geophysical Fluid Dynamics Laboratory and the National Center for Atmospheric Research.*"

The Accelerated Strategic Computing Initiative (ASCI) at Lawrence Livermore National Labs has historically been at the forefront of atmospheric supercomputing. With its roots in Lawrence Livermore's extensive history of computer simulated nuclear bomb detonations and artificial intelligence experimentation, the ASCI evolved from the Defense Advanced Research Projects Agency's Strategic Computing Initiative. The ASCI began operating in the early 1990s and their focus has been on climate change and the efficacy of climate change mitigation measures.

The National Center for Atmospheric Research (NCAR) has opened a special supercomputing center called the NCAR-Wyoming Supercomputing Center (NWSC). Approval for the NWSC was secured in 2007, construction began in 2010, and operations began in 2012. NCAR has transitioned their supercomputing efforts away from their home of Boulder, Colorado to this new facility. The NWSC's supercomputing power is, of course, accessed remotely. The NWSC is dedicated to the study of 'Earth system processes' including climate change and carbon sequestration. The NWSC is sponsored by the

National Science Foundation. This facility may be instrumental to the New Manhattan Project as well.

So what comprises today's New Manhattan Project supercomputing effort? As we have seen, NCAR and LLNL both have many of the world's fastest supercomputers and they have been deeply involved in atmospheric modeling. Today's New Manhattan Project computing effort most probably involves the supercomputers of NCAR and LLNL and possibly others. The only real question here is if today's NMP computing effort utilizes computing power from the larger Internet as well.

You see, most, if not all of the world's supercomputers have been connected via the Internet for a long time now and this allows for the computing power of these computers (and any computer for that matter) to be shared via the Internet. One might suggest then, that the computing power of today's NMP is drawn from not only the supercomputers at NCAR, LLNL, and possibly others, but also from the larger computing power of the Internet. This may be the case, but one must also take into consideration that sharing computing power might create unnecessary security and technical issues. My science advisor informs me of private, government networks, not connected to the standard Internet which may very well be used as part of the NMP.

Atmospheric tracers

The main purpose of all of the New Manhattan Project atmospheric monitoring operations described here is to track the movements of air masses. These movements of air masses are tracked with the use of atmospheric tracers. The best atmospheric tracers are man-made as well as naturally occurring radioactive particles. As noted in the previous chapter, coal fly ash can consist of radioactive molecules such as barium and strontium and, believe it or not, naturally occurring radioactive particles are common in our atmosphere. The historical literature pertaining to weather modification and the atmospheric sciences is replete with references to radioactive materials being used as atmospheric tracers.

As early as 1958, the chairman of President Eisenhower's Advisory Committee on Weather Control, the aforementioned Howard T. Orville, noted that radioactive particles have been used as atmospheric tracers. An article written by Orville and appearing in

Popular Science magazine states, "Sea water tagged with radioisotopes from H-bomb fallout has enabled us to learn that certain regions of the earth get most of their rain from specific parts of the ocean."

A 1962 report by the National Academy of Sciences speaks to injecting radioactive substances into the air. The report states:

"*Radioactive substances of suitable half lifes injected into the air are very useful as tags and may be used to study air motions on a variety of scales. Tracers used in sufficient amounts for this purpose could add immeasurably to our knowledge of the currents of the atmosphere and the dispersion within air masses.*"

A little later in "The Atmospheric Sciences 1961-1971," the authors expand further upon the usefulness of radioactive isotopes. Under the heading "Research on Trace Substances," the report's authors write:

"*Where radioactive compounds are involved, containing tritium or carbon 14, radioactive decay times lead to a calculation of the time elapsed since the formation of the water or carbon dioxide in the sample. Thus, it is possible to date the rain water and the water in wells, rivers, and oceans, and to use this information to study the exchange processes between the upper and lower atmosphere, between the atmosphere and the oceans, and between the atmosphere and the Earth. Research in this area should be vigorously pursued.*"

This passage suggests that radioactive particles are being used today as a way to trace the entire hydrological cycle.

In a 1972 Interdepartmental Committee for Atmospheric Sciences report, the authors write of spraying 'chemical tracers' developed by the Atomic Energy Commission into the atmosphere from aircraft. It reads:

"*Two types of tracers have been used, specific chemical elements rare in abundance in the atmosphere, and the cosmogenic radionuclides produced naturally in the atmosphere by cosmic ray actions with argon.* ***The chemical tracers are introduced into the storm as aerosols via aircraft and/or surface generators.*** *[author's emphasis] Analysis of the resulting precipitation for the tracer elements provides insight into the time scales and trajectories of the air motions within convective storms and into the hydrometeor growth rates and deposition patterns.*"

It continues:

"Currently, the AEC [Atomic Energy Commission] support of the Illinois State Water Survey in Metromex is directed at the use of chemical tracers to determine the dynamics and efficiency of urban modified severe storms to ingest and precipitate atmospheric aerosols. ***The tracers are released either by aircraft into the storm updraft or from the surface.*** *[author's emphasis]"*

A 1983 paper by Warren Johnson of SRI International goes over the state of the art in atmospheric tracer techniques. Johnson writes of particulate tracers, tracer smokes, and oil fogs (among other materials) being used for atmospheric tracer experiments. Johnson also writes of an ideal tracer having a small, finite, and predictable decay rate to be measured with lidar (laser radar) remote sensing as it floats through the air.

Barium is often found in coal fly ash and barium sulphate is commonly used in radar applications similar to those seen in the New Manhattan Project. It shows up on radar consistently well due to its opacity. Barium in particular is probably the predominant atmospheric tracer of the New Manhattan Project.

The MITRE Corporation

Evidence suggests that under the Defense Advanced Research Projects Agency (DARPA), a corporation called MITRE has been in charge of the New Manhattan Project's systems architecture. Systems architecture refers to the organization and harmonization of all the different, scientific New Manhattan Project sub-operations such as: operation of the ionospheric heaters, control of the aircraft, collection of atmospheric data, satellite communications, etc. MITRE can make all these things work together and create an elegant end-user experience to boot.

Along with designing and building technical systems architecture, the MITRE Corporation of Bedford, Massachusetts designs and helps build military command and control centers. In fact, they have designed and helped build every major command and control center in America. From the National Military Command Center under the Pentagon to Strategic Air Command (SAC) in Nebraska to the North American Aerospace Defense Command (NORAD) facility in Cheyenne Mountain, Colorado, MITRE has built them all. MITRE also worked with NASA to develop the Apollo Mission Control Center in Houston (now the Johnson Space Flight Center).

MITRE Corporation headquarters

The MITRE Corporation has extensive involvement in just about every sub-operation of the New Manhattan Project. From satellite communications to atmospheric monitoring to remotely controlling fleets of airplanes to supercomputing to operating ionospheric heaters, MITRE can do it all. The New Manhattan Project is operationally right up their alley. So much so that it makes one wonder if the MITRE Corporation was created specifically to be the New Manhattan Project's development coordinator and technical manager. As far as the author can tell, the MITRE Corporation is the only American corporation with the right skill set to build the systems architecture of the New Manhattan Project - probably the only one in the world.

Much of the information recounted here is contained within the pages of a 1979 book entitled *MITRE: The First Twenty Years*. It was written and published by the MITRE Corporation. There may not be a single book more important to the history of the New Manhattan Project.

MITRE was founded in 1958 with about 500 employees from the Massachusetts Institute of Technology's Lincoln Laboratory to resolve a technical mess made by the previous administrators of a system called the SAGE Air Defense System. The RAND Corporation, the Ford Foundation, and the Massachusetts Institute of Technology were disproportionally represented on MITRE's first board of trustees. Connections to the original Manhattan Project, Edward Teller, Rockefellers, and Rothschilds are readily found. Although they themselves do not specify exactly what their name means, the author speculates that MITRE is an acronym for the Massachusetts Institute of Technology Radiation Experiment.

The SAGE system was another tremendously huge, covert effort involving radar, computers, command and control centers, and fleets of unmanned, remotely controlled missile-like jet aircraft designed by Boeing to hit enemy bomber aircraft before they could reach American cities. Originally named Project Whirlwind, it was a direct spin-off of the nascent automated missile battery systems developed at the MIT Radiation Laboratory during WWII (ch 2). This was all, of course, to keep us safe from the Russian nuclear threat. It involved the first use of centralized computers for air traffic control. MIT and the Air Force Cambridge Research Laboratories founded the SAGE project. It dates back to a 1950 committee called Project Charles headed by an MIT professor named George E. Valley (1913-1999). John von Neumann was a consultant to Project Charles. The chief architect of this SAGE system was an MIT scientist by the name of Jay Forrester (1918-2016). The project was known originally as Project Lincoln and it was launched in 1951 at MIT's former Rad Lab in Kendall Square.

The earliest official mention of something like the SAGE system is found with Theodore Von Kármán and the other authors of "Where We Stand" (ch 3) who wrote of, "large numbers of long-range ground-to-ground pilotless aircraft" controlled by a central location. Nikola Tesla and Vannevar Bush theorized about such things as well.

Information gathered in the field from radar and other sensors was to be fed into a central computer which was designed to streamline and automate the process of launching the missiles. The central computer of the SAGE system was known as Whirlwind and later as the IBM AN/FSQ-7. Work on the SAGE system was originally conducted at MIT's Lincoln Laboratory, which was created for the purpose. Like the MITRE Corporation, Lincoln Laboratory's headquarters were and are in Bedford, Massachusetts.

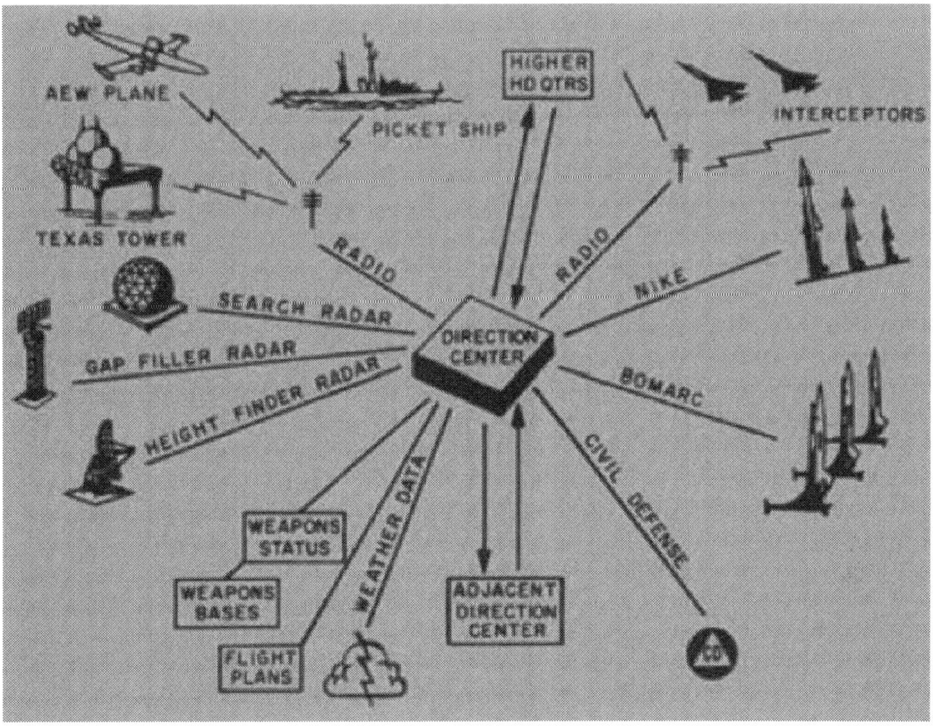

SAGE system schematic

Jay Forrester

The SAGE system never fully attained its planned capabilities. There were many delays and cost increases during the course of SAGE's development. On top of this, in the meantime, the enemy's offensive capabilities changed from bombers to intercontinental missiles which created a situation of moving goalposts. Due to these factors, in the late 1950s the Department of Defense recognized the realities of the situation and cut their losses by beginning to scuttle the project before it was completed.

Nevertheless, it is this SAGE system from which large portions of today's New Manhattan Project evolved. In fact, it is this SAGE system that has served as the nucleus of the New Manhattan Project - passed on throughout the decades. Most notably, the air traffic control systems of today's NMP evolved from SAGE.

Over the years, MITRE has consistently developed many different air traffic control systems capable of remotely controlling large numbers of aircraft. In 1959, MITRE partnered with the Federal Aviation Agency to develop something called SAGE Air Traffic Integration (SATIN). MITRE had been developing this system since their founding. According to a book commemorating their 50th anniversary, SATIN, "...was aimed at developing a single, unified system for managing all aircraft in the nation's airspace."

Later, something called Position Location, Reporting, and Control of Tactical Aircraft (PLRACTA) took shape. Let us reference a passage from *MITRE: The First Twenty Years*:

"In the late 1960s, the activities being pursued separately under CNI [Communications, Navigation, and Identification] and CASOFF [Control and Surveillance of Friendly Forces] were combined into a single advanced development program called PLRACTA, standing for Position Location, Reporting, and Control of Tactical Aircraft, headed by C. Eric Ellingson. PLRACTA's goal was to provide information exchange among a maximum of one thousand aircraft and other, ground elements of the control system."

PLRACTA later became known as Seek Bus, then later still as the Joint Tactical Information Distribution System (JTIDS). JTIDS is in use today. There is a JTIDS *Wikipedia* page. The New Manhattan Project probably utilizes an enhanced and customized version of the JTIDS system.

The NMP's air traffic control is most probably operated out of North American Aerospace Defense Command (NORAD) in Cheyenne Mountain, Colorado. This location is the most dominant North American air traffic controller and the site of intense MITRE activity over the years. NORAD is a joint U.S. and Canadian operation. Word is that NORAD's command center has actually been moved recently to underneath Denver International Airport, but at this time, this story cannot be substantiated.

NORAD

Separately, MITRE did the big study that provided a scientific basis for the Environmental Protection Agency's 1976 Toxic Controlled Substances Act. Before they produced that study, MITRE worked with the Department of the Interior's Office of Coal Research to produce a study about possible cleaner uses of coal. For the Interior Department, MITRE investigated methods for doubling or tripling coal production as well. These facts are interesting because, as detailed in the previous chapter, chemtrails usually consist of coal fly ash. Maybe MITRE and the Environmental Protection Agency (EPA) worked together to make sure that plenty of toxic and otherwise illegal coal fly ash was removed by government mandated coal fly ash scrubbers and therefore available as New Manhattan Project chemtrail spray.

Since their founding and throughout the production of the New Manhattan Project, MITRE Corporation's world headquarters has been in Bedford, Massachusetts. This is interesting because Bedford is in the Boston metropolitan area. As previously noted, the Boston metropolitan area has an abnormally high concentration of organizations involved in the development of the New Manhattan Project such as: the American Meteorological Society, Harvard University, the Massachusetts Institute of Technology (MIT), and many more. In order to produce history's largest scientific endeavor, it would be advantageous to have a community of some of the world's top academic, research, and development institutions located within a 15 mile radius; all answering to MITRE the coordinator. The Boston area was the New Manhattan Project's developmental epicenter and MITRE was most probably the NMP's development coordinator.

Once this Project became operational, it would be advantageous to retain the maintenance and support services of the people that designed and built it. It's like buying a brand new car from the dealer and getting premium factory service. Doing it this way makes things as easy as possible. As they are an ongoing concern, the MITRE Corporation probably continues to manage the systems of the New Manhattan Project to this day. They've been sucking up our taxpayer dollars this whole time. Now, where is our congressional investigation?

Gordon J.F. MacDonald

None other than our good friend Gordon J.F. 'How to Wreck the Environment' MacDonald was a MITRE Corporation chief scientist, served on their board of directors, and in 1966 became a trustee. He worked with the aforementioned Whirlwind computer at MIT. MacDonald was deeply involved in weather modification and the atmospheric sciences. He was involved in the wartime rainmaking operations over North Vietnam. He knew the famous Manhattan Project scientist James Conant well. He knew Edward Teller and Bernard Vonnegut and worked with many other scientists mentioned in the pages of this book. He worked for the Carnegie Institution as well as NASA. He got his PhD in geology at Harvard, taught geophysics at MIT and UCLA and had a hand in creating the EPA. He has connections to the Central Intelligence Agency and Al Gore. As a science advisor to President Johnson, MacDonald co-authored the aforementioned 1965 document titled "Restoring the Quality of Our Environment" which simultaneously established the modern theory of man-made climate change and the Solar Radiation Management geoengineering thesis. As one can see, it suffices to say that he was well connected to the New Manhattan Project.

It appears that MacDonald was the scientist who finally crystalized the NMP into a comprehensive project which takes into consideration not only the weather that we experience down here in the lowest level of the atmosphere known as the troposphere, but rather something that takes into consideration: all the geophysical aspects of terrestrial Earth, the Earth's lower and upper atmosphere, the moon, the sun, everything in between the sun and the Earth, other planets in our solar system, and outer space. All these things have varying degrees of effects upon our daily tropospheric weather and, although he was actually building upon a previously established school of thought known as 'cosmic physics,' MacDonald appears to have been the scientist who integrated this philosophy into the New Manhattan Project. He was probably responsible for popularizing the term 'space weather.'

In her book *The Pentagon's Brain*, Annie Jacobsen writes:

"By 1960, he [MacDonald] said, 'I was becoming more interested in the atmosphere, working on climate problems.' The University of California, Los Angeles, was developing a program in atmospheric science, and he accepted a position there as director of the Atmospheric Research Laboratory. At UCLA he found himself working on weather and the ionosphere. This led him to become interested in climate control."

Jacobsen continues:

"At the same time, in his classified work, Gordon MacDonald was becoming deeply interested in weather modification. He told the Journal of the American Statistical Association: 'I became increasingly convinced that scientists should be more actively engaged in questions of environmental modification, and that [the] federal government should have a more organized approach to the problem. While research could take place in both the public and the private sector, the government should take the lead in large-scale field experiments and

monitoring, and in establishing appropriate legal frameworks for private initiatives.'"

He said that he met with Al Gore during the Carter administration to discus what should be done about global warming. During a 1994 interview he said:

"Raith Pomerantz, who was then head of Friends of the Earth, saw it. He was an active environmental lobbyist. And called me up and said this was a problem that he felt was very serious. And first he wanted to understand the science, and we spent many, many days discussing the science. And then he said, 'What I want to do is for you to brief as many people in Washington as we can get to, to tell them about this problem.' And so as an environmental lobbyist he had many, many contacts on the Hill in the Carter Administration, with the Carter people. Now so early on I started to meet with people like Tim Wirth, Al Gore."

He says that Gore, "...quickly understood it [man-made climate change]." As if we could forget, Al Gore has been the most prominent proponent of the theory of man-made global warming and how we need to sacrifice our quality of life because of it.

MacDonald went on to testify before the Senate Energy Committee in November of 1987 about how America, as a global warming mitigation effort, needs to shift more to natural gas usage. Before the committee, he also advocated for a 'carbon tax' and conducted a detailed discussion about how that would work.

MacDonald also chaired the Central Intelligence Agency's MEDEA Committee which was the purported brainchild of former vice-president Al Gore. 'Purported' because it is hard to imagine anything being Al Gore's brainchild. Do you remember how he invented the Internet too? Nevertheless, the MEDEA (Measurements of Earth Data for Environmental Analysis) Committee existed between 1992 and 2001 and it went about using data collected from satellites for the purpose of solving Earth's environmental problems. Apparently the solution to our problems involves spraying megatons of toxic waste from supertanker aircraft. MacDonald's National Academy of Sciences obituary reads:

"Gordon chaired the MEDEA Committee (initially the Environmental Task Force) of the Central Intelligence Agency, a brainchild of Senator (later Vice-President) Gore for the application of

'overhead assets,' that is, information collected by intelligence satellites, to the solution of environmental problems. Here the intelligence and academic communities, two disparate communities meeting initially under conditions of mutual mistrust, developed over the years a feeling of trust and respect. Gordon was at his best, combining his unique environmental background with patience and perseverance. In 1994 the CIA presented Gordon with the Seal Medallion, the highest civilian honor of the agency."

This information suggests that MacDonald was instrumental in the development of the New Manhattan Project's satellite remote sensing needs.

The CIA's MEDEA Committee was terminated in 2001. According to a Jan. 4, 2010 *New York Times* article, the MEDEA Committee was resurrected as a CIA backed, National Academy of Sciences run, environmental remote sensing program. Five years later, the *Washington Times* and *Mother Jones* reported that this program had, once again, been shut down.

Gordon J.F. MacDonald

Today's New Manhattan Project is the pre-eminent masterwork of what is known as 'geophysical warfare.' Geophysical warfare, as it is synonymous with geoengineering, involves the manipulation of Earth's natural systems for strategic purposes. In the canon of publicly available literature pertaining to geophysical warfare (at least as far as sheer ostentatiousness is concerned) MacDonald's aptly titled 1968 essay "How to Wreck the Environment," as it appeared in the 1968 book *Unless Peace Comes*, stands out prominently. "How to Wreck the Environment" speaks to the use of directed electromagnetic energy in geophysical warfare with an emphasis upon weather and climate modification. It also asserts the theory of man-made global warming and the Solar Radiation Management (SRM) geoengineering thesis. This is why the document is so important to today's New Manhattan Project, which uses dispersed particles and directed electromagnetic energy in order to modify the weather and the climate. In 1968, the year 'How to Wreck the Environment" was published, Gordon MacDonald was the associate director of the Institute of Geophysics and Planetary Physics at the University of California, Los Angeles as well as a member of President Johnson's Science Advisory Committee. Now that we can see all of this in context, let us examine MacDonald's extraordinary specimen. He starts his paper with:

"Among future means of obtaining national objectives by force, one possibility hinges on man's ability to control and manipulate the environment of his planet. When achieved, this power over his environment will provide man with a new force capable of doing great and indiscriminate damage. Our present primitive understanding of deliberate environmental change makes it difficult to imagine a world in which geophysical warfare is practised. Such a world might be one in which nuclear weapons were effectively banned and the weapons of mass destruction were those of environmental catastrophe. Alternatively, I can envisage a world of nuclear stability resulting from parity in such weapons, rendered unstable by the development by one nation of an advanced technology capable of modifying the Earth's environment. Or geophysical weapons may be part of each nation's armoury. As I will argue, these weapons are peculiarly suited for covert or secret wars."

MacDonald continues:

"The key to geophysical warfare is the identification of the environmental instabilities to which the addition of a small amount of energy would release vastly greater amounts of energy. Environmental instability is a situation in which nature has stored energy in some part of the Earth or its surroundings far in excess of that which is usual. To trigger this instability, the required energy might be introduced violently by explosions or gently by small bits of material able to induce rapid changes by acting as catalysts or nucleating agents. The mechanism for energy storage might be the accumulation of strain over hundreds of millions of years in the solid Earth, or the super-cooling of water vapour in the atmosphere by updraughts taking place over a few tens of minutes. Effects of releasing this energy could be world-wide, as in the case of altering climate, or regional, as in the case of locally excited earthquakes or enhanced precipitation."

MacDonald is writing about the so-called 'butterfly effect' here. The notion that a relatively small climatic intervention can have a tremendous effect upon the weather later is central to the New Manhattan Project and has now been repeated many, many times throughout the pages of this book.

Next, under the heading of 'Weather Modification,' MacDonald makes the oft-repeated comparisons between the power of storms and atomic bombs (ch 2), writing, "The quantities of energy involved in weather systems exceed by a substantial margin the quantity of energy under man's direct control." That's enough to make your average paranoid power junkie begin to salivate.

MacDonald then goes over some atmospheric physics pertinent to some overt weather modification activities. Like this part, much of MacDonald's piece is pure Central Intelligence Agency (CIA) talking points and disinformation. The author has seen these types of CIA droppings hundreds of times. After all, as alluded to earlier, MacDonald was a CIA minion. He not only served as chairman of their aforementioned MEDEA Committee, he also served as the CIA liaison to the aforementioned JASON group, which we will discuss next.

In his piece, MacDonald doesn't mince words when he suggests that a hurricane might be used as a weapon. He writes that, "A controlled hurricane could be used as a weapon to terrorize opponents over substantial parts of the populated world."

Under the next heading of 'Climate Modification' MacDonald goes over some basics of atmospheric physics, then states the theory of man-made global warming and the Solar Radiation Management geoengineering thesis which, as we already know, involves scattering sunlight back into space with the use of dispersed atmospheric particles. This, MacDonald asserts, can save us from the dreaded catastrophic global warming. To accomplish this, MacDonald suggests using rockets, not airplanes, but the idea is the same. This is what the geoengineers do: they state the theory of man-made global warming and the SRM geoengineering thesis. Straight from the establishment, they've got your fix for their supposed problem. It's just going to cost you some money. It is reminiscent of the Black Hand telling a local shopkeeper that, unless he pays up, his building might burn down.

He then suggests that creating a hole in the Earth's ozone layer might be an effective weapon that could be 'fatal to all life.' Interesting. Outdoing even that, he then speculates that man might engineer the sun. He writes that, "advanced techniques of launching rockets and setting off large explosions" might do the trick. Then it's on to using atomic bomb blasts to make polar ice sheets slide into the ocean, thus causing a tremendous tidal wave. Rounding out MacDonald's cavalcade of mad science, the author writes of (among other things) nuclear bombs being used to trigger earthquakes, which, in turn, may create tidal waves that would be, "catastrophic to any coastal nation."

After speculating that lightning strikes might be geoengineered, MacDonald also famously writes about using the Earth's atmosphere as a transmission medium for the delivery of electromagnetic frequencies capable of altering the Human mind. Citing some of the current research of the time, he notes that this type of activity could, "seriously impair brain performance in very large populations in selected regions over an extended period." He writes that, "Perturbation of the environment can produce changes in behaviour patterns." Rounding out this section he boldly asserts that, "No matter how deeply disturbing the thought of using the environment to manipulate behaviour for national advantage is to some, the technology permitting such use will very probably develop within the next few decades."

Near the end, MacDonald writes:

"Deficiencies both in the basic understanding of the physical processes in the environment and in the technology of environmental change make it highly unlikely that environmental modification will be an attractive weapon system in any direct military confrontation in the near future. Man already possesses highly effective tools for destruction. Eventually, however, means other than open warfare may be used to secure national advantage. As economic competition among many advanced nations heightens, it may be to a country's advantage to ensure a peaceful natural environment for itself and a disturbed environment for its competitors. Operations producing such conditions might be carried out covertly, since nature's great irregularity permits storms, floods, droughts, earthquakes and tidal waves to be viewed as unusual but not unexpected. Such a 'secret war' need never be declared or even known by the affected populations. It could go on for years with only the security forces involved being aware of it. The years of drought and storm would be attributed to unkindly nature and only after a nation were thoroughly drained would an armed take-over be attempted."

The previous paragraph is about using geophysical warfare against foreign enemies. In his final sentence, MacDonald hints that these technologies might be used domestically. He writes that political issues arising from the use of geophysical warfare technologies, "deserve consideration by serious students throughout the world if society is to live comfortably in a controlled environment."

MacDonald's prophecies have come true.

Gordon J.F. 'How to Wreck the Environment' MacDonald is, by all indications, along with Teller, Vonnegut and others, one of the keys to all of the scientific aspects of today's New Manhattan Project. MacDonald may be the most important because of how he apparently brought everything together.

It is interesting to note that not only was Dr. MacDonald a science advisor to multiple presidents, a prominent member of the Council on Foreign Relations, and a member of a slew of top-shelf weather modification committees, but that he was also a very active member of, and CIA liaison to, something called JASON. Evidence suggests that JASON has worked on developing the New Manhattan Project under DARPA.

JASON

Jason (JASON) is an independent group of scientists that have historically done contract work mainly for DARPA, but have also done work for other New Manhattan Project progenitors such as: the CIA, the Navy, NASA, the National Science Foundation, the Department of Energy, and the MITRE Corporation. Jason has historically employed the best scientific minds. Jason has a history of solving the problems nobody else can. Starting in 1960, all the available Jason scientists have gathered annually (usually in the Summer) to tackle scientific problems brought to them by their sponsors.

The original Jason scientists and those that went on to run Jason for many years were comprised mostly of men whom had worked on the original Manhattan Project and their students. These were famous scientists like: Edward Teller (ch 2), John Wheeler (1911-2008), Hans Bethe (1906-2005), Murph Goldberger (1922-2014), Freeman Dyson (1923-2020), Val Fitch (1923-2015), Luis Alvarez (ch 2), Pief Panofsky (1919-2007), Bill Nierenberg (1919-2000), Ed Frieman (1926-2013), Murray Gell-Mann (1929-2019), Keith Brueckner (1924-2014), Marshal Rosenbluth (1927-2003), George Kistiakowsky (1900-1982), Charles Townes (1915-2015), and Richard Garwin. Jason's first sponsor was DARPA's first chief scientist and the first director of Lawrence Livermore National Labs: Herbert York (1921-2009).

York left work at the Pentagon in 1961. After that, he began a longtime, close working relationship with both the Institute for Defense Analyses (IDA) and something called the Aerospace Corporation. He describes the Aerospace Corporation as working almost exclusively for the Air Force and he describes the IDA as working almost exclusively for the Office of the Secretary of Defense, including the Office of the Joint Chiefs of Staff. York describes Jason as 'one of IDA's offspring.' York says that Jason stayed with IDA for its first 14 years, then went on to become a division of the Stanford Research Institute, then it became a division of the MITRE Corporation. York writes that, since he was on the board of IDA, that he had a hand in starting Jason and has participated in many of Jason's activities. He writes that Jason, "pioneered the work in beam weapons of all kinds." He writes that Jason reviewed the plans for the Star Wars program.

In the early 1970s Jason moved their operations to the headquarters of the Stanford Research Institute (SRI) in Menlo Park, CA. The Stanford Research Institute has contracted with Jason extensively and SRI has done much work in the vein of the New Manhattan Project.

In 1977 Jason formed a group to study Earth's climate. As described by the author of the definitive book on Jason, Ann Finkbeiner, our friend Gordon MacDonald and a man by the name of Henry Abarbanel served as, "...more or less the group's constant core." Other regular members included: William Nierenberg, Ed Frieman, and Freeman Dyson. Dyson reportedly says that the first climate study had no sponsor and that it was done with 'free money.' Finkbeiner writes that, "Jason decided to study the subject on its own." 'Free money' sounds like good work if you can get it.

Herb York

In the spring of 1977, the Jasons convened a two-day meeting at the National Center for Atmospheric Research in Boulder, Colorado. At this meeting, participants from government and academia had a fear mongering session about the non-existent threat of global warming due to man's carbon emissions. The product of this meeting was a climate model, made by Jason, which showed increased global temperatures correlated to increases in atmospheric CO2. Good work, men.

In 1979, Jason climate group regular Ed Frieman left his other job at Princeton and, according to Finkbeiner, "...became the head of the

Office of Energy Research at the newly created Department of Energy. For the next two years he commissioned Jason to do a series of studies on the relationship between carbon dioxide and global warming."

In 1981, Jason left their offices at the Stanford Research Institute and made a new home at facilities owned by none other than the aforementioned MITRE Corporation in McLean, VA. G.J.F. MacDonald himself said in an interview that MITRE provided the space, the salary of Jason's executive director, and money for additional costs.

After the initial flurry of Jason climate studies in the late 1970s, Jason did not give the issue much attention for many years. Then in 1986, Jason, led by Gordon J.F. MacDonald, began advising the Department of Energy in regards to their new, interagency Atmospheric Radiation Measurement (ARM) program. The Atmospheric Radiation Measurement program involved the establishment of a comprehensive array of atmospheric monitoring equipment and, secondly, bringing collected atmospheric measurements into a computer model. An ARM site was created on the North Slope of Alaska, where the HAARP antenna gets its power. This was all done, of course, to enhance our understanding of climate change, you understand. ARM was created in 1989 and continues to this day.

ARM's work has always centered on atmospheric aerosols. They usually dance around the issue by claiming to be talking about everything *but* man-made particles dispersed from planes for the purpose of weather modification. But sometimes they let one out in church. In an April 2002 ARM report by Jason and the MITRE Corporation, under the heading of 'Active Experiments,' the authors write of the atmospheric effects of man-made cloudiness such as particles emitted from an airplane used in the course of weather modification activities - 'active experiments' pertaining to climate change, indeed.

Command Centers

Like classical music conductors, today's top geoengineers must bring together all the members of their orchestra. This Project requires a small group (or groups) of people running it as a team. Any team such as this would necessarily require facilities capable of running the

New Manhattan Project. There aren't too many known locations in the country or in the world that might qualify. One can suppose what they would need.

Evidence and logic suggest that the command and control centers of the New Manhattan Project employ large holographic displays. These displays probably produce very realistic three-dimensional imagery in a theatre-in-the-round type of setting viewable by groups of lead scientists and others. This would be the best way to display the information and allow for discussion. Technicians can more easily respond to orders from the lead scientists and lead scientists can more easily respond to others when they are all looking at the same thing.

A NMP command center probably looks a little something like this:

Theater in-the-round

The background walls would be plastered with computer screens showing all kinds of relevant information. In the rows would be comfortable recliners with power and possibly data connections for laptop computers. In the center, holograms would appear.

It is not so far off to think that the command centers of the New Manhattan Project may be employing holograms. Holograms go all the way back to 1948 with Dr. Dennis Gabor (1900-1979) and his Nobel Prize. Holographic movies have been around since the 1960s. As early as 1964, our Air Force was recreating atmospheric volumes in a holographic fashion. In "The Electronics Research Center: NASA's Little Known Venture into Aerospace Electronics," a paper by Andrew Butrica, PhD, the author writes of holographic displays being developed at NASA's Cambridge, Massachusetts Electronics Research Center in the late 1960s. This facility today is known as the John A. Volpe National Transportation Systems Center. In the Summer of 1968, the NSF reported the Bureau of Reclamation's successful production of three-dimensional atmospheric models. Way back in 1975 the Army reported that they were collecting atmospheric data in a fashion that would lend itself to being displayed on a holographic display. They wrote:

"*...developments in microwave radar, laser radar, passive radiometry, and acoustic sounding have demonstrated that it is now feasible to consider real-time atmospheric sensing systems for the combat Army. Remote sensing permits the measurement of relevant parameters of the atmosphere in one, two, or three spatial dimensions, all as a function of time,...*"

The 1998 Army document "3D Holographic Display Using Strontium Barium Niobate" speaks to a breakthrough ability to both record and produce very realistic holographic imagery in real time and in free space. To produce such holographic imagery, a barium or strontium mist would not be necessary as the barium and strontium mentioned here actually refers to crystals used as part of the projector. This is probably the technology in use today as part of New Manhattan Project command and control centers.

Now we get down to the last locked dungeon door. It's dark and dank. Green slime and black grime covers the huge blocks of stone walls. A rat scurries along in the corner. The smell of rotting flesh permeates the air. We can hear the low growling, the howls of pain, and the insane, echoing laughter.

Where are they? Where have our tormentors been hiding? Who has been laughing at our pain?

Let us use our key that has been fashioned through so much scratching and clawing research. Now we slide the key into the lock, feeling the exact fit. It turns snugly. The portal is open. There is no going back now. The spirits beckon us. Let us pass from ignorance into wisdom.

Conclusions

When searching for the large scientific facilities necessarily housing the people and equipment running the New Manhattan Project, there aren't too many locations in the world that have anything resembling the massive technological resources necessary. As some of the very few locations capable, the evidence leads firstly to Lawrence Livermore National Laboratory (LLNL) in Livermore, California and secondly to the National Center for Atmospheric Research (NCAR) in Boulder, Colorado. There may be others, including a third command center in the East.

LLNL has a grand history of conducting classified scientific projects. LLNL has been deeply involved in the design and manufacture of nuclear weapons. Nuclear weapons were originally produced by the first Manhattan Project. Strangely enough, as we have seen and will continue to see, nuclear weapon development largely morphed into the atmospheric sciences.

The Program for Climate Model Diagnosis and Intercomparison (PCMDI) at Lawrence Livermore Labs is probably running the New Manhattan Project's atmospheric modeling efforts. The PCMDI was founded by the Department of Energy who has extensive ties to the New Manhattan Project.

The Accelerated Strategic Computing Initiative (ASCI) at Lawrence Livermore Labs is probably running the New Manhattan Project's supercomputers. The ASCI evolved from a Defense Advanced Research Projects Agency (DARPA) program. It looks like DARPA has been executively overseeing the scientific aspects of the New Manhattan Project all along.

MITRE probably designed and built LLNL's New Manhattan Project command center and probably manages it to this day.

LLNL's collaborations with NCAR are interesting because these investigations have shown that NCAR has been instrumental in the development of the New Manhattan Project as well. NCAR has always had similar equipment and conducted similar activities to those of LLNL. The main difference between LLNL and NCAR is that NCAR specializes in the atmospheric sciences only, while LLNL does not. NCAR's location near the geographical center of the contiguous United States would provide a more convenient location for geoengineers who live in the central regions of the country. NCAR is located on a mesa at the foot of the Rockies. The facility is protected by being deeply embedded in the Earth - much like NORAD which is further south and built into the same geological formation.

For this reason of convenience, there may also be a third C4 center in the East. The prime candidate is Wright-Patterson Air Force Base. As these investigations have shown, Wright-Patterson has some of the most extensive ties to the New Manhattan Project. My science advisor also suggests that Brookhaven National Laboratory on Long Island in the state of New York may house a command center. Some other interesting locations include Los Alamos National Laboratory in New Mexico as well as the NASA Ames Research Center near Mountain View, CA.

Now we have some idea of from whence the New Manhattan Project is being conducted. Somebody is doing it. At least one C4 center has to be somewhere. All will be revealed.

References

Planet Earth: The Latest Weapon of War a book by Rosalie Bertell, published by Black Rose Books, 2001

Atmospheric Sciences at Nasa: A History a book by Erik M. Conway, published by The Johns Hopkins University Press, 2008

Introduction to Remote Sensing: Fifth Edition a book by James B. Campbell and Randolph H. Wynne, published by The Guilford Press, 2011

"Government Weather Programs (Military and Civilian Operations and Research)" a report by the Committee on Government Operations, published by the U.S. Government Printing Office, 1965

"Analysis of Smoke Trail Photographs to Determine Stratospheric Winds and Shears" a report by the Air Force Geophysics Laboratory, Hanscom Air Force Base, 1976

Beyond the Atmosphere: Early Years of Space Science a book by Homer E. Newell, published by Dover Publications, 2010

"Chemical Releases in the Ionosphere" a paper by T. Neil Davis, published by the Institute of Physics, 1979

Rockets Over Alaska: The Genesis of the Poker Flat Research Range a book by Neil Davis, published by Alaska-Yukon Press, 2006

"Observations of the Development of Striations in Large Barium Ion Clouds" a report by T. N. Davis et al. for the Rome Air Development Center, Advanced Research Projects Agency, published by the National Technical Information Service, 1972

The Jasons: The Secret History of Science's Postwar Elite a book by Ann Finkbeiner, published by Penguin Books, 2007

"Report of Chief, AFSWP to ARPA" a film produced by the Armed Forces Special Weapons Project, 1958

The Interdepartmental Committee for Atmospheric Sciences (ICAS) reports 1960-1978, published by the Federal Council for Science and Technology

U.S. patent #4,402,480 "Atmosphere Modification Satellite" 1983

U.S. patent #5,762,298 "Use of Artificial Satellites in Earth Orbits Adaptively to Modify the Effect that Solar Radiation Would Otherwise Have on Earth's Weather" 1998

U.S. patent #5,984,239 "Weather Modification by Artificial Satellites" 1999

Satellite Technology: Principles and Applications, Second Edition a book by Anil K. Maini and Varsha Agrawal, published by John Wiley and Sons, 2011

The General Electric Story: A Heritage of Innovation 1876-1999 a book by the Hall of Electrical History, Schenectady Museum, published by Hall of Electrical History Publications, 1999

"Use of the ERTS-1 Satellite Data Collection System in Monitoring Weather Conditions for Control of Cloud Seeding Operations" a report by Archie M. Kahan, published by the Department of the Interior, Bureau of Reclamation, Division of Atmospheric Water Resources, 1974

"National Environmental Satellite, Data, and Information Service, 1996-97" a report by the United States National Environmental Satellite, Data, and Information Service, 1997

"Inquiry into the Feasibility of Weather Reconnaissance from a Satellite Vehicle" by S.M. Greenfield and W.W. Kellogg, published by the Rand Corporation, 1960

Operation Paperclip: The Secret Intelligence Program that Brought Nazi Scientists to America a book by Annie Jacobsen, published by Little, Brown and Company, 2014

Hughes After Howard: The Story of Hughes Aircraft Company a book by D. Kenneth Richardson, published by Sea Hill Press, 2012

"A Review of Federal Meteorological Programs for Fiscal Years 1965-1975" a report by Clayton E. Jensen, published by the Federal Coordinator for Meteorological Services, 1975

"On Geoengineering and the CO2 Problem" by Cesare Marchetti, as published in *Climatic Change*, 1977

"The Changing Atmosphere: An Integrated Global Atmospheric Chemistry Observation Theme" a report by the International Global Observing Strategy, 2004

The Pentagon's Brain: An Uncensored History of DARPA, America's Top Secret Military Research Agency a book by Annie Jacobsen, published by Little, Brown, and Company, 2015

National Science Foundation Weather Modification Annual Reports 1959-1968

"National Advisory Committee on Oceans and Atmosphere First Annual Report to the President and the Congress" 1972 as it appeared in the congressional record of testimony by Dr. Alfred J. Eggers Jr., NSF during hearings before the Subcommittee on the Environment and the Atmosphere of the Committee on Science and Technology, U.S. House of Representatives, Ninety-fourth Congress, second session, 1976

"Evaluation of Iridium Satellite Phone Voice Services for Military Applications" a report by Caroline Tom and Lyle Wagner, published by the Defence Research Establishment, Ottawa, 1999

U.S. patent #20090189802A1 "System and Method for Using Iridium Satellite Signals for Meteorological Measurements," 2009

U.S. patent #7,728,759 "System and Method for Using Iridium Satellite Signals for Meteorological Measurements," 2010

"Space Operations: Through the Looking Glass (Global Area Strike System)" a paper by Lt Col Jamie G.G. Varni, Mr. Gregory M. Powers, Maj Dan S. Crawford, Maj Craig E. Jordan, and Maj Douglas L. Kendall, published by the United States Air Force, 1996

"Exploring the Interface Between Changing Atmospheric Composition and Climate" a report by the Network for the Detection of Atmospheric Composition Change, 2011

"The Global Atmospheric Research Program: 1979-1982" a paper by Jay S. Fein and Pamela L. Stephens, as published in *Reviews of Geophysics and Space Physics*, vol. 21, no. 5, p 1076-1096, June, 1983

"Weather & Climate Modification: Problems and Progress" a report by the National Academy of Sciences, 1973

U.S. patent #7,365,674 "Airborne Weather Profiler Network," 2008

"The White House Conference on International Cooperation" a report by the National Citizens' Commission on International Cooperation, Committee on Meteorology, 1965

"Weather Modification: Programs, Problems, Policy, and Potential" a report by the Congressional Research Service, republished by the University Press of the Pacific, 1978

"Government Weather Programs: Military and Civilian Operations and Research" a report by the Committee on Government Operations, published by the U.S. Government Printing Office, 1965

Hearing before the Subcommittee on Oceans and Atmosphere of the Committee on Commerce, United States Senate, Ninety-fourth Congress, second session, Feb. 17, 1976

"RAIN MAKERS China 'will move clouds and make it rain in different places using satellites' in terrifying weather control plan" an article by Patrick Knox, published by *The Sun*, Nov. 7, 2018

"Conversations with Jule Charney" a report by George W. Platzman, published by the Massachusetts Institute of Technology and the National Center for Atmospheric Research, 1987

"Outline of Weather Proposal" a report by Vladimir Zworykin, published by RCA Laboratories, 1945

"Studies in Statistical Weather Prediction" a report by the Travelers Insurance Company and the Air Force Cambridge Research Center, 1958

"A Numerical Simulation of Warm Fog Dissipation by Electrically Enhanced Coalescence" a paper by Paul M. Tag as it appeared in the proceedings of the American Meteorological Association Fourth Conference on Weather Modification, 1974

"The PCMDI Software System: Status and Future Plans" a report by Dean N. Williams for the Program for Climate Model Diagnosis and Intercomparison, published by Lawrence Livermore National Laboratory, 1997

"AMIP: The Atmospheric Model Intercomparison Project" a report by W. Lawrence Gates for the Program for Climate Model Diagnosis and Intercomparison, published by Lawrence Livermore National Laboratory, 1992

"Long-Range Weather Prediction and Prevention of Climate Catastrophes: A Status Report" a report by E. Teller, K. Caldeira, G. Canavan, B. Govindasamy, A. Grossman, R. Hyde, M. Ishikawa, A. Ledebuhr, C. Leith, C. Molenkamp, J. Nuckolls, and L. Wood, published by Lawrence Livermore National Laboratory, 1999

"Global Biogeochemistry Models and Global Carbon Cycle Research at Lawrence Livermore National Laboratories" a report by C. Covey, K. Caldeira, T. Guilderson, P. Cameron- Smith, B. Govindasamy, C. Swanston, M. Wicjett, A. Mirin, and D. Bader, published by Lawrence Livermore National Laboratories, 2005

"Terrestrial Biogeochemistry in the Community Climate System Model (CCSM)" a report by Forrest Hoffman, Inez Fung, Jim Randerson, Peter Thornton, Jon Foley, Curtis Covey, Jasmin John, Samuel Levis, W. Mac Post, Mariana Vertenstein, Reto Stöckli, Steve Running, Faith Ann Heinsch, David Erikson, and John Drake as published in the *Journal of Physics*: Conference Series 46, 2006

Climategate: A Veteran Meteorologist Exposes the Global Warming Scam a book by Brian Sussman, published by WND Books, 2010

Lawrence Livermore National Laboratory: Webster's Timeline History 1941-2007 a book by Professor Philip M. Parker, PhD, published by ICON Group International, 2009

The Supermen: The Story of Seymour Cray and the Technical Wizards Behind the Supercomputer a book by Charles J. Murray, published by John Wiley & Sons, Inc., 1997
The ILLIAC IV: The First Supercomputer a book by R. Michael Hord, published by Computer Science Press, 1982

Hearings Before the Subcommittee on the Environment and the Atmosphere of the Committee on Science and Technology, U.S. House of Representatives, Ninety-fourth Congress, second session, 1976

National Center for Atmospheric Research annual report, 1988

"The History of the Accelerated Strategic Computing Initiative" a report by Alex R. Larzelere II, published by Lawrence Livermore Laboratory, 2009

The Pentagon's Brain: An Uncensored History of DARPA, America's Top-Secret Military Research Agency a book by Annie Jacobsen, published by Little, Brown and Company, 2015

"30 Years of Technical Excellence: 1952-1982" a report by Lawrence Livermore National Laboratory, published by Lawrence Livermore National Laboratory, 1982

"Weather as a Weapon" an article by Howard T. Orville as told to John Kord Lagemann, published by *Popular Science*, June 1958

"The Atmospheric Sciences 1961-1971" a report by the National Academy of Sciences, Committee on Atmospheric Sciences, published by the National Academy of Sciences and the National Research Council, 1962

The Interdepartmental Committee for Atmospheric Sciences reports 1960-1978, published by the Federal Council for Science and Technology

"Meteorological Tracer Techniques for Parameterizing Atmospheric Dispersion" a paper by Warren B. Johnson, published by the *Journal of Climate and Applied Meteorology*, May 1983, p931-946

From Whirlwind to MITRE: The R&D Story of the SAGE Air Defense Computer a book by Kent C. Redmond and Thomas M. Smith, published by The MIT Press, 2000

Architects of Information Advantage: The MITRE Corporation since 1958 a book by Davis Dyer and Michael Aaron Dennis, published by Community Communications, 1998

MITRE: The First Twenty Years a book by the MITRE Corporation, published by the MITRE Corporation, 1979

"Aluminum Poisoning of Humanity and Earth's Biota by Clandestine Geoengineering Activity: Implications for India" a paper by J. Marvin Herndon, PhD, published by *Current Science*, 2015

The Early History of the MITRE Corporation: Its Background, Inception, and First Five Years a book by Howard R. Murphy, 1972

The MITRE Corporation: Fifty Years of Service in the Public Interest a book by the MITRE corporation, published by the MITRE Corporation, 2008

"Pioneering MIT Lab May Be Best Argument for Keeping Massachusetts Base Open" an article by Robert Weisman, published in the *Boston Globe* newspaper, Sept. 27, 2004

"Where We Stand" a report by Dr. Theodore von Kármán, published by the Army Air Force Scientific Advisory Group as a volume in the series *Toward New Horizons*, 1946

"The SAGE/Bomarc Air Defense Weapons System: An Illustrated Explanation of What It Is and How It Works" a report by the IBM Military Products Division, 1958

Arms and the Physicist a book by Herbert F. York, published by the American Institute of Physics, 1995

"How to Wreck the Environment" a paper by Gordon J.F. MacDonald as published in the book *Unless Peace Comes: A Scientific Forecast of New Weapons* edited by Nigel Calder, published by The Viking Press, 1970

"Gordon James Fraser MacDonald 1930-2002" a memoir by Walter Munk, Naomi Oreskes, and Richard Muller, published by the National Academies Press, *Biographical Memoirs*, volume 84, 2004

"Restoring the Quality of Our Environment" a report by the Environmental Pollution Panel of the President's Science Advisory Committee, published by the U.S. Government Printing Office, 1965

An interview of Gordon J.F. MacDonald by James Roger Fleming conducted at the University of California, San Diego, published by the American Institute of Physics as part of their *Oral History Interviews*, March 21, 1994

"How to Wreck the Environment" an essay by Gordon James Fraser MacDonald as it appeared in the 1968 book *Unless Peace Comes*, edited by Nigel Calder, published by Viking Books, 1968

Area 51: An Uncensored History of America's Top Secret Military Base a book by Annie Jacobsen, published by Back Bay Books, 2011

Gordon J.F. MacDonald interviewed by Finn Aaserud on April 16, 1986, published by the American Institute of Physics Oral Histories Interviews

"Gordon James Fraser MacDonald 1930-2002" a memoir by Walter Munk, Naomi Oreskes, and Richard Muller, published by the National Academies Press, *Biographical Memoirs*, volume 84, 2004

Killing the Planet: How a Financial Cartel Doomed Mankind a book by Rodney Howard-Browne and Paul L. Williams, published by Republic, 2019

"C.I.A. Is Sharing Data with Climate Scientists" an article by William J. Broad, published by *The New York Times*, Jan. 4, 2010

"CIA Ends Climate Research Program After Obama Calls Climate Change a Security Risk" an article by Douglas Ernst, published by *The Washington Times*, May 22, 2015

"The CIA Is Shuttering a Secretive Climate Research Program" an article by Tim McDonnell, published by *Mother Jones*, May 21, 2015

Letter from Gordon J.F. MacDonald of the MITRE Corporation to Dr. Ari Patrinos, Director of Health and Environmental Research, U.S. Department of Energy, July 27, 1995

"ARM" report by N. Lewis, et al., published by the MITRE Corporation and the Jason Program Office, April 2002

The Complete Book of Holograms: How They Work and How to Make Them a book by Joseph E. Kasper and Steven A. Feller, published by Dover Publications, Inc., 2013

"3D Holographic Display Using Strontium Barium Niobate" a paper by Christy A. Heid, Brian P. Ketchel, Gary L. Wood, Richard J. Anderson, and Gregory J. Salamo, produced by the Army Research Laboratory, 1998

"The Electronics Research Center: NASA's Little Known Venture into Aerospace Electronics" a paper by Andrew Butrica, PhD, published by the American Institute of Aeronautics and Astronautics, Inc., 2002

"Space Weather Architecture Transition Plan" a report by the Office of the Assistant Secretary of Defense for Command, Control, Communications, and Intelligence, 2000

Chapter 8
MOTIVES

"We're in the gardening business now, damn it!"
-geoengineer David Keith

It is incredibly bold to routinely dump tens of thousands of megatons of toxic materials into our atmosphere. Committing crimes of this magnitude, openly in the skies above us, could very well lead to new Nuremberg trials. Considering the level of risk involved, what might be the rewards for these outrageous crimes against Humanity? A key question in many investigations is: Who benefits? In Latin we say, *"Cui bono?"* So as we continue to investigate chemtrails and geoengineering, let us enquire as to who benefits from this New Manhattan Project. What might be the advantages of conducting a global weather modification project such as this? For the worldly, the potential rewards are great.

Weather's effect upon business activity has been studied extensively at the highest levels since at least the mid-1950s. In 1971, an organization called Research Applied to National Needs (RANN) was created within the National Science Foundation to study the economic benefits of weather modification. The weather modification literature is replete with examples of favorable cost-to-benefit ratios. From agriculture to energy to transportation to leisure to construction to just about every business activity on the face of the planet, weather has a direct effect upon operations. American business activity sensitive to weather is well understood and measured in the trillions of dollars annually.

The multi-billion dollar weather derivatives and catastrophe reinsurance markets rise and fall with the weather. Outcomes of wars are heavily dependent upon weather. The politics of government water distribution has also been hotly contested over the years and directly affected by the weather. Don't forget about all the hydroelectric power generated by rainfall.

Maybe a better question is: Who doesn't benefit? Although we the people are surely not benefitting, this is the biggest scientific endeavor in Human history and weather influences just about every aspect of life. It is literally the air we breathe. The corporate beast wants to own it. Many stand to gain. This chapter examines the New Manhattan Project's motives.

An ancient motive

Man's desire and efforts to change the weather undoubtedly go back to before recorded history. Weather is fundamental to the existence of life, and therefore, weather is often a matter of life and death. Extreme heat, extreme cold, draught, floods, and myriad other types of extreme weather have killed innumerable people over the millennia. The historical record shows that it is during these periods of life-threatening weather that Man has often attempted to change the weather and therefore save lives. Most commonly, man has gone about making it rain - especially in times of draught. The earliest recorded efforts were those of mystics. Local shamans would be called upon to ingest a certain concoction in order to communicate with the weather gods and ask for help. In some cases, the sacrifice of certain animals in certain fashions may have been the thing to do. Sometimes a good old rain dance may have done the trick.

Some early Western efforts to stop destructive weather are outlined in Professor James Fleming's book *Fixing the Sky*. Fleming writes, "In ancient Greece, the official 'hail wardens' of Cleonae were appointed at public expense to watch for hail and then signal the farmers to offer blood sacrifices to protect their fields: a lamb, a chicken, or even a poor man drawing blood from his finger was deemed sufficient." A little later, Professor Fleming writes, "In Austria, it was traditional to ring 'thunder bells' or blow on huge 'weather horns' while herdsmen set up a terrific howl and women rattled chains and beat milk pails to scare away the destructive spirit of the storm." Navajo Indians of the Southwest have an ancient practice of causing precipitation by burning the charred bark of a tree that has been hit by lightning.

The geoengineers

The geoengineers want their rewards. They are participating in an industry set to become very prosperous if geoengineering becomes generally accepted. Their spokespeople are here to sell the spraying program which has already begun. This makes perfect sense in our upside-down world. It appears that there are at least hundreds of people calling themselves geoengineers. In one way or the other, they are heavily invested in geoengineering. Some may own their geoengineering companies outright. Some may own financial products like stock or stock options. Some may stand to gain from patent royalties. Some may just be interested in getting big paychecks, bonuses, and retirement benefits for the rest of their lives. If geoengineering becomes a generally accepted practice, the reputations and fortunes of all those involved will undoubtedly be positively reinforced.

Weather and war

It is far from a coincidence that the New Manhattan Project was built and has been carried out by military agencies. They have been there all along. They are there because weather has determined the outcomes of wars innumerable times. An author named Robert Child wrote a short 2012 ebook titled *Weather and Warfare*. For the opening section, Mr. Child writes a paragraph that gives many examples of weather affecting war. It reads:

"*In 1588 more than half of the Spanish Armada, on its way around Northern Britain, was destroyed by storms in retreat back to Spain. Napoleon's attack on Russia was stopped cold by winter weather, as was Hitler's siege of Leningrad. In WWI, it was all-quiet on the western front during mud season. The miracle at Dunkirk took place under cover of heavy fog, the Japanese attack on Pearl Harbor was launched under cover of Pacific storms and the D-day invasion of Normandy was launched based on the most critical weather forecast in history.*"

During the late 1960s and early 1970s, in southeast Asia our U.S. military conducted weather modification operations under something usually called Operation Popeye. It involved spraying silver and lead iodide out of aircraft (and thereby inducing rain) in order to flood out the enemy's transportation routes. After three years of dodging Senate inquiries, the Department of Defense, on March 20, 1974, coughed up a top-secret briefing on these extensive rainmaking operations in Southeast Asia. It was declassified and made public on May 19, 1974. This is how the American public first officially learned of these operations. As the report detailed, these rainmaking operations were performed between March of 1967 and July of 1972 in order to slow enemy troop movements along the Ho Chi Minh Trail. The silver iodide sprayed from Air Force aircraft caused increased rainfall which made the trail muddy and therefore less passable. The project was code-named differently at different times. The names Popeye, Compatriot, and Intermediary were used. It was the first known, deliberate use of modern weather modification as an offensive weapon of war. Gordon J.F. 'How to Wreck the Environment' MacDonald writes, "Because the program was considered so politically sensitive, responsibilities for it were lodged within that part of the Joint Chiefs of Staff responsible for covert operations." MacDonald continues:

"During the hearings it was revealed that because of the supposed political sensitivity of the operations the number of high officials in government who were aware of the activity was extremely limited. Indeed, Secretary of Defense Laird had to reverse his statement to the Foreign Relations Committee that rainmaking had not been used over North Vietnam. Even though Laird had been Secretary of Defense during the critical period of U.S. disengagement in Vietnam, he was not aware that the rainmaking activities were underway or that they had been initiated during the Johnson Administration."

Weather and war recalls Secretary of Defense Cohen's 1997 remarks mentioned in the first chapter where he spoke about how terrorists might, "...alter the climate, set off earthquakes, volcanoes remotely through the use of electromagnetic waves..."

Although today we have what appears to be military groups that have embedded themselves into our otherwise constitutional republic carrying out the New Manhattan Project domestically, weather control systems employing New Manhattan Project technology are probably also used against foreign enemies in foreign countries - most probably in the Middle East as that region has seen recent, large-scale military operations. As noted earlier, the 1996 Air Force document "Weather as a Force Multiplier: Owning the Weather in 2025" outlines such a program.

Geoengineering and the aforementioned planetary engineering are schools not only of weather modification, but of geophysical warfare as well. All of these embody the manipulation of Earth's natural systems. In the case of geophysical warfare, Earth's natural systems are manipulated for military advantage. Throughout the pages of this book, the reader has found our military writing about using Earth's natural systems for strategic advantage. As we have seen, Rear Admiral Luis de Florez, Vice Admiral William Raborn, Dr. G.J.F. MacDonald, and many others have written about it.

Unsurprisingly, when looking back in time, we find many early examples of geophysical warfare. We will now take a look at some of them. In 612 BC, a combined force of Medes, Persians, and Babylonians apparently diverted a tributary of the Tigris river in order to penetrate the stronghold of the ancient city of Nineveh. In her landmark book *The History of the Ancient World: From the Earliest Accounts to the Fall of Rome* eminent historian Susan Wise Bauer writes, "A tributary of the Tigris ran through the city beneath the walls, providing it with water and making it difficult to besiege. But it seems likely that the attackers built a dam to divert more of the Tigris into the city, carrying away the foundations of the walls and breaking them away." Bauer continues, "With the walls crumbling, the Babylonians stormed the city and sacked it."

A little later in 539 BC, geophysical warfare (also known as 'geoengineering') was used to penetrate the ancient city of Babylon. Once again it was the Persians, who this time caused a *lack* of water rather than a flood. The Persian king Cyrus the Great caused the level of the Tigris river to fall so that his troops could enter the fortified city and lay waste to it. Bauer writes:

"Cyrus, realizing that it would take months if not years to starve the defenders out of such a huge and well-supplied city, formed another plan. Xenophon [an early Greek historian] *explains it: the Tigris, which flowed right through the middle of Babylon, was deeper than two men's height. The city could not easily be flooded, thanks to Nebuchadnezzar's reinforcements, but Cyrus had another strategy in mind. He had trenches dug all along the Tigris, upstream from the city, and during one dark night he had his men open all the trenches simultaneously. Diverted away from its main stream in many directions, the level of the Tigris that ran through the city sank at once, enough that the Persian soldiers could march through the mud of the riverbed, under the walls of the city. The core assault unit climbed up out of the riverbed inside the city at night, covered with mud, and stumbled along through the streets, shouting as if they were drunken revelers, until they reached the palace and took it by storm... The gates were opened from the inside. The rest of the Persians came in, and the city fell."*

In the late first century, Eric the Red was apparently forced to leave his native Norway after his henchmen killed his rival by causing a landslide to fall on the man's farm.

Some geophysical warfare efforts have been more successful than others. Around 1210, the Mongol Genghis Khan attempted to flood out his enemy with disastrous results. Khan and his men were attempting to flood out the fortified Xia capital city of Chung-hsing by damming up a nearby branch of the Yellow river. It didn't work. The poorly built dam broke and flooded the Mongol camp instead. So much for Mongolian engineering.

Using geophysical warfare for three years, the ancient Greek scientist Archimedes (287-212 BC) apparently defended the city of Syracuse from the Roman fleet and army. Part of his effort, which was the city's official effort, involved the use of huge concave mirrors made of highly polished metal. These mirrors caught and focused the sun's rays onto the invading Roman ships, causing the ships to catch fire.

Interestingly, the author of *Archimedes and the Door of Science*, Jeanne Bendick writes that, "Archimedes began the science of mechanics, which deals with the actions of forces on things - solid things, like stones and people; liquid things, like water; gases, like air or clouds."

This example of early geophysical warfare regarding Archimedes is particularly relevant to the New Manhattan Project for a few reasons. For one, Archimedes used a mirror to focus the sun's rays which is akin to what today's ionospheric heaters can do, bending the ionosphere to produce a similar effect as part of the NMP. Secondly, Archimedes is famous for his quote, "Give me a lever long enough and a fulcrum on which to place it, and I shall move the world." That sounds like geoengineering. In fact, the logo of the Central Intelligence Agency's Directorate of Science and Technology is that of a man moving the world with a lever and a fulcrum.

Central Intelligence Agency's Directorate of Science and Technology logo

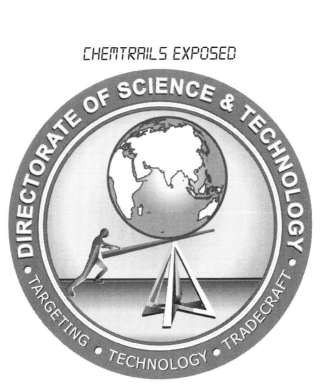

People involved in today's chemtrail spraying operations have probably been told that they, like Archimedes defending the city of Syracuse, are defending the planet from the scourge of global warming.

Agriculture

The link between weather and agricultural production is obvious. Production of crops and livestock was a primary motivation behind ancient weather modification activities and this motive continues to propel today's New Manhattan Project. There's nothing new under the sun. The relationships between weather modification and agricultural production have been well studied and well understood. The two main areas of overt public study, experiment, and implementation (precipitation enhancement and hail suppression) are directly relevant to agricultural production.

By 1968, the Department of Agriculture Joint Task Force on Weather Modification had issued a report (the first of many) stating that the federal Government should be concerned with the possible agricultural benefits from increased precipitation due to weather modification. They were talking about the conventional weather modification industry which, since 1947, has mostly concerned itself with precipitation (ch 3). The New Manhattan Project has since gone on to concentrate on many other geophysical conditions such as atmospheric temperature, but the principle is the same. Modification of many different weather phenomena can benefit agricultural production. The geophysical conditions previously achieved by the conventional weather modification industry may also today be achieved by the New Manhattan Project. Therefore, study applicable to the benefits of the conventional weather modification industry applies to the New Manhattan Project as well. That's why we look at these studies here.

The March, 1971 Interdepartmental Committee for Atmospheric Sciences Report found hail suppression to be quite cost effective. Their authors write, "Annual hail damage in the United States amounts to over $300 million. Where limited areas of certain crops have been 'protected' by commercial seeding operations, reported benefit-cost ratios have been generally better than 2 to 1."

In 1975, the Department of Atmospheric Science at Colorado State University produced a 200-plus page report assessing the present and potential role of weather modification in agricultural production. In this document, many renowned geoengineers statistically determined the relationships between precipitation and agricultural production and how said production might increase or decrease by weather modification.

In 1976 Congressional testimony, weather modifier Dr. D. Ray Booker wrote, "A study by the North Dakota State University indicates the gross state product of North Dakota would be increased by at least $300 million through the addition of one inch of additional rainfall during the growing season. This benefit would undoubtedly be several billion dollars if the same increase were spread over several states." $300M in 1976 is worth about $1.4B today. A billion dollars in 1976 works out to about $4.5 billion today.

The same organizations that have been bringing us the New Manhattan Project's technology have also been very interested in agricultural crop yields. The National Oceanic and Atmospheric Administration (NOAA), the National Aeronautics and Space

Administration (NASA), the World Meteorological Organization (WMO), and others came together in the mid-1970s to conduct something called the Large Area Crop Inventory Experiment. This experiment involved satellites using remote sensing technologies in order to determine crop growing conditions and thus predict forthcoming yields. Since then, this type of technology has become standard. There is a *Wikipedia* page for 'Satellite crop monitoring.'

If one can significantly influence crop yields with covert weather modification such as the New Manhattan Project, then one can make money in agricultural commodity financial markets. That's known as financial fraud.

Water

As we have seen, weather modification has historically most often been practiced in order to increase precipitation. In the case of today's New Manhattan Project though, it is also about the reduction of precipitation. What is done with water, whether it be abundant or lacking, is where your local water district comes in. The politics of water distribution, made famous in the 1974 movie *Chinatown*, have quite a reputation and, as we will see at the end of this section, strange occurrences continue.

It is interesting to note that in her book *The Brothers Vonnegut*, Ginger Strand writes that when Langmuir, Schaefer, and Vonnegut started the scientific era of weather modification in 1946 (ch 3), local water districts contacted General Electric wanting more rain for their districts. This was apparently when local water districts first got involved in weather modification. At the time, Senator Francis Case (1896-1962) from South Dakota and Senator Clinton Anderson (1895-1975) of New Mexico were big, public fans of weather modification. Case even did some of it himself.

Water resources management has been studied intensively at the highest levels. The July, 1977 Interdepartmental Committee for Atmospheric Sciences (ICAS) report states:

"*The importance of water resource management and planning has been recognized for some time, and related problems are being addressed at all levels of government in this country. Federal efforts include studies and analyses conducted by the Department of Agriculture (USDA), Department of Commerce (DOC), Department of*

Housing and Urban Development (HUD), Department of the Interior (DOI), and the U.S. Army Corps of Engineers, covering a broad spectrum of hydroclimatic impacts. Many of these activities are part of a program recommended by the Committee on Water Resources Research (COWRR) of the FCST [Federal Council for Science and Technology] directed toward specific needs and priorities identified by the COWRR."

In 1976, U.S. Congressman Bernice Sisk (D-CA) testified that, before any Federal funds arrived, local California water districts were paying for all weather modification activities. He said that after the Federal funds began flowing, local irrigation districts continued to provide a majority of the funding. Interestingly, he said the financing was done through California's college system.

Due to the 2015 draught in California, which was probably geoengineered, alfalfa farmers in the Palo Verde Valley went out of business. Swooping in to feast on their carcasses was the local water district. Of course, by 2016 the reservoirs were full again, so it looks like the water district got a good deal. Maybe this is the game plan going forward.

Energy

In addition to making plants and cities grow, rainwater can generate hydroelectric power. More rain and snowpack means more power. Over the years, this fact has fostered a cozy business relationship between weather modifiers and those responsible for the generation of hydroelectric power. As early as 1952, the Bonneville Power Administration of the Pacific Northwest was involved in cloud seeding operations. Other early participants include the California Electric Power Company (Southern California Edison) and the Pacific Gas and Electric Company (PG&E) of Northern California.

These electrical utilities have been participating in a conventional weather modification industry practice called 'orographic' cloud seeding. Orographic cloud seeding entails running ground-based silver iodide generators upwind from a mountain range. The silver iodide particles are taken by the naturally occurring wind and elevated as they race up the face of the mountains. The particles eventually reach an altitude at which they may nucleate, usually as snow, which increases the snowpack and thus, water available for hydroelectric power generation. This is much cheaper than operating aircraft and has been quite prevalent over the years. Electrical utilities engage in these activities to this day.

A paper by meteorologist Loren W. Crow appeared in one of the most referenced weather modification reports - 1958's *Final Report of the Advisory Committee on Weather Control*. In reference to money that might be made from additional hydroelectric power generation, her paper reads, "An annual [cloud] seeding project which covers 1,000 square miles might cost only $20,000 and have a reasonable chance of producing additional water worth $100,000. Another project with a target area of only 500 square miles might cost $50,000 per year to operate but the value of a small increase in additional water might be worth $1,000,000." She's writing about making, on a short-term basis, five to twenty times the money invested. Those are some healthy returns!

Fig. 17 -- Pacific Gas & Electric Company ground-based silver iodide generator.

A PG&E ground-based silver iodide generator

Pacific Gas and Electric (PG&E) did an initial study - the results of which were published in a 1966 paper. They were operating ground-based silver iodide generators in order to enhance mountain snowpack and thus increase hydroelectric power. PG&E found average cost/ return ratios of 1:5.56.

The Congressional Research Service's senior specialist on the subject wrote in a 1978 report:

"In a 1969 study, the Travelers Research Corp. estimated that run-off from the entire Connecticut River basin might be increased by about 2 million acre-feet (15 percent) per year through a weather modification program. It was calculated that this increment of water would cost $2.30 per acre-foot, or $4,600,000 annually. The report also stated that net benefits of $1,400,000 from municipal water supply, and $2,600,000 from supply of cooling water for thermal electric generating stations and increased flow for hydroelectric power generation might be realized by the 1980's."

Any way you slice it, if you're in the energy business, it makes plain old dollars and cents to modify the weather. Do you really think that all these companies would have been so invested in weather modification over the years if it didn't work? As far as the efficacy of conventional weather modification goes, that's about all you need to know.

The Travelers

Historically, when it comes to weather and the modification thereof, the Travelers Insurance Corporation of Hartford, Connecticut has been the most active insurance company. Their logo is an umbrella for crying out loud. We first learned about their participation in these areas in chapter 1 pertaining to their production of a paper for Luis de Florez's CIA science committee.

Weather related financial instruments have been pioneered by the Travelers. As early as 1954, in cooperation with the U.S. Weather Bureau, the Travelers simultaneously established a weather research organization called The Travelers Weather Research Center and a weather observing and forecasting station called The Travelers Weather Service. A radio station owned by the insurance company disseminated weather information to the public. Information was also shared with government groups. An associate professor of meteorology from the Massachusetts Institute of Technology named Thomas F. Malone headed these operations. At the time, the president of the company said, "This action is being taken after long and thorough investigation of the possibilities of weather study in relation to insurance underwriting." From 1968 to 1970, Malone was vice president of Travelers Insurance. The Travelers Weather Research Center was headed by J. Doyle Dewitt, president of the Travelers Insurance Company and a director of the Rockefeller-controlled Chase Manhattan Bank.

Throughout his life, Thomas F. Malone, PhD (1917-2013) of the Traveler's Research Department was very active in the atmospheric sciences. In the early 1950s, he served as a civilian consultant to the Office of Naval Research. Malone advised then President Kennedy on his aforementioned 1961 weather control speech before the UN (ch 1). He joined the U.S. national committee of the United Nations Educational Scientific and Cultural Organization (UNESCO) in 1965. He was a member of The Special Commission on Weather Modification of the National Science Foundation. The big 1966 National Academy of Sciences report "The Atmospheric Sciences 1961-1971" was produced under a committee chaired by Mr. Malone. Mr. Malone's committee included our good friend Gordon J.F. 'How to Wreck the Environment' MacDonald. Malone sat on the U.S. Department of Commerce's Weather Modification Advisory Board from 1977 to 1981. Mr. Malone was instrumental in the formation of the National Center for Atmospheric Research (NCAR) and later became Chairman of NCAR's Board of Trustees. Malone also served as the president of both the American Meteorological Society and the American Geophysical Union. Malone was an early, longtime, and steadfast promoter of the theory of man-made global warming. He helped establish the UN's Intergovernmental Panel on Climate Change (IPCC) in 1988.

Thomas F. Malone

In the early 1960s, the Travelers participated in the Bureau of Reclamation's massive conventional weather modification experiment Project Skywater. By 1967, the Travelers had studied the large-scale hydrology of North America east of the Rockies with the Environmental Science Services Administration (ESSA), the predecessor to today's National Oceanic and Atmospheric Administration (NOAA). In 1968 it was reported in the proceedings of the tenth NOAA weather modification conference that the Travelers was paid by the National Science Foundation (NSF) to review the NSF's weather modification documents in order to, "...investigate the possibility of abstracting useful scientific information...." The NSF reported:

"The Bureau of Reclamation has initiated a project with the Travelers Research Center in Hartford, Conn., to formulate a simple model to study the effects of artificial nucleation over relatively large

areas and the feasibility of surface or aircraft seeding. The qualitative value of precipitation augmentation over the Connecticut River Valley has now been established with respect to municipal water supply, pollution abatement, power generation, and recreation."

Do you think that a for-profit business like the Travelers would be involved in all these weather modification activities if there wasn't anything in it for them? They are an insurance company. As we have seen, weather can have a tremendous effect upon the health of people and business activities. The Travelers has been investing in weather modification for good reasons.

The icing on the cake here is that the Travelers insured the original Manhattan Project. Let us refer to a 1964 self-produced history. *The Travelers: 100 Years* reads:

"The Travelers was approached by the private contractors engaged in the project with the request that the Companies administer the forms of insurance which would be required to protect workers and the public. The Companies assented, on the condition that the government guarantee the payment of premiums. This done, a special underwriting formula was worked out."

Not only did the Travelers provide the Manhattan Project insurance, they also did a good job covering it up. *The Travelers: 100 Years* continues:

"The most ticklish phase of the adjustment of claims was the necessity of keeping them from reaching law courts, where publicity might have drawn undesired attention to the mysterious operations in process at Oak Ridge and Hanford."

Being that the New Manhattan Project is a gigantic secret as well, and taking into account all the other evidence presented here, The Travelers looks like the right company in the right line of work to be profiting handsomely from the NMP.

Weather derivatives

Since the heady days of The Travelers' early excursions into the study of weather-related financial products, other weather-related financial markets and products have arisen. The weather derivatives market is one of them.

In his paper "Why in the World Are They Spraying?," activist Michael Murphy floated the idea that chemtrails are sprayed (at least in part) in order to manipulate the weather derivatives market. He posted his story in October of 2011. He may not be too far off the mark as an investigation detailed here leads us to many questionable situations, strange bedfellows and none other than those legends of waste, fraud, and abuse: Enron. The thoroughly disgraced and vilified corporation was the founder of the weather derivatives market. Would you put it past Enron?

Weather derivatives are financial instruments (options, futures, and options on futures) anyone can buy that either pay off or don't pay off according to recorded atmospheric conditions such as temperature and rainfall. These instruments are mostly traded on the Chicago Mercantile Exchange (CME). They are also traded on smaller 'over the counter' (OTC) markets. Weather derivatives are usually structured as swaps, futures, and call/put options. Although they are available for sunshine hours, snowfall, rain, wind speed, and many other geophysical conditions, the most common type of weather derivative by far is based on temperature and this type is the focus of our investigation. According to industry experts, temperature-based weather derivatives account for anywhere from 75-99% of all weather derivatives sold. Observed atmospheric conditions are recorded and published by authorized organizations.

The weather derivatives market was created with the energy sector in mind. This is how temperature-based weather derivatives work. Indices take a location's average temperature, then the index value is determined by how much that average temperature deviates from 65 degrees Fahrenheit (or 18 degrees Celsius outside the U.S.). The number deduced determines a derivative's value and can be aggregated over a period of hours, days, weeks, months or seasons. This 65 degree baseline was chosen by the energy industry. The terms most commonly used to describe index values refer to a daily temperature average and are called Heating Degree Days (HDD) and Cooling Degree Days (CDD). Heating Degree Days refer to the number of degrees Fahrenheit below 65 the average temperature of a winter's day is, while Cooling Degree Days refer to the number of degrees Fahrenheit above 65 the average temperature of a summer's day is. It is this way because 65 degrees is about the temperature where, if it is warmer than that, people use more air conditioning (electricity) and, if it is cooler than that, people tend to use more heating (natural gas). In short, the average temperature recorded for a given time frame determines the derivative's value. One can bet that temperatures will be above or below the long term daily average for a particular hour, date, or group of dates. So, if one has insider knowledge about planned atmospheric temperature modifications, then one can make a killing in the weather derivatives market. The energy sector is the biggest buyer of weather derivatives.

The first weather derivative transactions were conducted over the counter in 1997 between Willis Group Holdings, Koch Industries, Pxre Reinsurance Company, and Enron. These transactions followed the deregulation of the energy market in the U.S. and shortly followed the commencement of large-scale domestic chemtrail spraying operations. The weather derivatives market was greatly expanded in 1999 when weather derivatives began trading on the Chicago Mercantile Exchange.

The leading weather derivatives industry association was founded in 1999 as well and is called the Weather Risk Management Association (WRMA). According to former Enron employee Lynda Clemmons, "The Weather Risk Management Association was launched by myself, and by Jim Gosselin of Castlebridge Partners, Darren Wilcox of Southern Co., Ravi Nathan of Aquila Energy and Jeff Porter of Koch Industries." Mrs. Clemmons neglected to mention another WRMA founder, Swiss Re. In 2011, the WRMA released the results of a survey which pegged the current global weather derivatives market value at about $12 billion.

USA Today says in their article "Weather Derivatives Becoming Hot Commodities" that the largest broker of weather derivatives in the world is TFS Energy. A man named Kendall Johnson, who is described as one of the industry's most powerful professionals, states, "Businesses in the U.S., Japan, London and Amsterdam are the most frequent users of weather risk management, though companies in emerging markets like India are beginning to trade weather derivatives."

Other big corporate players have included: British Gas, Hess Energy, ABN Amro, Merrill Lynch, AXA Re, RenRe Energy, Nephila Capital, Munich Re, Speedwell Weather Derivatives, Vyapar Capital Market Partners, Galileo Weather Risk Management, PCE Investors / Cumulus, EDF Trading Limited, Risk Solutions International, E.ON

Energy Trading, Mitsui Sumitomo Insurance Company, and Endurance Reinsurance Corporation of America. As one can see, re-insurers are some of the biggest market players.

Swiss Re is a name that comes up again and again, and just happens to be the insurer of the World Trade towers at the time of the 9/11 attacks. But, that's surely just a coincidence. There's nothing to see here. Move along. Swiss Re also happens to be the purchaser of General Electric's reinsurance company Employers RE. As we have seen in chapter 3, General Electric scientists famously kicked off the scientific era of weather modification in 1946 and they have been developing and producing many relevant technologies ever since.

Industry publications claim substantial non-financial or non-energy sector participation in the weather derivatives market. But, of businesses outside the finance and energy sectors, this investigation has revealed very little participation. It is unrealistic that an organizer of an outdoor event, let's say, would first of all even be aware of weather derivatives, much less use the time, energy, expertise, and money to buy such things. Businesses outside of finance and energy usually use more traditional forms of insurance or hedge with commodities contracts. Weather derivatives are almost entirely an energy and finance sector market. There is hardly any retail investor activity here, if at all.

Industry publications also often claim that weather derivatives are used by energy companies only as hedges against unforeseen demand lapses. If a particular winter is too warm, for example, an energy company would not make as much money selling fuel as they would in an abnormally cold winter. But, the reasoning goes, if they have purchased a hedge in the form of weather derivatives, they can make up those losses. But this investigation has found that weather derivatives are traded like any other Wall Street market. To make a buck, they are traded any way possible. Enron, the founder of the market, is famous for their trading desk which specialized in arbitrage. The 2007 *Bloomberg* article "Hedge Funds Pluck Money From Air in $19 Billion Weather Gamble" had it right. Nowhere in this article will you see any mention of non-financial or non-energy sector participation. In fact, industry professionals are quoted as saying they are, "...using weather as market intelligence." And that their business is, "...like playing poker."

Or consider Enron's John Sherriff, an originator of the weather derivatives market. Mr. Sherriff was quite the gambling man. The authors of the 2003 bestseller *The Smartest Guys in the Room* write, " The Sherriff legend was that he had made an enormous amount - tens of millions - in a single bet on short-term gas prices back in the mid-1990's, when the business was still in its early stages and such a windfall was not believed possible." Stories about the gamblin' John Sherriff seem to go on and on. Also from *The Smartest Guys in the Room* we find:

"*Every year on the annual retreat to the Hyatt Hill Country Resort in San Antonio for vice presidents and above, a group of traders would play a poker game called Omaha (where the lowest hand and the highest split the pot) at the same table in the lobby of the hotel. The pot was usually around $1,000, but in the final year it was played - 2000 - three players thought they had good odds of winning. The pot grew to $33,000, as the crowd gathered and the tension built into the early morning hours. One player had both the high hand and the low hand. He bought a new BMW. The other two - one of them London chief John Sherriff - were out $11,000 each.*"

Weather derivatives by themselves are big money gambles. They most probably contribute to making it worthwhile to put planes up in the sky spraying stuff on a daily basis. If one divides 2011's total market value ($12 billion) by the number of traded contracts (466,000), you get the average contract value which is $25,321. A matter of a few degrees on a given day or group of days could mean hundreds of thousands of dollars.

The weather derivatives market and other opportunities were made possible by the deregulation of the energy market. Deregulation requires government intervention. Was the Department of Energy in bed with Enron? As we will see in the next section, Enron CEO ol' Kenny Boy (as the Bushes called him) had extensive experience and connections with energy policy at the federal level.

The fact that Enron founded the weather derivatives market is very dubious. This is a company whose accounting firm, Arthur Andersen, shredded more than a ton of their documents in one day as Enron's chairman and CEO Ken Lay told everybody everything was fine. Enron's bankruptcy resulted in criminal charges against at least 21 former executives. People suffered under high power costs inflated by

Enron. When Enron and their cronies intentionally disrupted power service as they were known to do, people were injured and died. Who knows how many bodies they left behind? These guys were not playing Tiddlywinks. We shouldn't put anything past Enron.

A history of weather derivatives

Lots of information about the history of the weather derivatives market is available because Enron originated the weather derivatives market. The fall of Enron was one of the greatest corporate scandals in American history, so there has been much already investigated and exposed. This section examines this history of weather derivatives so inextricably connected to one of the most famously dishonorable corporations in history.

When you're talking about the history of weather derivatives, you're talking about Enron. They developed the most widely used early trading platforms, they were founding members of the leading industry association and were counterparties in the first known domestic and international transactions. For these reasons, in this section we will take a long look at Enron as well as the earliest beginnings of the weather derivatives market.

In their heyday, Enron received *Fortune* magazine's award for 'Most Innovative Company' six years in a row. They gave their Enron Prize for Distinguished Public Service to people like Nelson Mandela and Mikhail Gorbachev. Henry Kissinger and James Baker worked as Enron consultants, traveling to such far-flung destinations as Kuwait and China while preaching the Enron gospel. Stock analysts gushed over everything Enron did. Enron could seemingly do no wrong.

The problem was, Enron was cooking the books seven ways to Sunday. When the S.H.T.F., Enron's stock tanked like World Trade Center building 7 and eventually brought down one of the nation's oldest and largest accounting firms, Arthur Andersen. Although the company itself has long since been chopped up and sold off, former Enron employees now populate many other financial market trading establishments.

The particular division known to buy and sell weather derivatives was called 'Enron Weather.' Enron Weather started as a small, but promising bit of the company. By the time of Enron's demise in 2001, Enron Weather had grown to be a significant part of their business.

Although the author of the book *What Went Wrong at Enron* claims that the weather derivatives market probably came from an idea that just popped into the head of the TV weatherman brother of Enron executive Jeffery Skilling one day, there is quite a bit of documented history long before this. The weather derivatives market actually goes back to the late 1960s. From a 1968 example, we learn that the National Science Foundation (NSF) and the University of Missouri were laying the foundations of the weather derivatives market a long time ago and that its roots have everything to do with weather modification. The tenth annual NSF weather modification report reads:

"*Under NSF support, the University of Missouri is continuing its study of a method to determine the potential impact of weather and climate modification upon the social and economic structure of a sample State such as Missouri. The electrical power industry has been considered to be one which is both weather sensitive and one for which data useful for statistically isolating this relationship are available. For this reason it was decided to make study of the potential effects of weather modification on its operation. Daily electrical power demand data covering a number of midwestern states were supplied by the Edison Electric Institute.*

"*Using these data and Weather Bureau temperature records, it was possible to construct a series of power loads for each of a number of regions in the study area corresponding to the modified and nonmodified temperature series. Power generation costs and capacities for each of the regions in the study area were then entered into a linear programming model. With the modified and nonmodified load series, it was possible to construct two estimates of the cost of supplying power to the area. Each area power supply cost was a minimum given the loads and existing generation facilities. Comparisons of the costs of supplying power to the area during the summer months then provided an estimate of the direct effect of potential temperature modification on the electrical industry.*"

When one continues to dig a little deeper and finds further documentation of this work done at the University of Missouri, one finds the entire smoking gun. An April 1969 paper by M. Lawrence Nicodemus and James D. McQuigg titled "A Simulation Model for Studying Possible Modification of Surface Temperature" shows us that said temperature modification was to be accomplished by the generation of 'contrail cirrus' from jet aircraft. According to their model, this contrail cirrus was supposedly going to cause temperatures to fall in the warm summer months. The deceptive nature of this 1969 paper is simply astounding. First of all, when any type of cloud cover is present, surface temperatures go up, not down, because cloud cover traps heat. But the authors of this paper were apparently following the lead of the SRM geoengineering thesis which also involves such incorrect information. Secondly, the paper includes a photograph of chemtrails incorrectly labeled as contrails.

FIG. 3. Contrails over Columbia, Mo.

Chemtrails over Columbia, MO

Nicodemus and McQuigg write that their paper only pertains to *models* of possible future activities - while at the same time, they provide photographic evidence of these activities as being currently ongoing! Only the CIA could have come up with such convoluted nonsense. This paper provides the earliest example known to the author of the term 'contrail cirrus' being used as a misnomer for chemtrails. We should also note that modification of atmospheric temperature was, at the time, a whole new area of weather modification. Up until this point, the historical weather modification literature almost entirely concerned itself with the dispersal of silver iodide or dry ice for the purposes of precipitation enhancement or hail suppression.

Late in 1969, McQuigg along with two co-authors laid the foundations of the modern weather derivatives market by fusing the nascent field of atmospheric temperature modification with the generation of electrical power. In their seminal paper titled "Temperature Modification and Costs of Electric Power Generation," McQuigg once again writes of contrail cirrus from jet aircraft as being the means for modifying atmospheric temperature.

Are you ready? The earliest example known to the author of the aforementioned 'degree day' term and the 65 degree threshold appears in a 1975 paper titled "Economic Impacts of Weather Variability" also by James McQuigg. Boom. The weather derivatives market is directly related to weather modification. This is where the reader pauses for a moment to consider the magnitude of these findings.

McQuigg's work in these areas, going back to a 1961 report titled "Weather Variability and Economic Analysis" is most significant. McQuigg also did work pertaining to weather's economic impact upon the production of agriculture and livestock.

Expanding upon the degree day thesis, a 1976 ICAS report reads, "...CCEA [Center for Climatic and Environmental Assessment] is developing highly sophisticated mathematical models by which population-weighted degree days can be linked specifically to use of natural gas, to the exclusion of other fuels."

Although many others had previously laid the groundwork, it wasn't until the late 1990s that Enron entered the picture. Loren Fox, the author of *Enron: The Rise and Fall* tells the story of Enron Weather like this:

"*...weather derivatives came to be championed by an employee working at the grassroots level and seeing customers' daily needs. John Sherriff, who at the time managed gas trading for the western United States, began looking at derivatives linked to the weather in late 1995, and Vincent Kaminski's research group worked on the idea in 1996. A gas trader named Lynda Clemmons was very interested in the idea, based on her conversations with executives at electric* *utilities that used coal-fired power plants.... In 1997, Enron handed off its weather derivatives effort to Clemmons, who was only 27. She began a one-person weather-hedging department within ECT [Enron Commodities Trading]."*

Fox continues, "Clemmons built up Enron's weather business so that it did 350 transactions (hedging up to $400 million in potential revenues) in 1998, turning its first profit that year."

Lynda Clemmons

Enron initiated the weather derivatives market in Europe as well. Enron's Oslo office became the base of their European weather derivatives business. In 2000, Enron also introduced weather derivatives in Australia; offering temperature-based products for Sydney, Melbourne, Hong Kong, Tokyo, and Osaka.

In 2002, after the bankruptcy, the Enron trading desk (including Enron Weather) was bought by UBS Warburg. According to his websites, John Sherriff is now the owner of Lake Tahoe Financial and other Sherriff family businesses. Mrs. Clemmons left Enron in 2000. She took a number of her colleagues from Enron's weather team and set up a weather derivatives company called Element Reinsurance. After Enron, Lynda Clemmons also worked at XL Weather & Energy, The Storm Exchange Inc. and Vyapar Capital Market Partners. According to her Linkedin profile, Mrs. Clemmons is now an independent consultant.

The Enron financial market trading desk, just like the company itself, had a history of corruption. Back in 1987, the Enron trading operations were called 'Enron Oil.' When questions about an Enron account at New York's Apple Bank began surfacing, Enron management turned a blind eye. Money was pouring into this questionable account from a bank in the Channel Islands and flowing out to the account of a man named Tom Mastroeni, the treasurer of Enron Oil.

Enron management explained away these money flows and this questionable account as completely legal profit shifting. Although Mastroeni produced doctored bank statements and admitted a cover-up, Enron management didn't pursue the issue. Nobody involved was even reprimanded. Revelations continued to come out, but Enron management did nothing. Enron's accounting firm Arthur Andersen exhibited a similar disinterest. After management confirmed an official lack of responsible oversight, Enron Oil traders ignored position limits and got themselves in big trouble. Only then did Enron executives show interest. The authors of *The Smartest Guys in the Room* tell the story like this:

"*For months, Borget [the CEO of Enron Oil] had been betting that the price of oil was headed down, and for months, the market had stubbornly gone against him. As his losses had mounted, he had continually doubled down, ratcheting up the bet in the hope of recouping everything when prices ultimately turned in his direction. Finally, Borget had dug a hole so deep - and so potentially catastrophic - that there was virtually no hope of ever recovering.*"

Enron brass was in a panic. Enron was looking at a $1 billion loss - enough to bankrupt the company. Management sent in some expert traders who, over the course of a few weeks, managed to clear out these positions with only a $140 million loss. Enron had to tell people about this $140 million trading loss, though. Enron's stock slid 30%. The blame game began. Ken Lay, the affable Enron founder and CEO, denied any responsibility. News of the scandal conveniently came out right after a big bank loan approval. The U.S. attorney's office charged Borget and Mastroeni with fraud and personal income tax violations. In 1990, Borget pled guilty to three felonies and was sentenced to a year in jail and five years' probation. Mastroeni pled guilty to two felonies. He got a suspended sentence and two years probation.

Ken Lay

Ken Lay helped to create new financial markets (such as the weather derivatives market) and helped his company make more money through energy sector deregulation. Deregulation created a situation where Enron and others could more effectively manipulate and arbitrage (specifically, regulatory arbitrage) markets. Let us refer to a feature article published by industry publication *Risk.net*:

"*'Enron was the focal point of the deregulation agenda,' says Jonathan Whitehead, who started with Enron Europe in 1996 and was heading the liquefied natural gas (LNG) business in Houston at the time of Enron's demise. 'It was the most vocal when explaining to regulators and governments and customers the benefits of deregulated markets. I don't think deregulation in power and gas in Europe or the US would have come as far as it has without Enron,' he says.*

"Pushing for deregulation was very much a part of the company's strategy from the start. 'Ken Lay was the visionary at the time as far as seeing where deregulation could go and actually driving deregulation,' says Mark Frevert, who worked at one of Enron's predecessor

companies, Houston Natural Gas, from 1984 and stayed at Enron until it's demise."

Deregulation necessarily required help from the federal government. To make it happen, Enron and Ken Lay (who died in 2006) had the right federal connections in spades.

Ken Lay enlisted in the Navy in 1968. His friend pulled some strings and had him transferred to the Pentagon. According to the authors of *The Smartest Guys in the Room*, Lay spent his time at the Pentagon, "...conducting studies on the military-procurement process. The work provided the basis for his doctoral thesis on how defense spending affects the economy." Ken Lay had many high-level military connections. As we have seen, the evidence suggests that today's domestic chemtrail spraying operations are most probably carried out by what passes for our military.

Mr. Lay had the right political and, specifically, energy-related political connections. Under his former professor Pinkney Walker, he worked in the Nixon administration as a Federal Power Commission (FPC) aide. According to Russ Baker, "After less than two years [as an FPC aide], Lay was put in charge of coordinating government energy policies, as deputy undersecretary of Interior." Later as chairman and CEO of Enron, he sat on the boards of prestigious Washington DC think tanks and often travelled to Washington.

A revolving door existed between Enron and the federal government. Enron executive Tom White left the company to join the Bush Jr. administration as secretary of the army. Enron executive Herbert 'Pug' Winokur was Lay's old Pentagon friend. Robert Zoellick worked for Enron, then became a United States Trade Representative and later the Deputy Secretary of State. According to the authors of *The Smartest Guys in the Room*, "In 1993, Lay added Wendy Gramm [to the Enron board], who had just finished a stint as chairman of the Commodities Futures Trading Commission (CFTC) and was married to Texas Senator Phil Gramm." It continues, "Just after Wendy Gramm stepped down from the CFTC, that agency approved an exemption that limited the regulatory scrutiny of Enron's energy-derivatives trading business, a process she had set in motion."

When Enron galloped into Europe, they had an influential lord on their side. Journalist Greg Palast covers it like this:

"The fact that a truly free market [in electricity] didn't exist and cannot possibly work did not stop Britain's woman in authority, Prime Minister Margaret Thatcher, from adopting it. It was more than free market theories that convinced her. Whispering in her ear was one Lord Wakeham, then merely 'John' Wakeham, Thatcher's energy minister. Wakeham approved the first 'merchant' power station. It was owned by a company created only in 1985 - Enron. Lord Wakeham's decision meant that, for the first time in any nation, an electricity plant owner, namely Enron, could charge whatever the market could bear... or, more accurately, could not bear.

"It was this act in 1990 that launched Enron as the deregulated international power trader. Shortly thereafter, Enron named Wakeham to its board of directors and placed him on Enron's audit and compliance committee."

A scheme to manipulate financial markets by modifying the weather would necessarily involve intelligence agencies. According to the authors of *The Smartest Guys in the Room*, Enron had CIA connections:

"One of Enron's key advantages over its competitors was information: it simply had more of it than its competitors. Its physical assets provided information, of course. And Enron didn't stop there. It employed CIA agents who could find out anything about anyone. In stead of tracking the weather on the Weather Channel, the company had a meteorologist on staff. He'd arrive at the office at 4:30 A.M., download data from a satellite, and meet with the traders at 7:00 A.M. to share his insights."

It continues, "By the late 1990's, these research efforts were herded together into something called the fundamentals group - fundies in trader parlance. The fundies group produced intelligence reports and held morning briefings..."

Mr. Lay and Enron also had many connections to the Bush family and their cohorts. President George W. Bush (Bush Jr.) lovingly called Ken Lay 'Kenny Boy.' Mr. Lay was also close to his father, fellow Houstonian, George H.W. Bush (Bush Sr.). In 1991, Bush Sr. offered Mr. Lay the position of Commerce secretary. Mr. Lay turned him down. He wanted to be Treasury secretary. Greg Palast, in his book *The Best Democracy Money Can Buy*, characterizes the Bush/Enron relationship like this:

"But what about Pioneer Lay of Enron Corp? His company, America's number one power speculator, was also Dubya's number-one political career donor ($1.8 million to Republicans during the 2000 presidential campaign). Lay was personal advisor to Bush during the postelection 'transition.' And his company held secret meetings with the energy plan's drafters. Bush's protecting electricity deregulation meant a big payday for Enron - subsequent bankruptcy not withstanding - sending profits up $87 million in the first quarter of Bush's reign."

Other Bushes were apparently getting some, too. Palast writes:

"Two months after the bankruptcy, Governor Jeb Bush of Florida traveled to the Texas home of Enron's ex-president, Rich Kinder, to collect a stack of checks totaling $2 million at the power pillager's $500-per-plate fund-raising dinner. There are a lot of workers in Florida who will wish they had a chance to lick those plates, because that's all that's left of the one-third of a billion dollars Florida's state pension fund invested in Enron - three times as much as any other of the fifty states."

Mr. Palast continues:

"Governor Bush encouraged a scheme by a company called Azurix to repipe the entire Southern Florida water system with new reservoirs that would pump fresh water into the swamps. From the view of expert hydrologists, such a mega-project is a crackbrained and useless waste of gobs of money. As part of the deal, Azurix would be handed the right to sell the reservoir's water to six million Florida customers. Azurix was the wholly owned subsidiary of Enron that had recently been kicked out of Buenos Aires."

Enron was a business laboratory. In the new world of deregulation created by the Bushes, Enron, and others, the purpose of the company was to throw things against the wall and see what stuck. Enron was the perfect environment in which to try something new like weather derivatives. In light of the histories of corruption exhibited by those involved, it is understandable that the weather derivatives market they created was probably yet another murderous rip-off.

While most financial market trading operations are seen as unreliable, Enron's trading desk was often regarded as their most productive and stable business. Could this have been because they were getting inside information about weather produced by geoengineering activities? They had the ways and means necessary. Enron had motive and opportunity. Enron benefitted from weather derivatives.

Catastrophe reinsurance

Besides weather derivatives, there is also another, larger financial market which rises and falls with the weather: the catastrophe reinsurance market. In this market, the most commonly issued security is something called a 'catastrophe bond.' The issuance of catastrophe bonds themselves is a bespoke market. That means participants engage in direct and often lengthy negotiations to craft customized agreements. You may have heard of the catastrophe reinsurance market. It's home to the 'terrorism insurance' market. The catastrophe reinsurance market is also home to the Special Contingency Risk (Kidnap & Ransom) market. Don't forget the death bond market. In the death bond market you stand to gain healthy rates of interest if a certain number of people don't die. But if they do, you lose, and the banks get all your money, capiché? Catastrophe insurance derivatives market participants come together at the New York based Catastrophe Risk Exchange (CATEX). Once issued, these catastrophe bonds (or 'cat bonds' as they are called) are then cut into little pieces and sold in a process known as 'securitization.' Insurance Catastrophe Futures Contracts have been trading on the Chicago Board of Trade (CBOT) since 1992. If a bank or insurance company wants to position themselves to gain a whole lot of money in the case of a series of catastrophic hurricanes in a particular region, let's say, or a certain region suffering a sustained draught or floods or tornadoes or earthquakes, that's what the catastrophe reinsurance market is for and the NMP can produce all these phenomena.

Although catastrophe bonds have only been around since the early 1990's, the catastrophe reinsurance market has been around since the 1960's. The leading industry association was founded in 1968 and is called the Reinsurance Association of America (RAA). The RAA describes its priorities thusly:

"The RAA's public policy priorities include: federal and state financial role for natural disaster and terrorism catastrophe risk; regulatory reform efforts at the federal and state level; international trade, accounting and tax policy; accounting and financial reporting;

solvency oversight and reinsurance recoverables; and ***climate change and environmental risk*** *[author's emphasis]."*

The catastrophe reinsurance market involves much more money than the weather derivatives market. While the Weather Risk Management Association relatively recently pegged the value of the global weather derivatives market at about $12B, industry player Nephila Capital's website states, "The amount of notional exposure that trades in the catastrophe reinsurance market each year is approximately $200B." Now we're talking about your 'disaster capitalism' industry!

2011 was a big year for the catastrophe reinsurance market. In September of 2011, the industry's biggest broker Guy Carpenter released a report stating, "The devastating earthquakes in New Zealand and Japan, along with damaging tornadoes and floods in the United States and Australia, have resulted in insured losses of around USD70 billion so far this year." Munich Re put 2011 industry losses for the first half of the year at $265 billion; easily surpassing the previous full-year record amount of $220 billion set in 2007. According to Munich Re, "The 9.0 magnitude earthquake, the strongest ever registered in Japan, is also the costliest natural catastrophe on record."

Like the weather derivatives market, financial and energy market deregulation enabled the existence of today's catastrophe reinsurance market. The repeal of Glass-Steagall was key. Here is a passage from a book by industry insider Erik Banks, "In the US, product and market convergence has been aided by the passage of the 1999 Financial Modernization Act (i.e., the Gramm-Leach-Biley Act), which eliminated the 1933 Glass-Steagall Act and Depression-era legislation that prohibited banks and insurance companies from encroaching on each other's territory." When Mr. Banks refers to 'convergence,' he's writing about money pouring into the insurance industry from the banking industry.

According to Mr. Banks, for tax and regulatory reasons, many catastrophe reinsurance industry participants choose to domicile in countries such as: Bermuda, the Cayman Islands, the British Virgin Islands, Luxembourg and Ireland. The catastrophe reinsurance market consists of: related financial industry businesses, brokers, banks, hedge funds, and insurance companies. None of the material this author read mentioned any retail investor participation. This is a professional insurance market. Although industry consolidation remains a trend, businesses providing instruments such as Bermuda Transformers are needed to perform specific functions like converting derivative instruments into reinsurance contracts. Brokers negotiate deals. The biggest brokers are Willis Group Holdings and Guy Carpenter & Company. Banks provide the capital. Big banks have internal hedge funds which play the catastrophe reinsurance market. Independent hedge funds such as Nephila Capital are also here to play. Insurance companies are here because they originated the market. The terms 'insurance' and 'reinsurance' are used here interchangeably because they are both essentially the same thing. Insurance companies sell catastrophe bonds and insurance-linked securities. Swiss Reinsurance America Corporation and Munich Reinsurance America are the most prolific issuers of cat bonds and related securities. Many of the same companies which participate in the weather derivatives market also participate in the catastrophe reinsurance market. Since both markets rise and fall with the weather, it makes sense that they would. Here is a partial list of dual market players: Endurance Reinsurance Corporation of America, Nephila Capital, Swiss Re, Willis Group Holdings, and Munich Re.

On a completely unrelated note, catastrophe reinsurance industry heavyweight Marsh & McLennan (who owns the industry's biggest American broker Guy Carpenter) had all their World Trade Tower offices totally wiped out on 9/11. You see, Marsh & McLennan occupied floors 93-100. On the day of 9/11, the first plane completely gutted floors 93-99; killing every person on all of those floors. Maybe somebody wanted to make sure certain people were taken care of. Maybe the target was some people on the 96th floor, which was direct center. Maybe guys from Afghanistan who didn't know how to fly jumbo jets did it. Maybe it was Bigfoot, the Tooth Fairy, or a purple dinosaur. 9/11 was an outside job and chemtrails do not exist. Stop thinking and please return to watching television.

A history of catastrophe reinsurance

Following a spate of presumably natural disasters occurring from 1991-1994, climatologist Stanley Changnon (1928-2012) served as the lead author of a 1996 report advocating for today's modern catastrophe reinsurance market. Even though Changnon concluded that man-made global warming was probably not to blame, "Impacts and Responses of the Weather Insurance Industry to Recent Weather Extremes" suggests that finance and climatology should work with government in order to provide more protections for those affected by severe storms (catastrophes). This was when the big banks and the reinsurers initially converged and the modern catastrophe reinsurance market came into its own. The theory of man-made global warming served as a market catalyst. A passage from this report tells the story:

"The major storm losses of 1991-1993 created by 1993 a new level of interest built around a new theme: that the new extremes were a signal of a changed climate. Time [magazine] and some trade journals carried articles about the hazard upsurge and its possible connection to a greenhouse-induced climate change (Linden, 1994). This concept was actively promoted by environmental interests who had a stake in the climate change issue and who claimed that the recent upsurge in catastrophes and record losses were 'indicative of a change in climate due to the greenhouse effect' (Leggett, 1993). They were joined by some atmospheric scientists who also claimed climate change was responsible for the upswing in losses since their calculations indicated that the greenhouse change would produce a climate with more storms, higher winds, and longer storm tracks (Rountree, 1994).

"Leaders in the insurance industry began to consider the issue. Frank Nutter, President of the Reinsurance Association of America stated 'the insurance business is first in line to be affected by climate change' (Linden, 1994). Some insurance leaders, in assessing what factors had caused the recent (post 1990) upsurge in storm losses, included the possibility that climate change was a factor along with demographic shifts, growth of exposure, decaying infrastructure, and poor building codes (Lecomte, 1993; Berz, 1994). Certain insurance leaders claimed that climate change and the increase in storms were related, whereas others were not convinced (Viewpoint, 1995; Stix, 1996). Flavin (1994) addressed the key issues in a thorough analysis noting there had recently been a large number of storms, major recent increases in losses, and a vulnerable industry that became willing to believe that unique circumstance caused by a change in climate had occurred. Deering (1993) noted that those promoting concern over global climate change and a $1.8 billion U.S. research program, including environmental groups and atmospheric scientists, had acted

to get the insurance industry to believe in climate change and thus to become allies on their side in promoting research and policy development relating to climate change. Over the last few years, the greenhouse-induced climate change issue has become a national and international policy issue with scientists, government bureaucrats, and policymakers arguing all sides of the issue (Flavin, 1994; Glantz, 1995)."

This passage continues the story:

"Another financial result of the recent catastrophes was the expansion of the Bermuda-based reinsurance market (Jennings, 1995). The market was originally developed to create capacity and take advantage of the hard casualty market conditions of the 1980s, under the less restrictive regulations of the Bermuda government. After Hurricane Andrew, the Bermuda market saw the arrival of the 'super cat' reinsurers. These new companies were formed with different levels of capital, with most exceeding $300 million, and having the objective of providing a large participation in property catastrophe reinsurance programs. The capitalization of these companies was developed through public equity markets, private placements and investments by other insurance companies, reinsurance firms, investment banking firms, and insurance and reinsurance brokerage companies."

The report continues:

"CAT's [catastrophe bonds] first started trading in 1992. Initially the traditional insurance and reinsurance market did not know how to react to the product. Segments of the market saw this as an unneeded form of compensation. Many felt that the shortage in catastrophe capacity would be solved in the traditional market, as the pricing of that product improved, and at first, CAT's were minimally traded.

"However, over the last three years, open interest in these contracts has risen steadily. In April 1993, only 61 contracts traded. By January 1994, open interest was 4,800 contracts, and by August 1994, open interest exceeded 5,800 contracts. These futures contracts trade only on the floor of the Chicago Board of Trade (CBOT)..."

Lastly, the report adds:

"In response to the need for increased catastrophe risk capacity, an Exchange has been established in New York. In July 1995, the New York Insurance Department approved the licensing of a risk exchange facility to be domiciled and operated in that state. Catastrophe

Exchange (CATEX), licensed as a reinsurance intermediary, was brought into existence through the efforts of the former Insurance Commissioner of New Jersey. It reflects interest in increasing capacity for the Long Island shoreline to obtain coverage against windstorm and beach erosion damage."

For this new CATEX, Enron's now defunct accounting firm Arthur Anderson was slated to, "...establish an index that prices and rates exposures."

Crop insurance

In the 2012 documentary video *Why in the World Are they Spraying?*, an independent commodities trader at the Chicago Mercantile Exchange by the name of Michael Agne says that, in combination with weather derivatives and catastrophe reinsurance, traditional crop insurance may be gamed. With foreknowledge of the weather, it is as simple as insuring a crop for more than it would be worth at market, then collecting the insurance money when the crop fails.

Oddly enough, the biggest player in the space is the federal government. Something called the Federal Crop Insurance Corporation operates under the Department of Agriculture's Risk Management Agency. But, as we all know, our federal government would never engage in any such thing as insurance fraud associated with weather modification, so we should probably look elsewhere.

A comprehensive plan

A massive global weather modification endeavor such as the New Manhattan Project requires that all the disparate financial motives outlined here need necessarily be folded into a comprehensive plan to manage the Nation's and the world's weather resources. This is how every aspect of domestic weather-related activity can be managed comparatively and simultaneously. This would be the best way to manage the New Manhattan Project's socio-economic impacts. It is not a surprise that at least one such early study has been undertaken.

The 1966 book *Human Dimensions of Weather Modification* featured a paper by James Hibbs titled "Evaluation of Weather and Climate by Socio-Economic Sensitivity Indices." This paper outlines a comprehensive plan for managing the socio-economic impacts of a national weather modification program. Under this plan, all socio-economic impacts of weather modification activities are categorized as being applicable to either 'consumers' or 'producers.' Consumers are defined as using weather services to:

"*...enhance values generated outside the market place. Personal enjoyment of leisure time and other nonincome producing (in the dollar sense) activities are included. Also included are governmental actions using weather services to maintain or increase health, safety, general welfare, natural resource preservation, etc..*"

The paper classifies producers as using weather services:

"*...for purposes of enhancing their professional, commercial or industrial activities. Thus, specialized services are used by professional meteorologists, agriculturists, commercial airlines and private business pilots, and marine transportation.*"

Hibbs then goes on to describe a system of 'measures of benefit' applicable to all the defined groups potentially affected. His paper reads:

"*It appears feasible to develop indices suitable for use in decision-making based upon the aggregate of factors discussed in the previous section. A first effort has been made to use the concept of a generalized index of 'weather influence' combined with a dollar-weighted index of 'benefit potential' distributed among various weather influenced activities and within broad geographical areas.*"

The paper then goes on to quantify potential weather modification benefits by dividing the map of the United States into 8 regions. Each region is then assigned a numerical value representing four factors: the potential benefit for producers due to better weather forecasts, the potential benefit for producers due to weather modification, the potential benefit for consumers due to better weather forecasts, and the potential benefit for consumers due to weather modification. These values are to determine where and when different types of weather modification activities are to be conducted.

APPENDIX B

DERIVATION OF WEATHER INFORMATION BENEFIT POTENTIALS BY REGIONS

Benefit Measurements	Beneficiary Class												
	General Public	Land Transp.	Air Transp.	Water Transp.	Construction	Water Supply & Control[1]	Energy Prod. & Distrib.	Merchandising	Recreation[1]	Manufacturing	Fishing	Agriculture[2]	Forestry[3]	Communications[4]
1 Influence Index[5]	10.2	5.8	6.8	6.0	6.8	3.6	5.4	3.6	5.4	1.8	10.0	6.0	6.3	3.4
2 Adj. Factor[6]	1.5	1.5	1.5	1.5	1.5		2.6	1.9		1.3				1.5
SOUTHEAST (TABLE 33)[8]														
3 Wages & Salaries[7]	46.2	1.7	.4	.3	2.7		.6	8.1		12.5				1.4
4 Pers. Income (2)(3)	70.0	2.6	.6	.45	4.1	.23	1.6	15.4	.2	16.3	.03	5.5	.2	2.1
5 Benefit Potential (1)(4)	714.0	15.1	4.1	2.7	27.9	.8[1]	8.6	55.4	1.1[1]	29.3	.3[1]	33.0[2]	1.3[3]	5.1[4]
PLAINS (TABLE 25)														
6 Wages & Salaries[7]	20.9	1.2	.15	.1	1.4		.3	4.3		5.5				.7
7 Pers. Income (2)(6)	35.4	1.8	.23	.15	2.1	.1	.8	8.2	.1	7.2	.01	4.6	.1	1.1
8 Benefit Potential (1)(7)	361.1	10.4	1.7	.9	14.3	.4[1]	4.3	29.5	.5[1]	13.0	.1[1]	27.6[2]	.6[3]	3.7[4]
MID-EAST (TABLE 12)														
9 Wages & Salaries[7]	74.5	2.2	1.0	.6	4.0		1.0	13.5		24.7				2.6
10 Pers. Income (2)(9)	108.4	3.3	1.5	.9	6.0	.4	2.6	25.7	.5	32.1	.02	1.4	.3	3.9
11 Benefit Potential (1) (10)	1105.7	19.1	10.2	5.4	40.8	1.4[1]	14.0	92.5	2.7[1]	57.8	.2[1]	8.4[2]	1.9[3]	13.3[4]
GREAT LAKES (TABLE 19)														
12 Wages & Salaries[7]	63.9	2.6	.36	.24	3.1		.9	10.8		27.6				.9
13 Pers. Income (2)(12)	92.7	3.9	.5	.4	4.7	.3	2.3	20.5	.3	35.8	.02	3.1	.3	1.4
14 Benefit Potential (1) (13)	945.5	22.6	3.4	2.4	32.0	1.1[1]	12.4	73.8	1.6[1]	64.0	.2[1]	18.6[2]	1.9[3]	4.8[4]

Weather sensitivity numerical values

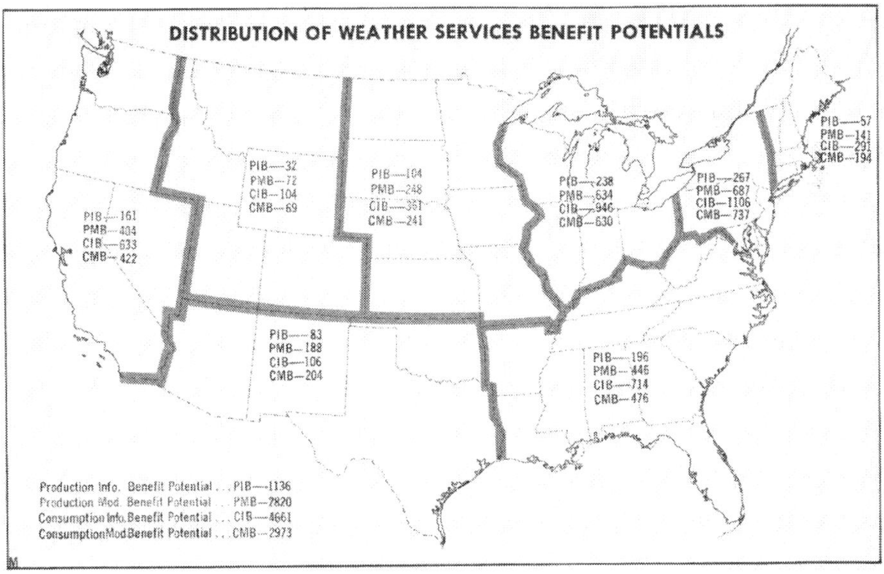

Map of U.S. w/ regions & weather sensitivity numerical values

"Evaluation of Weather and Climate by Socio-Economic Sensitivity Indices" discloses that the predecessor to NOAA, the Environmental Science Services Administration (ESSA) was developing these actuaries in order to aid ESSA policy and planning decisions. This suggests that NOAA manages the socio-economic impacts of today's New Manhattan Project. The paper concludes by noting:

"The program would coordinate not only activities and available resources of many ESSA components, but also capitalize upon available resources and experiences within TAD-NBS, Census Bureau, Bureau of Labor Statistics, and NREC-OEP."

Conclusions

Geoengineers argue that all the motives presented here are precisely why we need a global weather modification project. They argue that the Earth's resources need to be better managed and that a global weather control program, producing better weather forecasts, allows for that. The truth is that we don't need it or want it and we can't trust the people doing it anyway.

We don't need it. Humanity has been developing quite well for thousands of years without a global weather modification project, thank you. We need a clean environment, not one contaminated by geoengineering sprays and electrosmog.

We don't want it. We don't want to breathe in the chemtrail witches' brew and/or get zapped by this Project's electromagnetic rays. Whether or not one realizes what is going on, these things are bad for us and our environment. Stop assaulting us!

We can't trust the people behind it. Does a burglar ask for your permission before he robs your house? No. Does a mad military/industrial complex ask you for permission before they ruin your health and wreck your environment? No, they don't. The people responsible for this New Manhattan Project are not to be trusted. We don't want to do business with them. These claims of deleterious Human health impacts and environmental devastation will be discussed in the next chapter.

What we need is a system that does not allow the socio-economic elite to steal and hoard the Earth's wealth as they spray us with chemtrails. We need a system that disallows the type of psychopathic gamesmanship currently going on. We need a system of government and commerce that cuts out the middlemen and allows for the delivery of the wealth of the Earth *directly* to the people. We need the unequivocal enforcement of the American Constitution. If this is achieved, there will be abundance for all and no want or need for any global weather modification project.

References

How to Cool the Planet: Geoengineering and the Audacious Quest to Fix Earth's Climate a book by Jeff Goodell, published by Mariner Books, 2010

Fixing the Sky a book by James Roger Fleming, published by Columbia University Press, 2010 p112

Storms Above the Desert: Atmospheric Research in New Mexico 1935-1985 a book by Joe Chew with the assistance of Jim Corey, published by the University of New Mexico Press, 1987

"Economic Impacts of Weather Variability" a report by James McQuigg, published by the University of Missouri - Columbia, 1975

Interdepartmental Committee for Atmospheric Sciences report 18, published by the Federal Council for Science and Technology, May, 1974

"Design Study for Economic Analysis of Weather Modification" a paper by Edward A. Ackerman as it appeared in *Final Report of the Advisory Committee on Weather Control* by the Advisory Committee on Weather Control, 1958

Interdepartmental Committee for Atmospheric Sciences reports 1960-1978, published by the Federal Council for Science and Technology
"U. S. Economic Sensitivity to Weather Variability" a report by Jeffrey K. Lazo, Megan Lawson, Peter H. Larsen, and Donald M. Waldman, published by the American Meteorological Society, 2011

Weather and Warfare an ebook by Robert Child, published by Robert Child, 2012

"Weather as a Force Multiplier: Owning the Weather in 2025" a report by Col. Tamzy J. House, Lt. Col. James B. Near, Jr., LTC William B. Shields (USA), Maj. Ronald J. Celentano, Maj. David M. Husband, Maj. Ann E. Mercer, and Maj. James E. Pugh, published by the United States Air Force, 1996

"Weather Modification as a Weapon" an article by Gordon J.F. MacDonald, published by *Technology Review*, October/November, 1975

The History of the Ancient World: From the Earliest Accounts to the Fall of Rome a book by Susan Wise Bauer, published by W.W. Norton, 2007

The History of the Medieval World: From the Conversion of Constantine to the First Crusade a book by Susan Wise Bauer, published by W.W. Norton, 2010

The History of the Renaissance World: From the Rediscovery of Aristotle to the Conquest of Constantinople a book by Susan Wise Bauer, published by W.W. Norton, 2013

Archimedes and the Door of Science a book by Jeanne Bendick, published by Bethlehem Books, 1995

Prophecy Fulfilled: 'Toward New Horizons' and its legacy a book by the United States Air Force, published by Progressive Management, 1994

The Weather Changers: The Remarkable Story of How Man May Control and Change the Weather and Make More Accurate and Longer-Range Predictions a book by D.S. Halacy, Jr., published by Harper and Row, 1968

Climate and Food: Climatic Fluctuation and U.S. Agricultural Production a book by The National Research Council, published by the National Academy of Sciences, 1976

Weather Modification: Programs, Problems, Policy, and Potential a book by the Congressional Research Service, 1978

"Assessment of Weather Modification in Alleviating Agricultural Water Shortages During Droughts" a report by the Illinois State Water Survey at the University of Illinois, Urbana, produced by the National Science Foundation and Research Applied to National Needs, 1977

Tenth annual National Science Foundation weather modification report, 1968

The Interdepartmental Committee for Atmospheric Sciences Report No. 15, published by the Federal Council for Science and Technology, March, 1971

"Compilation of Workshop Materials: Workshop for An Assessment of the Present and Potential Role of Weather Modification in Agricultural Production" a report by the Department of Atmospheric Science at Colorado State University, compiled by Lewis O. Grant and John D. Reid, published by Colorado State University, 1975

Atmospheric Control Act hearings before the Subcommittee on Oceans and Atmosphere of the Committee on Commerce, United States Senate, Ninety-fourth Congress, second session, 1976

"The Large Area Crop Inventory Experiment" a report by R. B. MacDonald and the Earth Observations Division of the National Aeronautics and Space Administration, as it appeared in the 2nd Annual William T. Pecora Memorial Symposium, 1976

The Brothers Vonnegut: Science and Fiction in the House of Magic a book by Ginger Strand, published by Farrar, Straus and Giroux, 2015

Hearings Before the Subcommittee on the Environment and the Atmosphere of the Committee on Science and Technology, U.S. House of Representatives, 94th Congress, June 15, 16, 17, & 18, 1976

"Water Agency's Land Purchase Rattles California Farmers" an article by Elliot Spagat and Jae Hong, published by the *Los Angeles Daily News*, Nov. 21, 2015

"A United States Climate Program Plan" as it appeared in the Interdepartmental Committee for Atmospheric Sciences report 20b, published by the Federal Coordinating Council for Science and Technology, July, 1977

"Development of Desirable Operational Standards" a paper by Loren W. Crow as it appeared in *The Final Report of the Advisory Committee on Weather Control* by the Advisory Committee on Weather Control, 1958

"Weather Modification and the Operations of an Electric Power Utility: The Pacific Gas and Electric Company's Test Program" a paper by Donald L. Eberly as it appeared in the book *Human Dimensions of Weather Modification* edited by W.R. Derrick Sewell, published by the University of Chicago, 1966

"Economic Aspects of Weather Modification" a report by Warren Viessman, Jr. as it appeared in *Weather Modification: Programs, Problems, Policy, and Potential* a book by the Congressional Research Service, 1978

"The Travelers Weather Research Center" an article by Weatherwise, December, 1954, volume 7, issue 6, p159

Weather modification hearings before the Subcommittee on the Environment and the Atmosphere of the Committee on Science and Technology, U.S. House of Representatives, Ninety-fourth Congress, second session, published by the U.S. Government Printing Office, 1976

Weather Modification Ninth Annual Report a book by the National Science Foundation, published by the U.S. Government Printing Office, 1967

"National Science Foundation Program in Weather Modification for FY 1968" a report by P. H. Wyckoff as it appeared in *Proceedings of the Tenth Interagency Conference on Weather Modification* by the National Oceanic and Atmospheric Administration, published by the National Oceanic and Atmospheric Administration, 1968

"Thomas F. Malone 1917-2013 A Biographical Memoir" by David T. Malone, produced by the National Academy of Sciences, 2014

"Thomas F. Malone (1917-2013)" by Gregory A. Good, as it appeared in *Eos*, published by the American Geophysical Union, Vol. 94, No. 50, Dec., 10, 2013

Tenth annual National Science Foundation weather modification report, 1968

The Travelers: 100 Years a book by the Travelers, published by the Travelers, 1964

Killing the Planet: How a Financial Cartel Doomed Mankind a book by Rodney Howard-Browne and Paul L. Williams, published by Republic, 2019

"A Recommended National Program in Weather Modification: A Report to the Interdepartmental Committee for Atmospheric Sciences" a report by Homer E. Newell, published as Interdepartmental Committee for Atmospheric Sciences report number 10a, 1966, appendix II

The Smartest Guys in the Room a book by Bethany McLean and Peter Elkind, published by Portfolio Books, 2004

"Weather Products; Managing global weather exposures. Growing opportunities. Reducing Risks" a brochure by the Chicago Mercantile Exchange, 2009
"Weather Derivatives Instruments and Pricing Issues" a report by Financial Engineering Associates, 2000

"Weather Derivatives" a paper by Pauline Barrieu & Olivier Scaillet, London School of Economics, Swiss Finance Institute and University of Geneva, 2008

"Want a Weather Forecast? Ask Wall Street" an article by Alice Gomstyn, Rich Blake and Dalia Fahmy, *ABC News*, 2010

"General Electric Exits Insurance; Sells Insurance Solutions, Employers Re to Swiss Re" an article published by *Insurance Journal*, Nov. 18, 2005

"Weather derivatives becoming hot commodities" an article, *USA today*, 2008

"Firing Up the Market for Weather Contracts" an article by Antoine Gara, published by *Bloomberg Businessweek*, 2011

"OTC weather risk market grows 30% to $2.4bn" an article by Charlotte Dudley, EnvironmentalFinance.com, 2011

"Weather, Finance and Meteorology - forecasting and derivatives" a paper by the Samuel Randalls School of Geography, Earth and Environmental Sciences, University of Birmingham

"Energy Innovators: Ringing in an Age of Enlightenment" an article by Public Utilities Reports, Inc., 1999

"Enron: Charting the Legacy 10 Years on" an article by *Risk.net*, 2011

Family of Secrets: The Bush Dynasty, America's Invisible Government, and the Hidden History of the Last Fifty Years a book by Russ Baker, published by Bloomsbury Press, 2009

"On Modeling and Pricing Weather Derivatives" a paper by Peter Alaton and David Stillberger, 2002

"Impacts and Responses of the Weather Insurance Industry to Recent Weather Extremes" a report by Stanley A. Changnon, David Changnon, E. Ray Fosse, Donald C. Hoganson, Richard J. Roth, Sr., and James Totsch, published by *Changnon Climatologist*, May, 1996

"Hedge Funds Pluck Money From Air in $19 Billion Weather Gamble" an article by Peter Robison, published by *Bloomberg*, Aug 1, 2007

Enron annual report, 2000

The Smartest Guys in the Room a documentary produced by Magnolia Pictures, 2005

Conspiracy of Fools a book by Kurt Eichenwald, published by Broadway Books, 2005

What Went Wrong at Enron: Everyone's Guide to the Largest Bankruptcy in U.S. History a book by Peter C. Fusaro and Ross M. Miller, published by John Wiley & Sons, 2002

Tenth annual National Science Foundation weather modification report, 1968

“A Simulation Model for Studying Possible Modification of Surface Temperature” a paper by M. Lawrence Nicodemus and James D. McQuigg, published by the *Journal of Applied Meteorology*, April 1969

“Temperature Modification and Costs of Electric Power Generation” a paper by S.R. Johnson, James D. McQuigg, and Thomas P. Rothrock, published by the *Journal of Applied Meteorology*, December 1969

“Weather Variability and Economic Analysis” a report by James D. McQuigg and John P. Doll, published by the *Research Bulletin*, June 1961

Interdepartmental Committee for Atmospheric Sciences report 20, published by the Federal Council for Science and Technology, May, 1976

Enron: The Rise and Fall a book by Loren Fox, published by John Wiley & Sons, Inc., 2003

"Enron: Charting the Legacy 10 Years on" an article by *Risk.net*, 2011

The Best Democracy Money Can Buy a book by Greg Palast, published by Penguin Books, 2003

"Lynda Clemmons - Fast 50 2002" an article by *FastCompany.com*, 2002

"The Market for Catastrophe Risk: a Clinical Examination" a paper by Kenneth A. Froot, published by *The Journal of Financial Economics*, 2001

"A Buyer’s Guide for Options and Futures on a Catastrophe Index" a paper by Glenn G. Meyers, PhD, FCAS

"World Catastrophe Reinsurance Market Review" a report by Guy Carpenter & Co., 2011

Alternative Risk Transfer a book by Erik Banks, published by John Wiley & Sons, Ltd., 2004

"Marketplace Realities: Solid Footing and a Foundation for Growth" report by Willis Group, 2011

"Accumulation of Very Severe Natural Catastrophes Makes 2011 a Year of Unprecedented Losses" a report by Munich Re, 2011

Why in the World Are They Spraying? a documentary by Michael J. Murphy, produced by Truth Media Productions, 2012

"Evaluation of Weather and Climate by Socio-Economic Sensitivity Indices" a paper by James B. Hibbs as it appeared in the book *Human Dimensions of Weather Modification* edited by W.R. Derrick Sewell, published by the University of Chicago, 1966

Chapter 9
BIOLOGICAL IMPACTS

When airplanes routinely dump megatons of toxic garbage into our atmosphere as they have been doing for twenty years plus now, the most obvious question is: What are the biological impacts? What are the environmental implications of very small coal fly ash particles entering our bodies and fouling our biosphere? As one might guess, the implications are grave. Although the geoengineers will undoubtedly tell us that everything is fine, the best available evidence shows that the general population's health is being negatively impacted, at least hundreds of thousands of people are dying, and our environment is being summarily wrecked as well. These are the biological impacts of the New Manhattan Project.

Particulate matter

The inhalation of aerosolized particulate matter has generally harmful human health impacts. This is not a matter of debate. Common sense and many studies show this. A slew of studies referenced at the end of this chapter shows that inhalation of fine particulate matter is associated with: Alzheimer's disease, risk for stroke, risk for cardiovascular disease, lung inflammation and diabetes, reduced renal (kidney) function in older males, morbidity and premature mortality, decreased male fertility, low birth weight, onset of asthma, and increased hospital admissions.

Coal fly ash

As far back as October of 1979, a study was performed about the health effects of aerosolized coal fly ash. Unsurprisingly, the authors of the study found that exposure to aerosolized coal fly ash through the lungs causes harm. In other news, the geniuses at the World Health Organization found that bullets fired from guns can kill people.

We should be thankful that the good Dr. Marvin Herndon has recently produced a series of peer-reviewed, published journal articles detailing the health effects of exposure to that specific material being routinely pumped out of jet aircraft. His first paper in this area titled "Coal Fly Ash Aerosol: Risk Factor for Lung Cancer," published in February of 2018, was co-authored by Dr. Mark Whiteside, MD, MPH, the Medical Director of the Monroe County, Florida Department of Health. Herndon and Whiteside found that coal fly ash has lots of nasty, cancer-causing stuff in it. The authors write:

"CFA [coal fly ash] *contains a variety of potentially carcinogenic substances including aluminosilicates, an iron oxide-containing magnetic fraction, several toxic trace elements, nanoparticles, and alpha-particle-emitting radionuclides. Silica, arsenic, cadmium, and hexavalent chromium are found in CFA and all have been associated with increased lung cancer risk."*

Further, the authors write, "Chronic exposure to aerosolized CFA, emplaced in the atmosphere for climate intervention, may be an important, yet unrecognized, environmental risk factor for development of lung cancer."

Dr. Mark Whiteside

As we can see from the passage above and as many have feared, Dr.s Herndon and Whiteside have found that at least some of these atmospheric coal fly ash particles are nano-sized. This is a concern because when nano-sized particles are inhaled, they are so small that they go directly into the blood stream and right into the brain, often causing a host of neurological disorders. Nano-sized particles are so small that one ingests them through one's skin.

Herndon and Whiteside teamed up again for the March 2018 publication of their paper "Aerosolized Coal Fly Ash: Risk Factor for Neurodegenerative Disease." The authors write:

"The recent finding of spherical exogenous magnetite (Fe3O4) nanoparticles in the brain tissue of persons with dementia suggests an origin in air pollution produced by coal fly ash. The primary components of coal fly ash, iron oxides and aluminosilicates, are all found in the abnormal proteins that characterize Alzheimer's dementia. The presence of these substances in brain tissue leads to oxidative stress and chronic inflammation. Energy absorbed by magnetite from external electromagnetic fields may contribute to this neuropathology."

Later, in May of 2018, Herndon and Whiteside were published once again. This time, their paper titled "Aerosolized Coal Fly Ash: Risk Factor for COPD and Respiratory Disease" found that:

"Aerosolized CFA [coal fly ash] *is a particularly hazardous form of deliberate air pollution. Ultrafine particles and nanoparticles found in coal fly ash can be inhaled into the lungs and produce many toxic effects including decreased host defenses, tissue inflammation, altered cellular redox balance toward oxidation, and genotoxicity. Oxidative stress and chronic inflammation can predispose to chronic lung disease. Recognition and public disclosure of the adverse health effects of geoengineering projects taking place in our skies, and their concomitant cessation will be necessary to prevent an ever-widening epidemic of COPD and other respiratory illnesses."*

Rounding out this duo's series of papers on the Human health impacts of chemtrails, Herndon and Whiteside wrote a November 2019 paper titled "Geoengineering, Coal Fly Ash and the New Heart-Iron Connection: Universal Exposure to Iron Oxide Nanoparticulates." The authors write:

"Coal fly ash is a rich source of nano-sized metal, iron oxide, and carbonaceous particles. Previous findings revealed that coal fly ash is widely utilized in undisclosed tropospheric aerosol geoengineering. Proper iron balance is central to human health and disease, and the harmful effects of iron are normally prevented by tightly controlled

processes of systemic and cellular iron homeostasis. Altered iron balance is linked to the traditional risk factors for cardiovascular disease. The iron-heart hypothesis is supported by epidemiological, clinical, and experimental studies. Biogenic magnetite (Fe3O4) serves essential life functions, but iron oxide nanoparticles from anthropogenic sources cause disease. The recent finding of countless combustion-type magnetic nanoparticles in damaged hearts of persons from highly polluted areas is definitive evidence of the connection between the iron oxide fraction of air pollution and cardiovascular disease. Spherical magnetic iron oxide particles found in coal fly ash and certain vehicle emissions match the exogenous iron pollution particles found in the human heart. Iron oxide nanoparticles cross the placenta and may act as seed material for future cardiovascular disease. The pandemic of non-communicable diseases like cardiovascular disease and also rapid global warming can be alleviated by drastically reducing nanoparticulate air pollution. It is crucial to halt tropospheric aerosol geoengineering, and to curb fine particulate emissions from industrial and traffic sources to avoid further gross contamination of the human race by iron oxide-type nanoparticles."

Now that we have seen the Human health impacts of aerosolized coal fly ash, we will now take a look at the Human health impacts of some known constituents of coal fly ash.

Aluminum

As evidenced by voluminous rainwater sample lab reports (ch 1), chemtrails have been shown to consist significantly of aluminum oxide. Aluminum is a common component of coal fly ash. As we have learned from Dr.s Herndon and Whiteside, these particles can be nano-sized.

Aluminum nanoparticles are nasty stuff. A material safety data sheet (MSDS) produced by US Research Nanomaterials, Inc. says that they can cause: respiratory problems, skin irritation, eye irritation, tumors, Alzheimer's, pulmonary disease, neoplasms, and gastric or intestinal disorders. This MSDS also states that people coming in contact with aluminum nanoparticles should wear a respirator and a fully protective, impervious suit.

A 2016 paper titled "Assessing the Direct Occupational and Public Health Impacts of Solar Radiation Management with Stratospheric Aerosols" says that Aluminum aerosols will target these bodily systems: respiratory, cardiovascular, hematologic (blood), musculoskeletal (muscles & bones), endocrine (glands), immunologic, and neurologic (brain). They also say exposure to small atmospheric aluminum particles can cause cancer and death.

It appears coincidental that Wright-Patterson Air Force Base has studied the biological impacts of aerosolized aluminum. In March 2001, the Air Force Research Laboratory at Wright-Patterson published a study titled "In Vitro Toxicity of Aluminum Nanoparticles in Rat Alveolar Macrophages." Scientists exposed rats to airborne, nano-sized aluminum oxide particles. The authors concluded:

"Aluminum oxide nanoparticles displayed significant toxicity after 96 and 144 hours post exposure at high doses (100 and 250 µg/ml). Aluminum nanoparticles also showed slight toxicity after 24 hours at high doses (100 and 250 µg/ml). When these cells were dosed at lower non toxic levels (25 µg/ml) Al 50, 80, 120 nm caused a significant reduction in phagocytosis. Even at a dose as low as 5 µg/ml Al 50 nm still caused a significant reduction. None of these nanoparticles caused the induction of nitric oxide, TNF-alpha, or MIP-2, important components in inflammatory responses. In summary, based on viability aluminum nanoparticles appear to be slightly toxic to rat alveolar macrophages. However, there was a significant reduction in phagocytic function of macrophages."

In other words, they found that even at low doses, forcing rats to breathe in tiny aluminum particles screwed up their lungs. The induced lack of phagocytes means that the rats' immune systems (especially in the lungs) became unable to fight off invading harmful organisms.

"In Vitro Toxicity of Aluminum Nanoparticles in Rat Alveolar Macrophages" was but one of a series of studies produced by Wright-Patterson pertaining to aluminum nanoparticle exposure. Wright-Patterson also produced a 2010 study titled "Nanosized Aluminum Altered Immune Function" in which they found that inhaled aluminum nanoparticles impair human immune systems. The authors again noted that nanoparticles have more deleterious health effects than do larger sized particles. Curiously, "Nanosized Aluminum Altered Immune Function" also states that we are prone to inhale aluminum nanoparticles because they are used in jet fuels. This information, makes yet another case for aluminum-spiked jet fuels. All this is extremely interesting when one considers Wright-Patterson's involvement in the New Manhattan Project such as that which was documented in chapter 5.

A 2009 paper titled "Manufactured Aluminum Oxide Nanoparticles Decrease Expression of Tight Junction Proteins in Brain Vasculature" found that, due to brain cell death, aluminum exposure can cause: Alzheimer's, stroke, reperfusion, hypoxia, mitochondrial disease, and general vascular dysfunction.

In a 2012 paper written by one of the world's top neurosurgeons (now retired), many neurological diseases are linked to aluminum exposure. Russell Blaylock's "Aluminum Induced Immunoexcitotoxicity in Neurodevelopmental and Neurodegenerative Disorders" found a link between aluminum exposure and: Alzheimer's, Parkinson's, Huntington's, Pick's, HIV dementia, multiple sclerosis, viral encephalopathies, chronic traumatic encephalopathy, and amyotrophic lateral sclerosis (ALS / Lou Gehrig's disease). In this paper, Dr. Blaylock also found that aluminum exposure is linked to: impaired cognition, poor memory, impaired learning, poor attention, social withdrawal, irritability, reduced food and water intake and depression. Not only that, but Dr. Blaylock cites another paper here showing how extremely small aluminum particles like the ones used in today's New Manhattan Project can intensify adverse health reactions.

Dr. Blaylock has provided us with some impressive evidence for a causal relationship between chemtrails and Alzheimer's here. He tells us that the aluminum nanoparticles we constantly inhale are carried directly to the part of the brain that is first affected by Alzheimer's disease AND most severely affected by Alzheimer's disease. On March 28, 2013 Dr. Blaylock went on the *Linderman Unleashed* radio program. The host asked him how he became chemtrail aware and Dr. Blaylock said this:

Dr. Russell Blaylock, MD

"Well, you know, the connection has been the aluminum in the vaccines. I wrote several articles about the effects of the adjuvants in vaccines including the mercury and the aluminum effect.

"Then I found some articles about the chemtrails and there was a lot being said about it and I wasn't too sure whether it was true or not because in my state we rarely saw them. But as I started looking on the Internet and I would see these states in which there were these criss-cross patterns and they were very tight patterns and geometrical shapes where it was obvious that it was a purposeful covering of the atmosphere with these patterns and the trails were so long. Well now, you know, we're starting to see them in my state and as I look at them, they go from to horizon to horizon. Well, you know, I've been alive long enough to know that jets never did that in the past and I see the same patterning effect now where they're criss-crossing; it's an obvious pattern.

"And so I look into the literature and some of the reports and YouTube videos and they were saying that they were dropping as one of the ingredients, aluminum. Well, I had done a fair amount of writing and research on the effect of aerosolized chemicals in the nose when you breathe them. And what we knew was that these particles tend to

travel along the olfactory nerves which are the smell nerves in the nose. And it travels directly to the part of the brain that has to do with memory and emotions; the hippocampus, the interlinal area, and the prefrontal cortex. And that you can trace these chemicals traveling along that nerve and depositing in this area of the brain.

"The other thing that was known is that if you aerosolize aluminum, it's one of the metals that passes very easily along this track and directly into the brain. So it bypasses the blood-brain barrier and goes directly into the brain and accumulates. Well, if you do it in animals, it produces lesions, or damage in that area of the brain and the animal will begin to show changes of memory and learning and emotional changes.

"When we look at people who have Alzheimer's disease, ironically, the highest concentration of aluminum in the brain is that same entry point; what's called the interlinal cortex. And the levels continue to accumulate. So we have compelling evidence that aerosolized aluminum alone will enter the brain and produce damage to that critical area of the brain.

"The worst of all is the nano-sized. Nano-size means you make it such a small particle that it easily penetrates skin. It penetrates barriers in the body that normally metals cannot pass through. When you nano-size and produce these incredibly small particulate matter, it passes very easily. So when you nano-size aluminum and you use it in these aerosols through the nasal passages, it enters the brain in very high concentration and they find that the nano-sized aluminum in the brain is infinitely more toxic.

"Now one of the toxic reactions to aluminum is intense inflammation and activation of cells in the brain that are the immune cells called microglia. Aluminum is a very potent activator of these immune cells and that triggers the release of a powerful substance called glutamate which is an excitotoxin that causes cells to die from an excitatory mechanism. Kinda complex mechanism, but it is a combination of inflammation and excitotoxicity. And I coined the term in the medical literature called immunoexcitotoxicity to describe that process. So, we know that occurs. We know it occurs very easily.

"Now, the reports are coming out now that what they're spraying is nano-sized aluminum and the idea is the old concept of preventing global warming. And they nano-size the aluminum so it will stay in the

upper atmosphere longer; supposedly as a reflective compound metal. The problem with that, even from a climatological description is that if you make it into cirrus-like clouds rather than reflecting it upward and out of the atmosphere, it reflects the heat downward and actually causes global warming. So, you know, you could envision that they're doing this on purpose to make the atmosphere heat up so they can say, 'See, the atmosphere is warming up.'

"But what I'm concerned about mainly is the medical effect and that's because of these very strong connections between aluminum passing through this pathway into the brain [which] is so strongly connected with Alzheimer's disease and other diseases of memory.

"If you're aerosolizing this and spraying literally tons of it over the world, people are constantly breathing that aerosolized, nano-sized aluminum which will easily penetrate filters in your air-conditioning system [and] enter your home. So you're breathing it 24 hours a day; producing high levels of aluminum in this part of the brain. And the consequences could be absolutely devastating. It could cause a huge increase in Alzheimer's disease and inflammatory neurological disorders.

"I watched a YouTube which was a geoengineering conference that the government had put on. And in the conference, one of the questions somebody in the audience asked was: What is the medical effect of spraying aluminum in the atmosphere? And the speaker said, 'Well, uh, we don't really know. But we're in the process of researching that.' Well, of course that was an absolute lie. We do know what it does. But the fact that they were admitting that in fact they were going to spray, they gave it in the future tense that they were going to spray aluminum, the evidence now from the examination by biologists and scientists around the world is that the aluminum level in the lakes and streams and trees is increasing enormously. Some areas have incredible elevations of aluminum in the groundwater and in the vegetation. So if this indeed is happening, we're looking at a medical catastrophe that's worldwide."

There is lots of other highly credible evidence available linking aluminum exposure to the diseases mentioned here. If you want more information, please search the term 'aluminum toxicity.' Expediency demands that we move on.

Barium

Rainwater sample test results from around the world consistently show barium as well, and barium can also be a component of coal fly ash. Barium is highly toxic. Barium material safety data sheets (MSDS) readily available online will inform you that barium is extremely hazardous in case of inhalation. Severe exposure to barium can cause lung damage, choking, unconsciousness and death. Many other barium oxide MSDSs go on and on in a similar fashion. The Centers for Disease Control and Prevention (CDC) says that barium oxide reacts violently with water while the atmosphere has lots of water in it and our bodies consist mostly of water. My science advisor says that barium titanate and barium sulfate have been used in atmospheric dispersions as well.

The aforementioned paper "Assessing the Direct Occupational and Public Health Impacts of Solar Radiation Management with Stratospheric Aerosols" says that barium compounds used as atmospheric sprays target these Human bodily systems: respiratory, gastrointestinal, musculoskeletal, renal, metabolic, and neurologic. They also say barium compounds dispersed by aircraft as part of geoengineering programs can cause death.

Strontium

Rainwater sample test results as well as others such as ambient air sample test results collected by Dr. Herndon have also been showing a presence of strontium. Strontium can be a component of coal fly ash. It is not surprising, but, like aluminum and barium, strontium is highly toxic as well.

A strontium MSDS from Sigma-Aldrich states that it is corrosive. It causes burns when it comes in contact with the skin and can be absorbed through the skin. If one inhales it, the MSDS states that it is, "...extremely destructive to the tissue of the mucous membranes and upper respiratory tract." The MSDS continues:

"Inhalation may result in spasm, inflammation and edema of the larynx and bronchi, chemical pneumonitis, and pulmonary edema. Material is extremely destructive to tissue of the mucous membranes and upper respiratory tract, eyes, and skin."

The Sigma-Aldrich MSDS finishes up by noting that the chemical, physical, and toxicological properties of strontium have not been thoroughly investigated.

Strontium hydroxide is even worse. Being that there is lots of water in the atmosphere, the atmospheric strontium produced as part of the New Manhattan Project may react with it and form the extremely caustic strontium hydroxide. Not only that, but don't forget that our bodies are comprised of mostly H2O. Strontium in the atmosphere and inside of us has lots of opportunities to become strontium hydroxide. The Sigma-Aldrich MSDS cautions potential users to never expose strontium oxide to water because it reacts violently.

Because strontium can be a component of coal fly ash, it is interesting to note that studies have been done concerning exposure to the strontium found in 'fly ash.' The CDC writes:

"*Rats were exposed to aerosols of 85Sr [strontium] carbonate, phosphate, fluoride, oxide, or titanate (particle sizes and doses not specified) (Willard and Snyder 1966). Greater than 99% of the initial lung burden of 85Sr was cleared from the lung 5 days after inhalation of the carbonate, phosphate, fluoride, or oxide, whereas 60% of the 85Sr remained in the lung after inhalation of the more insoluble strontium titanate.*

"*In rats exposed to airborne fly ash (sieved to have a particle diameter of distribution of 90% less than 20 μm) for 6 hours, strontium was eliminated from the lung with a half-time of 23 days (observations were made for 30 days) (Srivastava et al. 1984b). One day after the exposure, the tissue: plasma strontium concentration ratios were 0.3–0.5 in the liver, kidney, small intestine, and heart. The report of this study does not indicate whether whole-body or nose-only exposures were utilized in the study; therefore, it is not possible to know for certain how much of the absorption may have resulted from ingestion of fly ash deposited on the animals. Furthermore, given the relatively large particle size of the fly ash, it is likely that deposition in the respiratory tract was largely in the tracheobronchial and nasopharyngeal region, from which the strontium may have been cleared mechanically to the esophagus and swallowed. Nevertheless, studies in which 89Sr-enriched fly ash was instilled into the trachea of rats indicate that strontium in this form was partly absorbed and appeared in plasma and other tissues within days of the exposure (Srivastava et al. 1984a)."*

The CDC goes on to note that the fly ash strontium administered to the lab rats ended up mostly in the bones. After that, it appeared in (in order of prevalence): muscle, skin, liver, and kidneys. Those heady days of just dumping dry ice into clouds are long gone.

Mercury

Dr. J. Marvin Herndon produced a December 2017 paper co-authored by Mark Whiteside, MD in which the authors write specifically of the Human health impacts of mercury. It has been well known for a long time now that mercury is one of the most toxic substances on the planet and we now know that mercury is a common constituent of the coal fly ash currently being sprayed by the megaton. The authors write:

"Despite strengthened mercury emission regulations, mercury measured in rainwater is increasing. Since it is known that the upper troposphere contains oxidized, particle-bound mercury, it is likely that covert aerosolized coal fly ash sprayed into this region is a major source of mercury pollution. Mercury affects multiple systems in the body, potentially causing neurological, cardiovascular, genitourinary, reproductive, immunological, and even genetic disease."

CDC rates of associated diseases

As this chapter has explained, chemtrails are associated with many diseases. As we have been assaulted by this New Manhattan Project for over twenty years now, it is no surprise that the best available data shows rates of the associated diseases going up significantly. Historical rates of every disease associated with chemtrail spray are not presented here due to a lack of CDC data. Every associated disease *with* available CDC data is presented.

Let's start with the most strongly correlated disease: Alzheimer's. According to the latest data from the CDC, from 1999 to 2014, age-adjusted rates of death from Alzheimer's increased 54.5% with the 2014 number of total deaths at 93,541. That means that in 2014 alone we saw tens of thousands of additional American deaths from Alzheimer's. If one adds up all the additional deaths from Alzheimer's between 1999 and 2014, we're talking about hundreds of thousands of additional deaths. Let us recall that large-scale domestic spraying operations began in 1996.

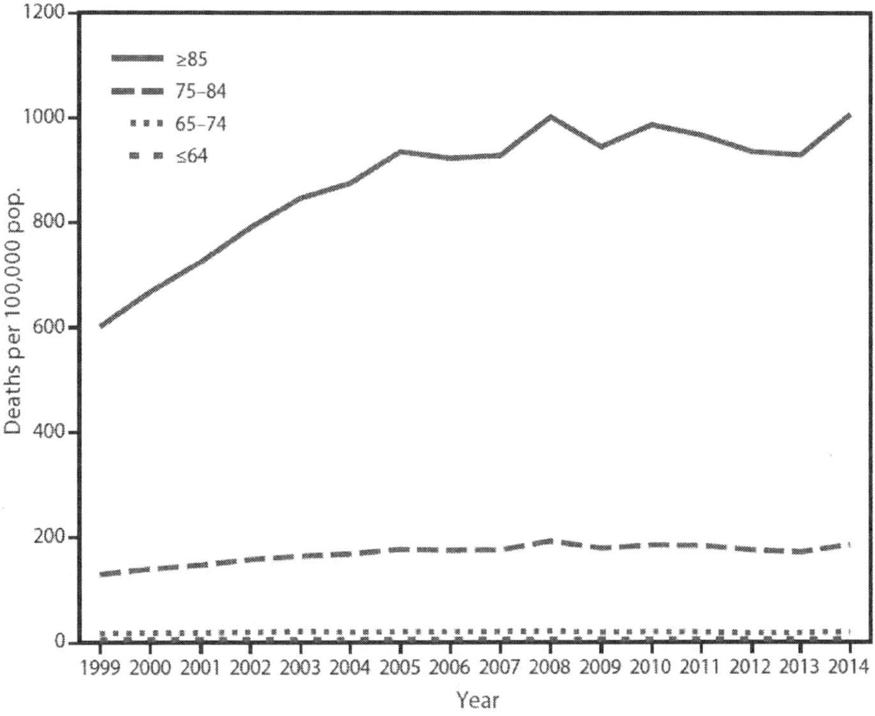

Age adjusted rates of Alzheimer's disease 1999-2014

In a 2013 report, the CDC found that while deaths from other diseases such as cancer, heart disease, and stroke decreased significantly, Alzheimer's deaths increased 39%. They write, "Mortality from Alzheimer's disease has steadily increased during the last 30 years." Knowing what we now know, it is reasonable to assume that chemtrails have contributed greatly to this.

Not only have the rates of adult Alzheimer's disease been increasing, but a disease that used to be relegated to the elderly is now showing up in children. Reports have been pouring in from around the world documenting research into Niemann Pick Type C disease, also known as 'childhood Alzheimer's.' As previously mentioned, Dr. Blaylock has seen this phenomenon as well.

~ ~ ~

Dr. Blaylock says that there is also a correlation between aluminum exposure and Parkinson's. The latest data from the CDC shows that between 1999 and 2017, the age-adjusted rate of Parkinson's disease in people aged 65 or older went from 41.7 per 100,000 to 65.3 per 100,000.

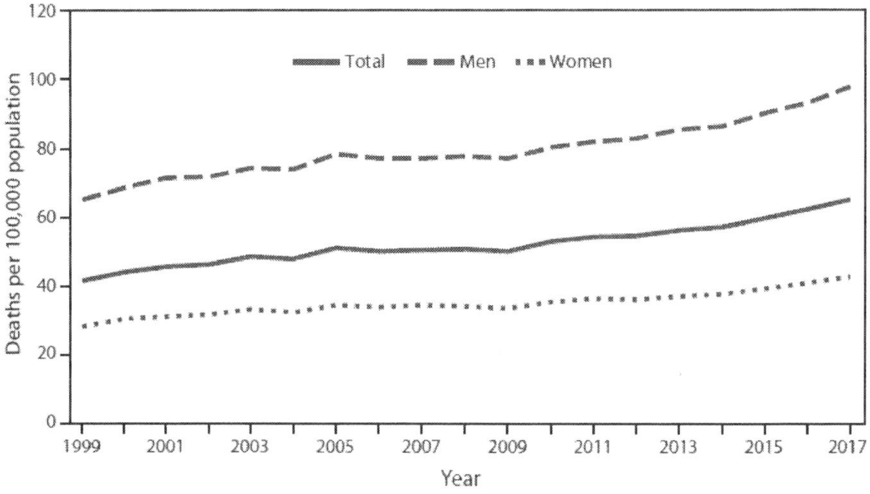

CDC rates of Parkinson’s disease 1999-2017

~ ~ ~

Despite what the tobacco companies said in the 1950s, routinely breathing in particulate matter is bad for your lungs. It is for this reason that we now take a look at the CDC data pertaining to diseases associated with the routine inhalation of particulate matter such as COPD, asthma, and lung cancer. Although the CDC found that the rate of chronic pulmonary disease (COPD) was stable between 1998 and 2009, they also found that the prevalence of asthma rose during a similar period (between 2001 and 2010). The CDC also reports that between 1995 and 2011, smoking went from 35% among students and 25% among adults to 18% and 19% respectively. Concurrently, the

CDC reports significant drops in the rate of lung cancer between 2002 and 2011.

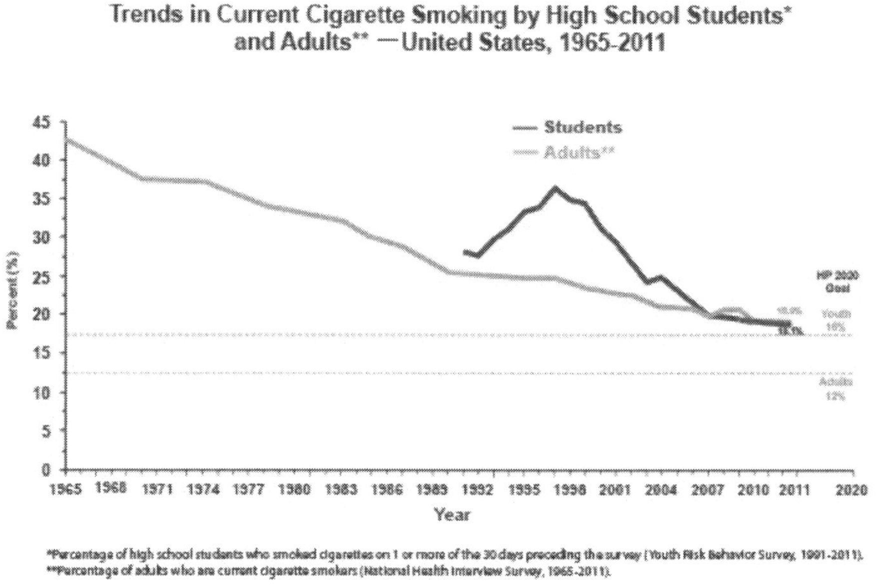

Rate of smoking from 1965-2011

With these big drops in the rate of smoking, one might assume that the rate of COPD and asthma would go down as well, instead of remaining stable. Chemtrails probably kept the rate of COPD stable while contributing to the prevalence of asthma. Lung cancer probably decreased because chemtrail exposure has not been as carcinogenic for your lungs as smoking. It's good to know that there are more carcinogenic things for your lungs than routine chemtrail exposure. Smoking cigarettes apparently fits that category. Moderate chemtrail exposure is probably better for you than inhaling burning plutonium too, but that doesn't mean it's ok.

Overall life expectancy

Very recently, we here in America have seen a slight reduction in our life expectancies. According to CDC data, for the first time in many decades, between 2016 and 2017 overall life expectancy at birth fell by .1 years.

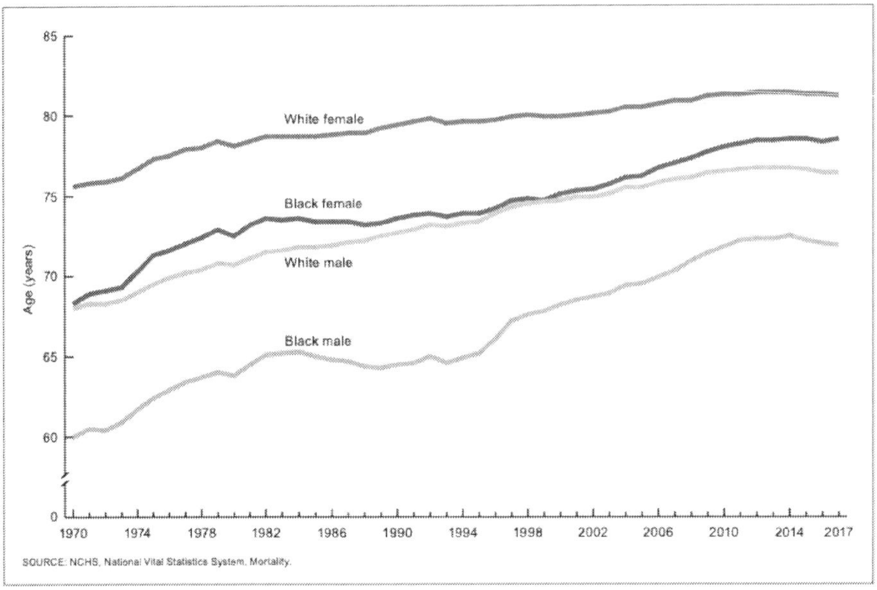

CDC life expectancy 1970-2017

One might think, with all the much-touted breakthroughs in medicine, a growing health care industry, expanded access to better nutrition such as organic foods and supplements, and the like, that we would be experiencing *longer* average life expectancies, not shorter. Might chemtrails have something to do with it?

Early exposures

Although it appears that our bodies have been finding ways to better cope with this daily onslaught of aerosolized toxic waste, around the times when people were first exposed, emergency rooms filled up. William Thomas' aforementioned 2004 book *Chemtrails Confirmed* chronicles many of these examples. Thomas recounts the words of a registered nurse:

"*Approximately December 16th or the 17th, while traveling north, I could see 'stripes' in the sky. It appeared as if someone took white paint on their fingers and from north to south ran their fingers through the sky. These contrails were evenly spaced and covered the whole sky! They covered it completely! When I was finished with the next visit, approximately 45 minutes, I came out of the house and found the whole sky was white. There was no definition in cloud pattern.*

"*Within the 24 hours I became very weak, feverish, and my asthma began to act up. I didn't think too much about it, until my boyfriend told me that many in his family started coming down with the same complaints. I also started noticing a lot of my patients and their family members were coming down with these symptoms at the same time. In our area we have one main hospital which I was the Supervisor of for four years. I worked there a total of six years. I stay in close contact with the nurses and physicians and am planning on investigating into this more. At that time, they complained of being extremely busy with respiratory diagnoses.*"

Another passage from *Chemtrails Confirmed* recounts the experiences of a restaurant owner from Oklahoma. The passage reads:

"*On January 24, 1999 [Pat] Edgar reported that on, 'Monday, Tuesday and Wednesday and Thursday of last week, we were really hit hard with the contrails. I mean real bad. Everybody in this town is sick right now; sicker than a damn dog. It's all in their head and their sinuses, and it hangs in the throat, (sore necks), ears ringing.'*

"*Edgar added: 'Some customers that frequent our business have stated that they have been to the doctor and the offices have been full of sick people. Same thing at the Indian clinic.'*

"'*People have to wait for hours because the waiting room is full. Some people have reported being on their third and fourth round of antibiotics and they are still ill. We noticed excessive contrails Thursday, Feb. 11th.'*

"*Edgar became ill the following day, and visited a doctor. From a friend he learned that Sparks regional hospital had over 500 people seeking medical attention at the emergency room for flu, or flu-like symptoms.*"

Others appearing in Thomas' book tell similar stories.

Bodily contamination testing

When we ingest aluminum, some of it eventually comes out in our hair and fingernails. There are many reports online of people finding high levels of chemtrail toxins in their hair and fingernails. Certain laboratories can analyze hair and fingernail samples for aluminum and other substances. If you are curious about your bodily contamination, one may get their hair tested by the Great Plains Laboratory or Analytical Research Labs. One can find their websites online.

Biospheric implications

There is evidence that chemtrails are changing soil pH. This could be very bad for our biosphere. As mentioned in the first chapter, anti-geoengineering activist Francis Mangels has a Bachelor of Science in forestry from the International School of Forestry at Missoula, spent 35 years with the U.S. Forest Service as a wildlife biologist and worked several years with the USDA Soil Conservation Service as a soil conservationist. In order to document the effect of chemtrail spray upon soil pH, Mr. Mangels wrote on Oct. 30, 2009:

"The soil scientists from the USDA Soil Conservation Department visited private property east of Shasta Lake, California, on Oct. 27, 2009. Mr. Bailey, Komar, and Owens tested the pH with standard federal meters. All agreed the pH should be 5.5.

"Under Douglas fir, the ph was 7.4, astoundingly basic for that habitat.

"Under Poderosa pine, at the precise soil-needle interface, I would expect a pH of 5. At that point, Bailey's meter showed 6.5. This is high for a microhabitat that should be very acid. Old soil surveys indicate this soil should be very acid, around pH of 5.5.

"I bought a house in Mt. Shasta old black oak/pine pasture in 2002, tested the pH at below 6, good for vegetable gardening. It was a major reason for purchase, and proceeded with highly acid composting of leaves and grass to drive the pH down or at least keep it low, as every master gardener knows. I added a touch of sulphur and avoided wood ash to insure acidity, and proceeded to teach organic gardening courses out of my yard through COS. The pH tests were an embarrassment because now my garden is pH 7, sometimes higher. This is the opposite of what should happen.

"The pH meter of Jon McClellan proceeded to show pH in McCloud gardens also running close to 7 or 8, which is too high for heavy organic mulch with no ashes. General lawns were also running over pH 7 under oaks and pines and fir trees. This is contrary to everything I learned in college and the Soil Conservation Service for 35 years. The old data sheets say these soils should be running at a pH of 5-6."

Francis Mangels

In the movie *What In the World Are They Spraying?*, Mr. Mangels says that when soil pH changes, soil arthropods (a vital link in our ecosystem) start to go away. This type of disruption could have negative effects up and down the food chain.

Reports of massive plant and animal die-offs potentially due to chemtrails are widespread. Spraying vast regions of the Earth with tens of thousands of megatons of toxic waste is probably contributing to the alarming rate of animal species extinction as well. Although many other factors are in play here, the chemtrails surely don't help. The Center for Biological Diversity reports that:

"*Scientists estimate we're now losing species at 1,000 to 10,000 times the background rate, with literally dozens going extinct every day. It could be a scary future indeed, with as many as 30 to 50 percent of all species possibly heading toward extinction by mid-century.*"

Once again, our Sparticus with the dragon energy, Dr. J. Marvin Herndon, PhD has been on the case. Dr. Herndon has again teamed up with the Medical Director of the Monroe County, Florida Department of Health, Dr. Mark Whiteside, MD, MPH to publish a series of peer-reviewed, published journal articles addressing the biospheric implications of the ongoing and uncontrolled geoengineering experimentation and we will go over them here.

Let's start at the bottom of the food chain. In June of 2019 Herndon and Whiteside published a paper titled "Role of Aerosolized Coal Fly Ash in the Global Plankton Imbalance: Case of Florida's Toxic Algae Crisis." In this paper, the authors provide evidence for the assertion that the coal fly ash sprayed by the megaton into our biosphere is causing, among other things, an overabundance of harmful plankton blooms which, in turn, has more harmful effects. The authors write:

"Our objective is to review the effects the multifold components of aerosolized coal fly ash as they relate to the increasing occurrences of HABs [harmful algal blooms]. *Aerosolized coal fly ash (CFA) pollutants from non-sequestered coal-fired power plant emissions and from undisclosed, although 'hidden in plain sight,' tropospheric particulate geoengineering operations are inflicting irreparable damage to the world's surface water-bodies and causing great harm to human health (including lung cancer, respiratory and neurodegenerative diseases) and environmental health (including major die-offs of insects, birds and trees). Florida's ever-growing toxic nightmare of red tides and blue-green algae is a microcosm of similar activity globally. Atmospheric deposition of aerosol particulates, most importantly bioavailable iron, has drastically shifted the global plankton community balance in the direction of harmful algae and cyanobacterial blooms in fresh and salt water."*

A little further up the food chain we find insects. Herndon and Whiteside have been working in this area as well. In August of 2018 their paper titled "Previously Unacknowledged Potential Factors in Catastrophic Bee and Insect Die-off Arising from Coal Fly Ash Geoengineering" was published. In this paper, the authors substantiate a multitude of harmful, observed effects upon insects from chemtrail spray. We can stop wondering why bee populations are being decimated. The authors write:

"The primary components of CFA [coal fly ash], *silicon, aluminum, and iron, consisting in part of magnetite (Fe3O4), all have important potential toxicities to insects. Many of the trace elements in CFA are injurious to insects; several of them (e.g., arsenic, mercury, and cadmium) are used as insecticides. Toxic particulates and heavy metals in CFA contaminate air, water, and soil and thus impact the entire biosphere. Components of CFA, including aluminum extractable in a chemically-mobile form, have been shown to adversely affect insects in terrestrial, aquatic, and aerial environments. Both the primary and trace elements in CFA have been found on, in, and around insects and the plants they feed on in polluted regions around the world. Magnetite from CFA may potentially disrupt insect magnetoreception. Chlorine and certain other constituents of aerosolized CFA potentially destroy atmospheric ozone thus exposing insects to elevated mutagenicity and lethality levels of UV-B and UV-C solar radiation."*

This information goes a long way towards explaining the tremendous drops in global insect populations lately. It's almost too scary to look into, but an Internet search of the term 'insect populations' will bring pages and pages of relevant results. Of course, many are blaming it on the dreaded global warming/climate change, but insect populations have done just fine throughout previous fluctuations in Earth's average temperature. In fact, insect populations have most probably done better in warmer climates, so maybe we should look instead at the gigantic aircraft routinely dumping megatons of toxic waste into our biosphere.

As noted, Francis Mangels has been observing a lack of insects as well. He logically attributes it to geoengineering. On July 19, 2017, Francis emailed to the author the following:

"Several streams were sampled for aquatic insects, and I likewise fished them hard to get stomach samples of trout. Total sample over 1000, lately around 400 stomach samples. Methods used were fairly casual, using typical nets for streams in gravel substrates that appeared similar. Standard data was orders of aquatics per square foot, accuracy about 80% due to equipment. It was very easy to see which streams would have the most trout.

"The bottom fell out of the sampling from 2000 to 2008, and it continues today. All major orders of bugs took severe hits from an unknown source. Then I was contacted by Dane [Wigington], and logic said only sky pollution could hit all the streams at once in the same way.

"Likewise, the trout I caught before then always were loaded with bugs and etc. food both terra and aquatic. Ever since about 2006, the trout stomachs were almost empty, and I quit taking data because there was no data to take, for the most part. A bug here and there, mostly terrestrials, very small amounts and the trout got skinny over the years (except for those freshly planted, that soon lost the fat and got skinny too, as we say, poor condition factors). Very clear streams went almost barren, no bugs or trout either.

"Net sweeps in lots showed plenty of earwigs, pill bugs, ants, aphids, box elder bugs, any SUCKING types. However, the caterpillar types for the most part became very scarce, as did moths and butterflies as you would expect (leaf eaters eat the aluminum). I turned in a huge collection to the American butterfly association of CA, but damned if I could do it now....Lepidoptera are around, but rare now except for the cabbage butterfly and a few swallowtails. Point is, this distribution showed in the trout stomachs, which caused me to do the sweeps."

Further up the food chain we find birds. Dr.s Herndon and Whiteside published a paper in November of 2018 titled "Aerosolized Coal Fly Ash: A Previously Unrecognized Primary Factor in the Catastrophic Global Demise of Bird Populations and Species." In this paper, the authors find that coal fly ash is causing unprecedented bird die-offs.

The authors write, "Bird populations and species world-wide are experiencing die-offs on an unprecedented scale." A little later, the authors continue, "Aerosolized CFA [coal fly ash], a particularly toxic form of air pollution, contains multiple metals and elements well-known to adversely affect all portions of the avian life cycle, in aerial, terrestrial, and marine environments. Studies from around the globe reveal systemic contamination of birds by these elements." The authors conclude that, "Coal fly ash, including its use in ongoing atmospheric geoengineering operations, is a major factor in global bird die-off. The accelerating decline of birds parallels the catastrophic decline of insects, due in part to the same type of aerial pollution."

Doctors Herndon and Whiteside have also looked at the biological impacts of chemtrails upon bat populations. In January of this year (2020), they published a paper titled "Unacknowledged Potential Factors in Catastrophic Bat Die-off Arising from Coal Fly Ash Geoengineering." In this paper, the authors find that bat populations worldwide are suffering a precipitous decline. The authors write:

"Bats are excellent mammalian bioindicators of environmental contaminants and it is known that their tissue contains high levels of metals and persistent organic pollutants. From a review of the literature, we show that the pollutant element ratios in bat tissue and bat guano are consistent with an origin in CFA-type air pollution. These findings suggest that CFA [coal fly ash], *including its use in covert climate engineering operations, is an unacknowledged factor in the morbidity and mortality of bats. Bats, therefore, are an important 'canary in the coal mine' pointing to the urgency of halting covert climate engineering and greatly reducing ultrafine particulate air pollution."*

~ ~ ~

As we saw at the beginning of this section, with all the professionally observed soil pH anomalies, plants are not doing very well under this New Manhattan Project either. Doctors Herndon, Whiteside and other co-authors have been doing work in these areas as well. In a series of published, peer-reviewed journal articles, they have found that a combination of factors, all caused by the spraying of coal fly ash, are causing mass die-offs of global vegetation. They found that trees, in particular, are weakened by increased UV radiation, desiccation, and toxicity - all caused by chemtrails. Once a tree is weakened by this trifecta, it becomes susceptible to insect infestations, fungal infections, and other biotic factors such as bacteria and viruses.

The result of all this is dry, dead and dying vegetation. An abundance of dry, dead and dying vegetation makes forest fires occur more often and burn more furiously. Herndon *et al.* find that this is most probably why we have seen such tremendously large forest fires lately. The increased levels of UV radiation noted by Herndon *et al.* as being harmful to vegetation, are also harmful to Humans as well as phytoplankton, coral, and insects.

Silver iodide

The conventional weather modification industry has been openly spraying vast areas of the United States with silver iodide since 1947. The super-secret New Manhattan Project only started spraying us with coal fly ash in 1996. Hence, the vast majority of the weather modification and atmospheric sciences literature is geared towards the dispersion of silver iodide. Although silver iodide is not what is used in today's New Manhattan Project, as a side issue, let's take a look at the scientific evidence (or lack thereof) concerning the biological impacts of silver iodide. Past is prelude.

Considering that this issue is the most obvious question and of grave importance, the lack of publicly available research pertaining to the biological impacts of silver iodide dispersion is quite shocking. You may read the 746 page, 1978 Congressional Research Service report on weather modification. You may read all 21 of the Interdepartmental Committee for Atmospheric Sciences reports or all of the National Science Foundation annual weather modification reports. You may read scores of weather modification reports, book after book, and myriad reports and papers about weather modification and the atmospheric sciences. But nowhere in any of these documents may you find an adequate examination of biological impacts and specifically human health impacts caused by exposure to atmospheric silver iodide. Only after reading a stack of documents about a yard high, did your author finally find a report containing an adequate discussion of this topic.

A popular silver iodide material safety data sheet describes silver iodide as, "Hazardous in case of skin contact (irritant), of eye contact (irritant), of ingestion, [and] of inhalation." Unbelievably, the authors of this data sheet write that much of the toxicology information is NOT AVAILABLE. They've been spraying us with this stuff since 1947 and the toxicology information is not available?! Equally as unbelievable, to date, no publicly available, long-term studies have been done.

It is widely suggested that exposure to silver iodide causes argyria - characterized by a blue-grey discoloration of the eyes, skin, mucous membranes, and internal organs. Does that sound healthy? Another MSDS produced by Fisher Scientific reads:

"*Chronic ingestion of iodides during pregnancy has resulted in fetal death, severe goiter, and cretinoid appearance of the newborn. Prolonged exposure to iodides may produce iodism in sensitive individuals. Symptoms could include skin rash, running nose and headache.*"

In spite of this information, the historical weather modification literature notes a lack of data. A 1966 National Science Foundation report stated, "The present state of knowledge places uncomfortable limits on the prediction of the biological consequences of modifying the weather." A 1969 Bureau of Reclamation report noted, "There has so far not been a single biological field study completed and reported in the literature specifically designed to identify any aspect of the ecological effects of weather modification." A 1972 study conducted by the Council on Environmental Quality stated, "Projects may have significant adverse environmental effects, ranging from immediate hazards to life and property to long-term alterations in land use patterns and threats to ecological systems."

Weather modifiers have exhibited a pattern of dismissing the potentially harmful effects of substances used in weather modification activities. In 1967 weather modifier Archie Kahan, writing for the Bureau of Reclamation, dismissed concerns about the use of silver iodide as he conflated the biological impacts of silver iodide with its efficacy as a nucleant and any possible hazardous weather that might arise from its use.

In 1972, decades after silver iodide was first used as a nucleant, Bernard Vonnegut and another atmospheric scientist by the name of Ronald Standler wrote a biology paper published in the *Journal of Applied Meteorology* that mollified concerns about their activities. Although the biological impacts of prolonged silver iodide dispersion has implications not only for Human health, but also for the health of the entire biosphere, the paper concerns itself almost exclusively with impacts upon Human health. The questionable biological impacts of their activities pertaining to plant and animal life is glossed over only briefly. They note that prolonged exposure to silver iodide has been known to cause Humans to exhibit an ashen appearance, but they claim that this is not of particular concern. They also dismiss concerns about silver iodide's ability to cause a yellowing of the skin when exposure is topical. They even dismiss two examples of individuals having been significantly harmed by exposure to silver iodide. Their paper is full of phrases like 'seem to be' and 'we do not expect' because much of what is presented in the paper is assumptions and extrapolations based on other people's work rather than any scientific findings of their own.

The vast majority of research done in this area does not even concern itself with Human health impacts or biospheric contamination. Rather, it focuses on the ancillary issue of how plants and animals may be affected by either more or less rainfall. The work that is publicly available is mostly cursory. In the vast majority of cases where the subject is even so much as broached, the literature quickly follows with assurances that there are probably no adverse effects and that further study is not necessary.

Thankfully, some research indicating silver iodide's negative biological impacts has surfaced. It is not good news, but we need to hear it. Evidence suggests that it is exceptionally bad for organisms further down the food chain. The aforementioned 1969 Bureau of Reclamation report also noted:

"Silver compounds are much more toxic to fish than to terrestrial vertebrates. Some of the higher concentrations of Ag recorded in precipitation from seeded storms are comparable to the lowest concentrations lethal to fish in the short run. In one set of experiments, sticklebacks were able to withstand no more than 0.003 ppm Ag in water at 15-18° C. The fish survived one week at 0.004 ppm, four days at 0.01 ppm, and but one day at 0.1 ppm."

This 1969 report also found silver to be, "...highly toxic to microorganisms...." The report continues:

"Many investigators have placed Ag at or near the top of the list among heavy metals in toxicity to fungi, slime molds, and bacteria. Water containing 0.015 ppm Ag from contact with specially prepared metal has exhibited bacteriocidal activity. 0.006 ppm Ag has killed E. coli in 2 to 24 hours, depending on numbers of bacteria. Bacteriocidal activity in this context usually implies death of 9.99% or so of the cells present."

Killing fungi, E. coli, and slime molds may sound like a good thing. But in the context of our complex and interdependent biosphere, it is not. Our overall ecosystem needs slime molds and the like. These things are vital links in the food chain.

Why does the conventional weather modification and atmospheric sciences literature not sufficiently address the issue of silver iodide's biological impacts? They wouldn't have anything to hide, would they? That which is not disclosed is often more incriminating than that which is. Although today's Weather Modification Association claims it is completely safe, they have a conflict of interest and they do not have enough data to sufficiently back up their claims.

The bottom line is that there is evidence showing that silver iodide has negative biological impacts. We cannot know for sure that spraying this stuff is safe if no public long term studies have been done. But they have been going ahead and doing it anyway - just like today's geoengineers.

Conclusions

Although it is currently not feasible to completely assess the damage to Earth's biosphere caused by this New Manhattan Project, the available evidence does not paint a pretty picture. This is an area of study and body of work which should be vastly expanded and updated in the coming years and decades. We already know that massive quantities of atmospheric coal fly ash are bad for Humans, animals, insects, plants, and the overall environment. In Humans, the rates of diseases linked to exposure are on the rise. Many people became very sick when first exposed. The historical precedent set by the conventional weather modification industry mandates irresponsibility. When geoengineers say that their activities are harmless, we have plenty of good reasons to not believe them.

References

"An Open Letter to Members of AGU, EGU, and IPCC Alleging Promotion of Fake Science at the Expense of Human and Environmental Health and Comments on AGU Draft Geoengineering Position Statement" a paper by J. Marvin Herndon, published by *New Concepts in Global Tectonics Journal*, September 2017

Kampa, M.; Castanas, E. Human health effects of air pollution *Environmental Pollution* 2008, 151, 362-367.

Calderon-Garciduenas, L.; Franko-Lira, M.; Mora-Tiscareno, A.; Medina-Cortina, H.; Torres-Jardon, R.; et al. Early alzheimer's and parkinson's diese pathology in urban children: Friend verses foe response - it's time to face the evidence. *BioMed Research International* 2013, 32, 650-658.

Moulton, P.V.; Yang, W. Air pollution, oxidative stress, and alzheimer's disease. *Journal of Environmental and Public Health* 2012, 109, 1004-1011.

Beeson, W.L.; Abbey, D.E.; Knutsen, S.F. Long-term concentrations of ambient air pollutants and incident lung cancer in california adults: Results from the ahsmog study. *Environ. Health Perspect.* 1998, 106, 813-822.

Hong, Y.C.; Lee, J.T.; Kim, H.; Kwon, H.J. Air pollution: A new risk factor in ischemic stroke mortality. *Stroke* 2002, 33, 2165-2169.

Haberzetti, P.; Lee, J.; Duggineni, D.; McCracken, J.; Bolanowski, D.; O'Toole, T.E.; Bhatnagar, A.; Conklin, D., J. Exposure to ambient air fine particulate matter prevents vegf-induced mobilization of endothelial progenitor cells from bone matter. *Environ. Health Perspect.* 2012, 120, 848-856.

Potera, C. Toxicity beyond the lung: Connecting pm2.5, inflammation, and diabetes. *Environ. Health Perspect.* 2014, 122, A29

Mehta, A.J.; Zanobetti, A.; Bind, M.-A., C.; Kloog, I.; Koutrakis, P.; Sparrow, D.; Vokonas, P.S.; Schwartz, J.D. Long-term exposure to ambient fine particulate matter and renal function in older men: The va normative aging study. *Environ. Health Perspect.* 2016, 124(9), 1353-1360.

Dai, L.; Zanobetti, A.; Koutrakis, P.; Schwartz, J.D. Associations of fine particulate matter species with mortality in the united states: A multicity time-series analysis. *Environ. Health Perspect.* 2014, 122, 837-842.

Dockery, D.W.; Pope, C.A.I.; Xu, X.P.; Spengler, J.D.; Ware, J.H.; et al. An association between air pollution and mortality in six U. S. Cities. *N. Eng. J. Med.* 1993, 329, 1753-1759.

Pope, C.A.I.; Ezzati, M.; Dockery, D.W. Fine-particulate air pollution and life expectancy in the united states. *N. Eng. J. Med.* 2009, 360, 376-386.

Pires, A.; de Melo, E.N.; Mauad, T.; Saldiva, P.H.N.; Bueno, H.M.d.S. Pre- and postnatal exposure to ambient levels of urban particulate matter (pm2.5) affects mice spermatogenesis. *Inhalation Toxicology: International Forum for Respiratory Research*: DOI: 10.3109/08958378.2011.563508 2011, 23.

Ebisu, K.; Bell, M.L. Airborne pm2.5 chemical components and low birth weight in the northeastern and midatlantic regions of the united states. *Environ. Health Perspect.* 2012, 120, 1746-1752.

Tetreault, L.-F.; Doucet, M.; Gamache, P.; Fournier, M.; Brand, A.; Kosatsky, T.; Smargiassi, A. Childhood exposure to ambient air pollutants and the onset of asthma: An administrative cohort study in quebec. *Environ. Health Perspect.* 2016, 124(8), 1276.

Bell, M.L.; Ebisu, K.; Leaderer, B.P.; Gent, J.F.; Lee, H.J.; Koutrakis, P.; Wang, Y.; Dominici, F.; Peng, R.D. Associations of pm2.5 constituents and sources with hospital admissions: Analysis of four counties in connecticut and massachusetts (USA). *Environ. Health Perspect.* 2014, 122, 138-144.

"The Effect of Reaerosolized Fly Ash from an Atmospheric Fluidized Bed Combustor on Murine Alveolar Macrophages" a paper by Patricia C. Brennan, Frederick R. Kirchner, and William P. Norris, published by Argonne National Laboratory, 1979

"Coal Fly Ash Aerosol: Risk Factor for Lung Cancer" a paper by Dr. Mark Whiteside and J. Marvin Herndon, PhD, published by the

Journal of Advances in Medicine and Medical Research, February 2018

"Aerosolized Coal Fly Ash: Risk Factor for Neurodegenerative Disease" a paper by Dr. Mark Whiteside and J. Marvin Herndon, PhD, published by the *Journal of Advances in Medicine and Medical Research*, March 2018

"Aerosolized Coal Fly Ash: Risk Factor for COPD and Respiratory Disease" a paper by Dr. Mark Whiteside and J. Marvin Herndon, PhD, published by the *Journal of Advances in Medicine and Medical Research*, May 2018

"Geoengineering, Coal Fly Ash and the New Heart-Iron Connection: Universal Exposure to Iron Oxide Nanoparticulates" a paper by Dr. Mark Whiteside and J. Marvin Herndon, PhD, published by the *Journal of Advances in Medicine and Medical Research*, November 2019

"Weather and Climate Modification: Report of the Special Commission on Weather Modification" by the National Science Foundation, 1965
Aluminum oxide material safety data sheet by US Research Nanomaterials, Inc., 2013

"Assessing the Direct Occupational and Public Health Impacts of Solar Radiation Management with Stratospheric Aerosols" a paper by Utibe Effiong and Richard L. Neitzel, published in Environmental Health, 2016

"In Vitro Toxicity of Aluminum Nanoparticles in Rat Alveolar Macrophages" a report by Andrew Wagner, Charles Bleckmann, and E. England of the Air Force Institute of Technology, Krista Hess of Geo-Centers, Inc., Dayton, Ohio, and Saber Hussain and John J. Schlager of the Air Force Research Laboratory, Human Effectiveness Directorate, Applied Biotechnology Branch, Wright-Patterson AFB, published by the Air Force Research Laboratory, Human Effectiveness

Directorate, Applied Biotechnology Branch, Wright-Patterson AFB, 2001

"Nanosized Aluminum Altered Immune Function" a paper by Laura K. Braydich-Stolle, Janice L. Speshock, Alicia Castle, Marcus Smith, Richard C. Murdock, and Saber M. Hussain, published by the American Chemical Society, 2010

"Manufactured Aluminum Oxide Nanoparticles Decrease Expression of Tight Junction Proteins in Brain Vasculature" a paper by Lei Chen, Robert A. Yokel, Bernhard Henning, and Michal Toborek, published by the *Journal of Neuroimmune Pharmacology*, December, 2008

"Aluminum Induced Immunoexcitotoxicity in Neurodevelopmental and Neurodegenerative Disorders" a paper by Dr. Russell L. Blaylock, as published in *Current Inorganic Chemistry*, 2012

"Gila Activation Induced by Peripheral Administration of Aluminum Oxide Nanoparticles in Rat Brains" a paper by X. Li, H. Zheng, Z. Zhang, M. Li, Z. Huang, H.J. Schluesener, Y. Li, and S. Xu, published in *Nanomed*, 2009, 5, (4), 473-479

Strontium oxide material safety data sheet by Sigma- Aldrich, 2007

"Aluminum Poisoning of Humanity and Earth's Biota by Clandestine Geoengineering Activity: Implications for India" a paper by J. Marvin Herndon, PhD, published by *Current Science*, 2015

"Strontium" a report by the Centers for Disease Control and Prevention

"Contamination of the Biosphere with Mercury: Another Potential Consequence of On-going Climate Manipulation Using Aerosolized Coal Fly Ash" a paper by Dr. Mark Whiteside and J. Marvin Herndon, PhD, published by the *Journal of Geography, Environment and Earth Science International*, December 2017

Inhaled Particles and Vapours a book edited by C.N. Davies, published by Pergamon Press, 1961

"Fine Particulate Air Pollution and Mortality in 20 U.S. Cities, 1987-1994" a report by Jonathan M. Samet, MD, Francesca Dominici, PhD, Frank C. Curriero, PhD, Ivan Coursac, MS, and Scott L. Zeger, PhD, published by the *New England Journal of Medicine*, volume 343, number 24, 2000

Pulmonary Deposition and Retention of Inhaled Aerosols a book by Theodore F. Hatch, Paul Gross, the American Industrial Hygiene Association, and the United States Atomic Energy Commission, published by Academic Press, 1964

"Mortality from Alzheimer's Disease in the United States: Data for 2000 and 2010" a report by Betzaida Tejada-Vera, M.S., published by the U.S. Department of Health and Human Services, 2013

"Deaths from Alzheimer's Disease - United States, 1999-2014" an article by Christopher A. Taylor, PhD, Sujay F. Greenlund, Lisa C. McGuire, PhD, Hua Lu, MS, and Janet B. Croft, PhD, published in the *Morbidity and Mortality Weekly Report* of the Centers for Disease Control and Prevention, May 26, 2017

"Age-Adjusted Death Rates for Parkinson's Disease Among Adults Aged ≥65 Years - National Vital Statistics System, United States, 1999-2017" an article by Nancy Han, MS and Barnali Das, PhD, published by the *Morbidity and Mortality Weekly Report* of the Centers for Disease Control and Prevention, Sept. 6, 2019

"Chronic Obstructive Pulmonary Disease Among Adults Aged 18 and Over in the United States, 1998–2009" a report by Lara J. Akinbami, MD; and Xiang Liu, MSc, published by the U.S. Department of Health and Human Services, 2011

"United States Life Tables, 2017" an article by Elizabeth Arias, PhD and Jiaquan Xu, MD, published by *National Vital Statistics Reports*, June 24, 2019

"National Surveillance of Asthma: United States, 2001-2010" a report by the Centers for Disease Control, U.S. Department of Health and Human Services, November, 2012

Chemtrails Confirmed a book by William Thomas, published by Bridger House Publishers, 2004

What In the World Are They Spraying? a documentary film by Michael Murphy, Paul Wittenberger, and Edward G. Griffin, produced by Truth Media Productions, 2010

"Role of Aerosolized Coal Fly Ash in the Global Plankton Imbalance: Case of Florida's Toxic Algae Crisis" a paper by Dr. J. Marvin Herndon, PhD and Dr. Mark Whiteside, MD, published by the *Asian Journal of Biology*, June 2019

"Previously Unacknowledged Potential Factors in Catastrophic Bee and Insect Die-off Arising from Coal Fly Ash Geoengineering" a paper by Dr. J. Marvin Herndon, PhD and Dr. Mark Whiteside, MD, published by the *Asian Journal of Biology*, August 2018

"Aerosolized Coal Fly Ash: A Previously Unrecognized Primary Factor in the Catastrophic Global Demise of Bird Populations and Species" a paper by Dr. J. Marvin Herndon, PhD and Dr. Mark Whiteside, MD, published by the *Asian Journal of Biology*, November 2018

"Unacknowledged Potential Factors in Catastrophic Bat Die-off Arising from Coal Fly Ash Geoengineering" a paper by Dr. J. Marvin Herndon, PhD and Dr. Mark Whiteside, MD, published by the *Asian Journal of Biology*, January 2020

"Previously Unrecognized Primary Factors in the Demise of Endangered Torrey Pines: A Microcosm of Global Forest Die-offs" a paper by J. Marvin Herndon, PhD, Dale D. Williams, and Dr. Mark Whiteside, MD, published by the *Journal of Geography, Environment and Earth Science International*, August 2018

"California Wildfires: Role of Undisclosed Atmospheric Manipulation and Geoengineering" a paper by J. Marvin Herndon and Dr. Mark Whiteside, MD, published by the *Journal of Geography, Environment and Earth Science International*, October 2018

"Deadly Ultraviolet UV-C and UV-B Penetration to Earth's Surface: Human and Environmental Health Implications" a paper by J. Marvin Herndon, PhD, Raymond D. Hoisington and Dr. Mark Whiteside, MD, published by the *Journal of Geography, Environment and Earth Science International*, March 2018

Silver iodide material safety data sheet produced by ScienceLab.com, 2010

Silver iodide material safety data sheet produced by Fisher Scientific, 2009

National Science Foundation Report No. 66-3 as it appeared in a hearing before the Subcommittee on Oceans and Atmosphere of the Committee on Commerce, United States Senate, Ninety-fourth Congress, second session, Feb. 17, 1976

"Ecological Effects of Weather Modification: A Problem Analysis" a report by Charles F. Cooper and William C. Jolly, produced by the U.S. Department of the Interior, Bureau of Reclamation, Office of Atmospheric Water Resources, published by the University of Michigan, 1969

"Some Comments About Weather Modification Affects on Man's Environment" by Archie M. Kahan, Office of Atmospheric Water Resources, Office of Chief Engineer, Bureau of Reclamation, Department of the Interior, published by the Department of the Interior, 1967

"Federal Regulation of Weather Modification" a report by the Council on Environmental Quality, Washington, D.C., 1972 as it appeared in a hearing before the Subcommittee on Oceans and Atmosphere of the

Committee on Commerce, United States Senate, Ninety-fourth Congress, second session, Feb. 17, 1976

Environmental Impacts of Artificial Ice Nucleating Agents a book edited and co-written by Donald A. Klein, published by Dowden, Hutchinson & Ross, 1978

"Weather Modification Association Position Statement on the Environmental Impact of Using Silver Iodide as a Cloud Seeding Agent" a paper by the Weather Modification Association, published by the Weather Modification Association, 2009

"Estimated Possible Effects of AgI Cloud Seeding on Human Health" a paper by Ronald B. Standler and Bernard Vonnegut, published by the *Journal of Applied Meteorology*, Volume 11, August 11, 1972

Chapter 10
OTHER AGENDAS

So far in this book we have only discussed the New Manhattan Project in the context of its most obvious purpose: weather control. In that purpose alone, the New Manhattan Project is easily the biggest scientific endeavor in Human history. Believe it or not, it appears that quite a few other massive projects have been piggybacked upon this global weather modification scheme.

They are already spraying us with tens of thousands of megatons of toxic waste. That by itself is enough to put people away in prison for the rest of their lives, or worse. So why stop there? The people responsible for today's New Manhattan Project (NMP) have apparently gone for broke. What else might one accomplish by spraying stuff all over God's creation and/or shooting electromagnetic energy all over the place?

What else have these mad scientists been up to? As opposed to the more understandable weather modification aspects of the NMP, this chapter examines the NMP's more unbelievable purposes. From mind control to Morgellons to Nikola Tesla's death ray, and more, this chapter examines the New Manhattan Project's other agendas.

Psychotronic weaponry

Other than weather modification, the most probable purpose of the New Manhattan Project is the remote control of our minds and bodies. It is well documented that certain types of electromagnetic energy (EM) waves can remotely manipulate people's moods, thoughts, and bodily functions. The aforementioned ionospheric heaters as well as NEXRAD machines (ch 3), and, for that matter, the whole panoply of EM emitting devices (including cell phones) can produce these types of signals. The relevant fields here are known as 'psychotronics,' 'psychotronic weaponry,' 'bioelectromagnetics,' and 'biophysics.' An overview of the principles, scientists, effects, and technologies applicable to these fields is presented here with an emphasis upon their potential use as part of the New Manhattan Project.

Machines that produce electromagnetic waves designed to have an effect upon our brain functions can accomplish these objectives with a couple of different methods. Firstly, these machines often operate upon the basic principle of brain entrainment. Here is a simple explanation of how brain entrainment works. The targeted individual's brain waves are firstly scanned to determine the frequency of the EM his or her brain is currently producing. Then, that same frequency currently being produced by the targeted individual's brain is also produced by the psychotronic weapon and directed at the targeted individual. This produces synchronization between the targeted individual's brain and the psychotronic weapon's directed EM. Once this synchronization occurs, the frequency produced by the psychotronic weapon is then gradually altered toward a desired frequency. Due to the already established synchronization, as the EM produced by the psychotronic weapon is gradually altered, the targeted individual's brain waves will follow. Other times, the targeted individual(s) will simply be hammered with powerful EM in the desired range, which can force the targeted individual's brain to comply.

Psychotronic weaponry that affects brain functions utilizes electromagnetic energy frequencies primarily in the Extremely Low Frequency (ELF) range (3-30 Hz) because ELF is primarily the frequency range of the electromagnetic energy that the Human brain naturally sends and receives. Electromagnetic fields in the Very Low Frequency (VLF / 3-30 kHz) and Super Low Frequency (SLF / 30-300 Hz) ranges can also have psychotronic brain effects.

For a full explanation of how all of this works, one simply must read a book written by the same man that wrote the definitive book about the world's most powerful and versatile ionospheric heater. That is the aforementioned Dr. Nick Begich and his book is titled *Controlling the Human Mind*. Dr. Begich is able to easily transition from the subject of HAARP to mind control because, as already noted, ionospheric heaters such as the HAARP antenna can produce the necessary frequencies and apply them over large areas and vast distances.

Not only can electromagnetic frequencies influence one's mind, they can also affect the rest of our bodily functions. Research by Dr. Robert Becker, Dr. Thomas Valone, and many others have strongly correlated all of the Human body's functions to electromagnetic fields. This correlation is due to the fact that the Human body is a complex bioelectric organism highly susceptible to electromagnetic energies and fields. The electromagnetic frequencies that can affect our bodily functions span the entire electromagnetic spectrum.

In the historical literature, it is interesting to see that three of the most important New Manhattan Project scientists: Alfred Lee Loomis, Bernard Vonnegut, and Gordon J.F. MacDonald all worked in the area of psychotronic weaponry.

Way back in the 1930s, the aforementioned Alfred Lee Loomis was studying and writing about brain waves as he helped to develop the field of electroencephalography - which is the field of brain wave measurement. This is the same Alfred Lee Loomis mentioned in chapter 2 who made his fortune in the electrical power industry and later went on to help develop technologies capable of bouncing electromagnetic communication signals off the ionosphere - transmitting them over thousands of miles. These facts are also significant because America's electrical power grid produces EM fields within the frequency range of the Human brain, and the electrical power industry has incentives such as the generation of hydroelectric power to motivate their interests in a global weather modification endeavor such as the New Manhattan Project (ch 8). Also, today's New Manhattan Project chemtrail spray has been found to be coal fly ash (ch 6). As we have learned, coal fly ash is a byproduct of the electrical power industry's coal-fired power plants. As if that wasn't enough, today's ionospheric heaters of the New Manhattan Project (with fundamental technologies developed by Loomis), as we have noted multiple times, are capable of producing tremendously strong EM signals and fields that can control the weather AND influence our moods, thoughts, and bodily functions. It truly makes one's head spin.

In the mid-1930s Loomis *et al.* wrote about their work in studying different types of brain waves produced by different types of individuals in different situations. They noted that certain brain wave patterns are consistently produced among individuals depending upon that individual's current activity. They noted six different encephalogram brain wave patterns: a 'saw toothed' pattern common among children, alpha waves, 'spindles' which appear during sleep, a compressed pattern rarely observed, beta waves, and random activity with no measurable frequency.

Loomis' work in the field of electroencephalography involved electrical diodes placed directly on a subject's skin. So, although his work is notable here as a precursor, it was a little bit different than modern psychotronics in that today's psychotronic weaponry is administered remotely, through the air. What we are examining here in this chapter is more akin to the later field of *magneto*electroencephalography, but many of the principles are the same.

Project Bassoon/Shelf/Sanguine/Seafarer/ELF at Clam Lake (ch 4) was probably the beginning of the development of these psychotronic technologies as part of the NMP. These projects and facilities in Wisconsin and Michigan were officially not about mind control; only naval communications. But, because they involved the appropriate frequencies, these projects were directly applicable to the field of psychotronic weaponry and were an early example of what was to come. In fact, in a 1981 paper titled "ELF Electric Field Coupling to Dielectric Spheroidal Models of Biological Objects" researchers who had previously worked for Boeing, the Hughes Aircraft Company, and the Navy's ELF Communications Project wrote of coupling ELF electrical fields with the Human body. They mention that the Navy's Seafarer antenna at Clam Lake would be suitable for this purpose. By 1989, the Navy described the Clam Lake facilities as "fully operational." In 2004, the project was technologically obsolete, so it was subsequently defunded and reportedly shut down.

Some years after the beginnings of project Bassoon/Shelf/Sanguine/Seafarer/ELF, along with two other scientists, that great pioneer of the New Manhattan Project's weather modification aspects, Bernard Vonnegut (ch 3) also laid the groundwork for future mind control experiments. He conducted experiments pertaining to the effects of electromagnetic energy, propagated through an atmospheric aerosol, upon the effigy of a man's head. In their 1961 report titled "Research in Electrical Phenomena Associated with Aerosols," the trio wrote:

"*One can make an estimate of the concentration of the field around an object or man by considering the potential in space in which it exists. For example, in a field of 100 volts cm−1 a man's head is in a region where the potential is about 20,000 V with respect to ground. Because of the relatively high conductivity of the body the man's head is at ground potential and therefore a corresponding amount of charge has passed from the ground up to his head. If we approximate the head as a sphere of 10 cm radius, its capacity is about 10−11 farads so the induced charge on it is about 2 x 10−7 coulombs and the field at its surface is approximately 2000 volts cm−1. Accordingly, we see that the field and hence the rate of aerosol deposition should be about 20 fold greater on the man's head than on the ground.*"

Seven years later, as alluded to in chapter 7, another great scientific champion of the New Manhattan Project, Dr. Gordon J.F. 'How to Wreck the Environment' MacDonald wrote:

"*The enhanced low-frequency electrical oscillations in the earth-ionosphere cavity relate to possible weapons systems through a little-understood aspect of brain physiology. Electrical activity in the brain is concentrated at certain frequencies, some of it extremely slow, a little around five cycles per second, and very conspicuous activity (the so-called alpha rhythm) around ten cycles per second. Some experiments have been done in the use of a flickering light to pull the brain's alpha rhythm into unnatural synchrony with it; the visual stimulation leads to electrical stimulation. There has also been work on direct electrical driving of the brain. In experiments discussed by Norbert Wiener, a sheet of tin is suspended from the ceiling and connected to a generator working at ten cycles per second. With large field strengths of one or two volts per centimeter oscillating at the alpha-rhythm frequency, decidedly unpleasant sensations are noted by human subjects.*

"*The Brain Research Institute of the University of California is investigating the effect of weak oscillating fields on human behavior. The field strengths in these experiments are of the order of a few hundredths of a volt per centimeter. Subjects show small but measurable degradation in performance when exposed to oscillating fields for periods of up to fifteen minutes.*

"*The field strengths in these experiments are still much stronger, by a factor of about 1000, than the observed natural oscillations in the earth-ionosphere cavity. However, as previously noted, the intensity of the natural fluctuations could be increased substantially and in principle could be maintained for a long time, as tropical thunderstorms are always available for manipulation. The proper geographical location of the source of lightning, coupled with accurately timed, artificially excited strokes, could lead to a pattern of oscillations that produced relatively high power levels over certain regions of the earth and substantially lower levels over other regions. In this way, one could develop a system that would seriously impair brain performance in very large populations in selected regions over an extended period.*

"The scheme I have suggested is admittedly far-fetched, but I have used it to indicate the rather subtle connections between variations in man's environmental conditions and his behaviour. Perturbation of the environment can produce changes in behavior patterns. Since our understanding of both behavioral and environmental manipulation is rudimentary, schemes of behavioral alteration on the surface seem unrealistic. No matter how deeply disturbing the thought of using the environment to manipulate behavior for national advantage is to some, the technology permitting such use will very probably develop within the next few decades."

The mind control technologies detailed here by MacDonald have been in development by the Stanford VLF Group.

In 1974 a book edited by Dr. Michael A. Persinger (1945-2018) titled *ELF and VLF Electromagnetic Field Effects* was published. This book references and details many studies conducted with animals and Humans pertaining to physiological and biochemical changes associated with EM fields in the ELF and VLF ranges. In the preface, Dr. Persinger writes that ELF signals are associated with, "weather changes, solar disturbances and geophysical-ionospheric perturbations." That is very convenient considering that this New Manhattan Project is primarily a global weather modification project. In fact, ELF and VLF EM is fundamental to natural life processes here on Earth. Dr. Persinger also writes that ELF signals are preferable because they, "…have the capacity to penetrate structures which house living organisms." He's writing about weaponized EM like MacDonald did about 10 years earlier. Persinger is writing here about psychotronic weaponry mind controlling people in their houses from outer space or other vast distances.

In 1980, Persinger's book *The Weather Matrix and Human Behavior* was published. *The Weather Matrix and Human Behavior* is a work of what is called 'human biometeorology.' Human biometeorology refers to the biological effects of weather upon Humans. Persinger's book features discussions of the biological and behavioral effects of naturally occurring ELF fields as well as the effects of ionized atmospheric particulate matter upon Humans. Describing Humans as 'a large antenna' and noting that ELF fields have 'a 99 percent capacity to penetrate most surface dwellings,' Persinger writes about the ability to effectively apply ELF fields from 'hundreds of kilometers' away. He then proceeds to write about the effect of ELF fields upon specific organs and Human bodily systems such as: the hypothalamus region of the brain, the blood, the bones, the central nervous system, the cardiovascular system, and cell membranes. Persinger describes the Human body as a 'biological homeostat' that has predictable responses to different inputs.

Persinger's tome also addresses the Human health effects of ionized particulate matter. Late in the book, under the heading of 'SMALL AIR IONS: PARTICULATE ELECTRIC FIELDS,' Persinger references the work of former Nazi scientist Hans Dolezalek as he notes the vulnerability of the Human lungs to ionized particulate matter. Persinger writes, "Inhalation of air ions is like the sudden injection of a chemical into the body volume. Considering the area of the lung lining, the more than 10,000 liters of air breathed daily, and the passage of all the body's blood through the lungs every 30 seconds, absorption of any bioeffective chemical through the respiratory system should have significant consequences."

You probably thought it ended there, right? Apparently Dr. Persinger never slept. In a 2011 video posted on *YouTube*, he talked about individuals having access to the information stored in every other Human brain on the planet, which, as he describes it, would create an environment of 'no more secrets.' No More Secrets was the title of his lecture. He talks about 7 billion Human brains all sharing the Earth's conductive field which has, "enough energy to store the experiences of every Human being who has ever lived." He says that individuals can purposefully access information from other Human brains at a distance. Not to be confused with the remote sensing discussed in chapter 7, this is known as 'remote viewing.' He says that the Central Intelligence Agency (CIA) has successfully worked on remote viewing.

Dr. Michael A. Persinger

Dr. Michael Persinger was a very public figure for many decades. He just went on and on about all this stuff. He taught it at Laurentian University in Ontario, Canada. He is most well known for inventing what is called the 'God Helmet.' The God Helmet was a device placed on a subject's head which could apparently produce in the subject the feeling of having a religious experience. He has a big *Wikipedia* listing. He even had a *YouTube* Channel.

Most relevant to our discussion here is the CIA's Operation PIQUE. As noted in *Controlling the Human Mind*, by 1978 the CIA had developed a psychotronic weapon that was capable of bouncing, "high powered radio signals off the ionosphere to affect the mental functions of people in selected areas, including Eastern European nuclear installations."

Some mainstream media articles have recently confirmed the reality of psychotronic weaponry. In October of 2019, *The Guardian* published an article about how individuals' minds can be read electronically. Also, in late 2018, in many articles, the use of psychotronic weapons was suspected in a series of attacks upon the United States Embassy in Cuba.

Voice to Skull

Making people hear things that no one else around them can hear through the use of atmospheric radio waves is a branch of psychotronic weaponry known as 'voice to skull' technology. This technology involves beaming sounds and speech directly into a subject's head. This technology is well documented and well known. It has been demonstrated in many movies and television shows such as the movie *Kingsman: The Secret Service* and the TV show *The X-Files*. The New Manhattan Project's ionospheric heaters can do this. An early book titled *Microwave Auditory Effects and Applications* by James C. Lin, PhD revealed the method. 1989's U.S. patent #4,877,027 "Hearing System" defines this technology. Many scientific papers and other patents on the subject are available. Voice to skull technology has been weaponized into the Navy's MEDUSA program. Subjects assaulted with this weapon are temporarily incapacitated by extremely uncomfortable sounds others nearby cannot hear. For the MEDUSA program there is a *Wikipedia* listing. Most recently, in a January 2019 press release, Massachusetts Institute of Technology researchers revealed that they had the technological means to send audible messages directly into people's skulls with the use of lasers.

Light from the gods

My science advisor Ginny Silcox mentions current research into how multiple ionospheric heaters such as the HAARP antenna could be used in combination to produce lightning-like energy beams capable of completely destroying individuals, buildings or large areas. This type of technology was first hypothesized by Nikola Tesla in 1934 and has been referred to as his 'death ray.' Tesla's 'death ray' was his vision of a powerful laser or particle beam used as a weapon of war. Tesla produced an unpublished paper titled "The New Art of Projecting Concentrated Non-Dispersive Energy Through Natural Media" which was the first technical description of a charged-particle beam weapon. The aforementioned Jason group was working on something similar in 1963. Their 'charged particle beam' program was called Seesaw.

There are a few ways Tesla's death ray might be made to appear like a lightning bolt. This may be accomplished by tapping into the ionosphere, generating it independently, or a combination of both. The ionosphere (about 50-375 miles above Earth's surface) is home to Earth's aforementioned auroral electrojet (ch 3). The auroral electrojet is a river of highly ionized, electrified particles making up a plasma encircling our planet. If lightning taps into that power by creating a circuit from the auroral electrojet down to the Earth, it will be incredibly powerful. This may happen naturally. A 1982 article co-written by the great New Manhattan Project progenitor Bernard Vonnegut recounts the experiences of pilots who have witnessed lightning extending upwards, above storm clouds, presumably to the ionosphere. You see, the earth is negatively charged and the ionosphere is positively charged. Opposites attract. This is why lightning between the two would be so powerful. Man might also make this happen. It is also conceivable that the New Manhattan Project's ground-based ionospheric heaters (possibly in conjunction with other, similar space-based antennas) could induce lightning from the ionosphere. These antenna arrays could create a conductive path known as a 'waveguide' which would cause the ionosphere's energy to flow down to Earth. These waveguides can also direct the energy more accurately. The metals found in today's chemtrails may allow for better propagation of these energy waves. The Stanford VLF Group has been active in this area as well.

Tesla's death ray might also be entirely man-made. As part of the proposed overhaul of Air Force (USAF) operations outlined in the aforementioned 1996 *Air Force 2025* documents, the USAF produced a paper titled "Space Operations: Through the Looking Glass (Global Area Strike System)." If you will recall, *Air Force 2025* is the set of documents containing the infamous "Weather as a Force Multiplier: Owning the Weather in 2025" paper. The authors describe Nikola Tesla's death ray as a probable future Air Force capability. The report reads:

"*Once again a small but capably armed country is threatening to seize its smaller but resource-rich neighbor. The Global News Network reports that the border has been violated. The same old story? No, the plot twists as a sophisticated satellite surveillance and reconnaissance system tracks the belligerent nation's leader. As he steps to the podium to incite his troops to greater violence, a blinding light from above vaporizes him and his podium leaving even his bodyguards untouched. His smarter brother, the second in command, countermands the invasion orders and in 12 hours the borders are restored. Stability, if not peace, reigns again.*

"This is not science fiction, but a mission well within the capabilities of Space Operations in 2025."

A little later in the piece, the authors expound upon the usefulness of such a weapon. The authors of "Space Operations: Through the Looking Glass" explain the incredible destructive power of such activities:

"At slightly higher powers, the enhanced heating produced by the laser can be used to upset sensitive electronics (temporarily or permanently), damage sensor and antenna arrays, ignite some containerized flammable and explosive materials, and sever exposed power and communications lines. The full power beam can melt or vaporize virtually any target, given enough exposure time. With precise targeting information (accuracy of inches) and beam pointing and tracking stability of 10 to 100 nanoradians, a full-power beam can successfully attack ground or airborne targets by melting or cracking cockpit canopies, burning through control cables, exploding fuel tanks, melting or burning sensor assemblies and antenna arrays, exploding or melting munitions pods, destroying ground communications and power grids, and melting or burning a large variety of strategic targets (e.g., dams, industrial and defense facilities, and munitions factories) —all in a fraction of a second."

The authors of "Space Operations: Through the Looking Glass" specifically mention the Iridium satellite constellation as one that would be suitable for producing high-power microwaves. The Iridium satellite constellation was noted in chapter 7 as particularly suitable for the New Manhattan Project's remote sensing needs.

Electromagnetic energy, like that which is employed in the New Manhattan Project, has also been weaponized into systems like Raytheon's 'heat ray' also known as their Active Denial System. This technology produces in subjects a burning sensation. If left exposed to the directed energy, subjects may have their skin burned at distances of 700 yards. There is a *Wikipedia* listing.

Judy Wood, PhD in her book *Where Did the Towers Go?*, documents 1400 burned cars up to one half mile away from ground zero. Nothing else near them was damaged. Trees still had leaves on them. There were reams of unburned sheets of paper. Though they were covered with ash, no humans were burned. The exceptional thing about the cars, she notes, was the fact that they were not grounded while everything else was. Was some sort of electromagnetic weapon used on 9/11? If so, how did cave dwellers from Afghanistan get one of those?

Maybe the original purpose of Raytheon is to come up with just such a weapon. Raytheon means 'light from the gods.' Raytheon has been all about these types of technologies all along. Their 2009 patent titled "Multifunctional Radio Frequency Directed Energy System" describes this type of weapon. As noted earlier, Raytheon acquired E-Systems; the company that built HAARP.

Genetic Modification

Somebody has apparently been planning ahead. If, from the megatons of aluminum-heavy toxins released into our biosphere, our environment becomes wrecked beyond being able to sustain life as we know it, don't worry! People have been genetically modifying plants to be able to thrive in a ravaged biosphere. A 2008 paper by geoengineer Alan Robock and two co-authors, published in the *Journal of Geophysical Research*, notes that, "...genetic modification of ecosystem productivity" has been suggested. They don't note who suggested it, but the rest of this section validates their declaration.

In U.S. patent #7,582,809 "Sorghum Aluminum Tolerance Gene, SBMATE," we learn that a genetically modified sorghum seed has been developed to thrive in aluminum contaminated soils. The abstract of the patent reads, "SbMATE can work across species to enhance tolerance to Al in other important crops grown in localities worldwide where Al3+ cations are present in acid soils and are toxic to plants." The makers of this sorghum seed assure us that this type of genetic manipulation may also work well with many other food crops such as: wheat, barley, rice, maize, cotton, peanut, sunflower, tobacco, rye, alfalfa, tomato, cucumber, soya, sweet potato, grapes, rapeseed, sugar beet, tea, strawberry, rose, chrysanthemum, poplar, eggplant, sweet pepper, walnut, pistachio, mango, banana, and potato. It appears that sorghum is just the proverbial camel's nose under the tent. If you will recall, Aluminum is a major constituent of coal fly ash (ch 6).

It is curious that this sorghum patent is assigned to (along with Brazilian officialdom) the United States of America as represented by the Secretary of Agriculture. One might think that this would be right up Monsanto's alley. There may be some sort of deception going on here. We do know that historically the Department of Agriculture has acted like it works for Monsanto. As described in Marie-Monique Robin's 2010 masterpiece *The World According to Monsanto*, the collusion between our Department of Agriculture and Monsanto has been pervasive. Maybe some sort of back-room deal has been struck between Monsanto and corrupt elements of our federal government enabling Monsanto to profit from the rollout of these aluminum resistant seeds - a rollout designed to appear as a benevolent government action.

If you noticed, one of the plants mentioned in patent #7,582,809 is poplar - a non-food producing tree. If food crops don't grow in a ravaged environment, one can expect that trees won't either. This is why we can thank our lucky stars that there is a corporation called ArborGen producing and selling genetically modified trees. They aren't saying that their trees are specifically resistant to aluminum contaminated soils yet, but it's good to know that they might help us in our future hour of need.

Regions continually desiccated by chemtrail spray or deluged by geoengineered storms might not be able to sustain trees. Of course, the geoengineers are here to help too. Our friends have developed wholly mechanical trees that can do what natural trees have traditionally done: remove CO2 from the atmosphere. It's a branch (pun intended) of geoengineering. For confirmation, search 'carbon capture trees.' Who needs Mother Nature? We can have giant multinational corporations that will take care of everything!

Other sprays

We are all but guinea pigs in their Human experimentations. While today's super-secret New Manhattan Project saturates our atmosphere with coal fly ash, conventional weather modifiers have also admittedly been spraying us with myriad substances. As noted in chapter 1, the books *Clouds of Secrecy* by Leonard A. Cole and *In the Name of Science* by Andrew Goliszek document hundreds of open air tests of all types of chemicals and biologicals on unsuspecting American citizens. For many decades now, we have been getting hit with everything under the sun. Documents show that many other substances have been released in open air experiments right here in America including: sarin nerve agent, sulfur dioxide, methylacetoacetate, lead iodide, urea, lewisite, mustard gas, carbon black, calcium chloride, lithium chloride, silicon, sulfur chloride, phloroglucinol, propane, metaldehyde, dihydroxynaphthalene, sodium chloride, polyelectrolytes, ammonium nitrate, lead diiodide, sodium iodide, potassium iodide, hydroxyquinol, pyrogallol, isopropylamine, acetone, aluminum oxide, and barium. Most of these substances have been used in weather modification experiments. Reports of all types of strange substances apparently being sprayed from aircraft are abundant and readily available. William Thomas' groundbreaking 2004 book *Chemtrails Confirmed* provides early examples. The book reads:

"*Human blood cells were found by a hospital lab technician in samples of gel-like material dropped over the tiny town of Oakville, Washington (pop. 665), covering a 20 square-mile area three times within a six-week period.*"

It continues:

"*Officer David Lacey was on patrol with a civilian friend at 3 am when the sticky downpour began. 'We turned our windshield wipers on, and it just started smearing to the point where we could almost not see,' Lacy said. 'We both looked at each other and we said, 'Gee this isn't right. We're out in the middle of nowhere, basically, and where did this come from?'*

"*Pulling into a gas station, officer Lacey pulled on a pair of latex gloves to clean his windshield. Lacey: 'The substance was very mushy, almost like if you had Jell-O in your hand.' Within hours, Lacey was in hospital unable to breathe.*"

Chemtrails Confirmed goes on to recount multiple other instances of this human blood cell goo (presumably dropped from airplanes) splattering all over people, property, and the Earth.

Thomas' book also references a 2001 *World Net Daily* article titled "Californians Confused Over Chemtrails." In this article, as reproduced in *Chemtrails Confirmed*, reporter Lance Lindsay recounts the experience of a former military radar technician named Gene Shimer. It reads:

"*One morning, about two years ago, Shimer spotted a long, thin string resembling a spider web floating in the sky, he said. It was about 20 feet in length and could only be seen in the reflection of the sun. He looked up, saw trails, and then saw something that appeared to be a glob of foam falling from the sky. 'It came floating down, free-floating,' Shimer says. 'I caught it with a spatula, scooped it off the ground, and I watched it as it shrank. It was about the size of my fist when it first started. It looked like a cross between soap bubbles and cotton candy.'"*

What Mr. Shimer described is not unlike something that was captured on video. It is video of a mass of foam about the size of a large man falling from the sky. The source of this foam is unknown. The title of the *YouTube* video reads: "Weird Cloud Falls to Earth and Begins to Make Shapes!! MUST SEE TO BELIEVE!" What was captured on video looks like what might be causing one of the most widely documented 'other' sprays: the so-called 'chem-webs.' As noted in William Thomas' book and in myriad online reports (the author has even been told this by a person with first-hand experience), a spider web-like substance is sometimes found covering grass, bushes, and trees. It looks like the area has been covered in a foam that has since deflated. The substance is usually described as completely foreign to the particular environments - i.e. it is not spider webs. Despite what the shills claim, this stuff was probably sprayed from an airplane. There is a United States patent describing a method of seeding clouds that produces foam which floats down to the ground. U.S. patent number 6,315,213 "Method of Modifying Weather" speaks to the use of a 'cross-linked aqueous polymer' which, once dispersed into a cloud, is agitated by the wind to produce a 'gelatinous substance' which then falls to the ground. The patent describes this method as a way to inhibit rainfall.

If all this isn't weird enough, our United States Air Force has written about spraying what is commonly referred to as 'smart dust.' The seminal 1996 Air Force document "Weather as a Force Multiplier: Owning the Weather 2025" mentions using smart materials for the purpose of weather modification. On page 17 it reads:

"*With regard to seeding techniques, improvements in the materials and delivery methods are not only plausible but likely. Smart materials based on nanotechnology are currently being developed with gigaops computer capability at their core. They could adjust their size to optimal dimensions for a given fog seeding situation and even make adjustments throughout the process. They might also enhance their dispersal qualities by adjusting their buoyancy, by communicating with each other, and by steering themselves within the fog. They will be able to provide immediate and continuous effectiveness feedback by integrating with a larger sensor network and can also change their temperature and polarity to improve their seeding effects. As mentioned above, UAVs [unmanned aerial vehicles] could be used to deliver and distribute these smart materials.*"

A document titled "Operational Defenses Through Weather Control in 2030" by the Air Command and Staff College Air University also speaks of nanobots capable of different types of weather modification. Anything that contains semi-conductor material is laced with dangerously toxic elements like cadmium and germanium. If these particles are small enough to float in the air, they will be inhaled by people and animals. This type of technology is also known as 'microelectromechanical systems' or MEMS and in 2002 a PhD scientist claimed a breakthrough. In his report titled "Global Environmental MEMS Sensors (GEMS): A Revolutionary Observing System for the 21st Century," John Manobianco, PhD writes:

"*Technological advancements in MicroElectroMechanical Systems (MEMS) have inspired a revolutionary observing system known as Global Environmental MEMS Sensors (GEMS). The GEMS concept features an integrated system of MEMS-based airborne probes that will be designed to remain suspended in the atmosphere for hours to days taking in situ measurements over all regions of the Earth with unprecedented spatial and temporal resolution. The GEMS concept is revolutionary because it foresees the future integration of evolving technologies to realize an observing system with scalability and applicability over a broad range of weather and climate phenomena. GEMS have the potential to expand our understanding of the Earth and improve weather forecast accuracy and efficiency well beyond current capability. Resulting improvements in forecast accuracy will translate directly into cost benefits for weather-sensitive industries worldwide, and mitigate the risk factors associated with life-threatening weather phenomena.*"

A year prior to Manobianco's paper, a UCLA scientist by the name of Jack Judy was claiming that MEMS can be brought down in size to the micrometer scale.

A 2016 *PRNewswire* article describes Manobianco's latest work. According to the article, Manobianco and his company Mano NanoTechnologies have received funding from the National Oceanic and Atmospheric Administration (NOAA) to develop MEMS on the micrometer *and* nanometer scales for the purpose of atmospheric observations. Their program is known as GlobalSense and the MEMS are being designed to be deployed from balloons or aircraft. One of the many uses outlined is the monitoring of man-made climate change.

Micro-electromechanical systems are developed at the aforementioned MITRE Corporation.

Morgellons

Reports of fibers growing out of people's skin along with sores, subdermal crawling sensations and a wide variety of other very strange symptoms have been pouring in. According to the Mayo Clinic, these symptoms generally occur along with severe fatigue, difficulty concentrating and short-term memory loss. The Mayo Clinic also notes that this, "is a relatively rare condition that most frequently affects middle-aged white women." Although the Centers for Disease Control (CDC) officially categorizes it as an 'Unexplained Dermopathy,' this disease is being called Morgellons.

Many people are claiming that Morgellons disease is caused by chemtrails. Whether or not Morgellons is caused by the usual chemtrails consisting of coal fly ash, is caused by other sprays, or whether or not chemtrails is a cause of it at all is unclear. But people have been drawing connections between heavy chemtrail spraying and Morgellons. There is an overwhelming amount of information available online.

A woman by the name of Dr. Ginger Savely was the first medical professional to investigate Morgellons disease extensively and her efforts did a lot to gain the medical establishment's acknowledgement of this newly recognized disease. Over the course of a decade, Dr. Savely has treated hundreds of Morgellons sufferers and had many journal articles on the subject published. She has since written a 2016 self-published book titled *Morgellons: The Legitimization of a Disease*. In her book, Dr. Savely writes that Morgellons is not a new disease. One of her patients has been exhibiting symptoms since 1954. This evidence does not tend to support the notion that chemtrails are causing Morgellons because chemtrail spraying was not prevalent in the 1950s, but it does not eliminate the possibility, either. In fact, in her book, Dr. Savely directly addresses the potential chemtrails/ Morgellons connection. Savely writes:

"According to the chemtrail theory, long-lasting trails behind high-flying aircraft contain aerosolized particulate used for geo-engineering, weather-modification and military purposes. It is proposed that this spray causes the general population to be over-exposed to aluminum oxide and other toxic metals and chemicals. Heavy metal toxicity is known to burden the immune system making those exposed more vulnerable to all kinds of disease. Furthermore, studies have shown specifically that Lyme patients with a heavy aluminum load have more difficult-to-treat infections.

"So, if in fact there were a relationship between MD [Morgellons disease] *and chemtrails, it would simply be that those with immune systems burdened by metal toxicity are more susceptible to it or to any other disease. Furthermore, we run into the same conundrum that we face with the GMO theory: the chemtrail discussion started appearing on the Internet in the 1990s. Morgellons disease has been around much longer than that. In summary, chemtrails and GMO's do* ***not cause*** *Morgellons disease. However, they* ***may*** *make certain susceptible individuals more vulnerable to any number of immune challenges, one of which could be Morgellons disease."*

Dr. Ginger Savely

In apparent, partial conflict with Dr. Savely is a scientist named Clifford Carnicom. He has been investigating Morgellons in connection with chemtrails for many years now. He suggests that a substance at least similar if not identical to the aforementioned goo consisting of Human blood cells may be causing Morgellons. He suggests that Morgellons is not a skin condition, but rather a systemic condition not limited to our bodies. He suggests that every living thing on this planet is currently being affected by whatever is causing Morgellons - which would speak to the substance being distributed in the fashion of chemtrails. The possibilities suggested by Carnicam are absolutely terrifying, mind-boggling, and largely beyond the scope of this investigation. Regrettably, this area of research promises to be greatly expanded in years to come. This may have been what George Bush, Sr. was talking about when he said, "If the American people ever find out what we have done, they would chase us down the street and lynch us."

Clifford Carnicom

My friend and fellow geoengineering researcher and writer Elana Freeland is finishing her third book on geoengineering. In another apparent, partial rebuke of Dr. Savely's assertions, Freeland's upcoming book ties geoengineering to the synthetic biology and nanotechnology (including Morgellons) that she says will wirelessly plug us into artificial intelligence systems that will make us brain-computer interface controlled cyborgs. The book will be out later this year (2020) and is not to be missed.

As we can see, the research pertaining to Morgellons and chemtrails is in a state of flux. In the months and years to come we should get more clarity here.

Extreme weather

As noted in the first chapter, U.S. Secretary of Defense William S. Cohen said:

"Others are engaging even in an eco-type of terrorism whereby they can alter the climate, set off earthquakes, volcanoes remotely through the use of electromagnetic waves... So there are plenty of ingenious minds out there that are at work finding ways in which they can wreak terror upon other nations. It's real, and that's the reason why we have to intensify our efforts, and that's why this is so important."

Knowing what we know now about the New Manhattan Project, it is reasonable to assert that the NMP is capable of producing the types of extreme weather noted by Sec Def Cohen. Providing further support for this assertion are the types of extreme weather written about by Gordon J.F. 'How to Wreck the Environment' MacDonald in chapter 7.

MacDonald wrote further about extreme weather in a 1975 piece titled "Weather Modification as a Weapon." Similar to what he wrote in "How to Wreck the Environment," MacDonald writes, "It may be possible in the future to trigger earthquakes with devastating results from a great distance, or to bring about major climatic changes by triggering the instabilities inherent in the Antarctic icecap."

People familiar with the types of technologies employed in today's New Manhattan Project, such as the author's science advisor, tell the author that powerful electromagnetic frequencies are absolutely capable of producing the types of extreme weather noted by SecDef Cohen and Dr. MacDonald.

A 2011 paper posted on Cornell University's *arXiv.org* website provided information showing how, in the days before the big Japan earthquake of that same year, the ionosphere above the area where the earthquake struck was manipulated. This ionospheric manipulation was probably the cause of the gigantic earthquake and tsunami that, in turn, caused the worst nuclear disaster in history.

Holography

New Manhattan Project operations create a situation conducive to the production of holograms. The substances sprayed into our atmosphere (particularly barium) can act as a three dimensional reflective screen onto which beams of electromagnetic energy can be projected. This is how holograms are often made. As weapons of deception, the aforementioned 1996 document "Space Operations: Through the Looking Glass (Global Area Strike System)" has a small section devoted to holography. The authors write:

"*It is certainly possible to make holograms of troop concentrations, military platforms, or other useful objects, although the larger the scene the more difficult it is to produce the proper conditions to create a convincing hologram. No credible approach has been suggested for projecting holograms over long distances under real-world conditions, although the Massachusetts Institute of Technology's Media Lab believes holographic color projection may be possible within 10 years. Holographic and other, less high-technology forms of illusion may became [sic] a potent tool in the hands of the information warriors...*"

Not long before his death, Canadian journalist Serge Monast made public something he had uncovered called Project Blue Beam. He said it was an establishment plan to destroy all national sovereignty and religions - replacing them with a dictatorial one world government and a satanic one world religion. This was largely to be accomplished, he claimed, by the use of incredibly realistic holographic imagery, projected in the sky and on the Earth, depicting an alien invasion. This fake alien invasion was to be used as a catalyst for all the people of the world to give up their countries and religions in favor of only one religion and only one government in order to unite against the perceived alien threat.

The chemtrails sprayed as part of today's New Manhattan Project have the potential to serve as the screen upon which the Project Blue Beam movie is projected. The aforementioned voice to skull technology could be broadcast over the audience as well to create a show with both images and sound. Ionospheric heater scientists have long been exploring the use of magnetosonic and ion-acoustic waves - both of which could provide the necessary elements of a full-blown 3D movie in the sky in tandem with the holography.

Depopulation

We know that mass murder is taking place. Just in America alone, the evidence from the previous chapter shows that at least hundreds of thousands of people have been murdered by the ongoing geoengineering activities. This is probably a very conservative estimate. This estimate is simply what the best available information shows. This is simply what can be proven at this time. We also know that two of the world's most prominent geoengineering advocates are population reduction advocates as well. Further, there are abundant calls for depopulation from many high-ranking establishment types. Considering all this in total, it is reasonable to assert that the many deaths caused by geoengineering are not simply an unfortunate side-effect, but rather part of an organized global depopulation agenda.

John P. Holdren was a Presidential science advisor to Barak Obama. He is also a geoengineering advocate. He regurgitates the same talking points that the geoengineers do. Without admitting that Solar Radiation Management geoengineering (chemtrails) is currently happening, he says that it might be a good thing to do because global warming is such a big problem. He, of course, couches this assertion in all sorts of disclaimers just like the geoengineers do. But the fact remains that, under certain circumstances, he recommends it. There are many reports of Mr. Holdren's position available online. John Holdren is also a population reduction advocate. In fact, he is arguably the most prominent population reduction advocate today. Back in 1977, Holdren co-authored a book called *Ecoscience*. In this book, he and his co-authors advocate for population reduction - a.k.a. mass murder. Holdren and his cohorts call for forced abortions, mass sterilizations, and a United Nations 'Planetary Regime' with the power of life and death over American citizens.

John P. Holdren

William Gates III, the co-founder of Microsoft, is another geoengineering advocate of the highest profile also advocating for population reduction. Bill Gates funds geoengineering. On Stanford University's website, it says that Bill Gates funds solar radiation management geoengineering research through something called the Fund for Innovative Climate and Energy Research (FICER). His geoengineering advocacy is further detailed in a 2012 *Guardian* article titled "Bill Gates backs climate scientists lobbying for large-scale geoengineering." Like John Holdren, Bill Gates also advocates for population reduction. He has stated this position many times in many venues such as on *CNN* and during a TED talk. He says vaccines and other so-called 'health care services' such as those embodied in Obamacare can help in this area. For confirmation, go to *YouTube* and search 'Bill Gates population reduction.' His father was the head of Planned Parenthood - an organization that has been reducing the population quite effectively for many years now. Now that Mr. Gates has conquered the world of business, he has apparently become a 'philanthropist.' He wants to help all right. He wants to help us into our graves.

Bill Gates

Bill Gates isn't the only top-level crony capitalist calling for population reduction. He's got plenty of company. Prince Philip, Duke of Edinburgh, leader of the World Wildlife Fund said, "In the event that I am reincarnated, I would like to return as a deadly virus, in order to contribute something to solve overpopulation … We need to 'cull' the surplus population." Ted Turner, the founder of CNN said, "A total population of 250-300 million people, a 95% decline from present levels, would be ideal." As reported in a 2013 *Telegraph* article Sir David Attenborough has said that Human beings are 'a plague on the Earth.' As documented in a 2017 *Telegraph* article, Prince William has warned us that there are too many people in the world.

Providing further evidence here, a 1974 National Security Council report titled "Implications of Worldwide Population Growth For U.S. Security and Overseas Interests," also known as the Kissinger Report, makes a case for the depopulation of foreign countries. The report and its suggested means for accomplishing these ends were subsequently endorsed and slated for implementation by President Gerald Ford.

Depopulation is a theme common among the socio-economic elite as well as their brainwashed sycophants. What most of these people fail to understand is that the problem is not overpopulation. The problem is the establishment's extremely poor management. It is a known fact that better education results in fewer births and smaller family sizes while the establishment promotes dumbed-down garbage like Common Core. Clean, free energy technologies have been brutally suppressed for over 100 years in favor of their pollution-spewing multinational oil and gas conglomerates while they tell us that we have to give up our quality of life because of pollution and the supposedly resultant global warming. Many other examples of the establishment's extreme mismanagement abound. Another book could be written.

In fact, something called the Demographic Transition Model (DTM) disproves the establishment's overpopulation thesis entirely. The DTM proves that populations actually, in the long-term, *decline* with economic development, not increase. Boom. All the fear-mongering about man-made climate change is directed towards shutting down industrial activity and economic development because they say it creates a situation where the world becomes overpopulated and Earth's resources become quickly depleted. The DTM and other supporting data prove that this predicted sequence of events is 180 degrees wrong. They have absolutely no reason to spray us *whatsoever* other than the consolidation of their own power and wealth.

Relevant to our discussion here, the Georgia Guidestones is a massive monument located in Elbert County, Georgia. Consisting of 5 granite slabs (plus a capstone) over 19 feet high and weighing over 21 tons each, the same message to Humanity is carved in 8 different languages. Along with some other rubbish, the stones read, "MAINTAIN HUMANITY UNDER 500,000,000 IN PERPETUAL BALANCE WITH NATURE." The current population of the Earth is over 7,000,000,000. Maybe if the makers of the Georgia Guidestones do a really good job with vaccines and other 'health care services' like Bill Gates suggests, they can kill 6.5 billion people.

The Georgia Guidestones

Geoengineering is not the only way the establishment is going about mass murdering us, but it sure is an effective one. In fact, short of thermonuclear war, the author cannot think of a more effective way to mass murder a population other than spraying megatons of toxic garbage all over them and their land. This is what geoengineers are doing, and probably a lot more.

References

The Body Electric: Electromagnetism and the Foundation of Life a book by Robert Becker and Gary Selden, published by William Morrow, 1998

Bioelectromagnetic Healing: A Rationale for its Use a book by Thomas Valone, published by Integrity Research Institution, 2000

"Bioeffects of Selected Non-lethal Weapons" an addendum to "Nonlethal Technologies - Worldwide" (NGIC-1147-101- 98) by the Department of Defense

Mega Brain: New Tools and Techniques for Brain Growth and Mind Expansion a book by Michael Hutchison, self-published, 2013

Angels Don't Play this HAARP: Advances in Tesla Technology a book by Dr. Nick Begich and Jeane Manning, published by Earthpulse Press, 1995

ELF: Webster's Timeline History 1590-2007 a book edited by Philip M. Parker, PhD, published by ICON Group International, 2009

VLF: Webster's Timeline History 1923-2006 a book edited by Philip M. Parker, PhD, published by ICON Group International, 2009

"Biological Effects of Extremely Low Frequency Electric and Magnetic Fields: A Review" a report by A.A. Marino and R.O. Becker, published by *Physiological Chemistry and Physics*, Volume 9, Number 2, 1977

"Biological and Human Health Effects of Extremely Low Frequency Electromagnetic Fields" a report by the Committee on Biological and Human Health Effects of Extremely Low Frequency Electromagnetic Fields, published by the American Institute of Biological Sciences, 1985

Controlling the Human Mind: The Technologies of Political Control or Tools for Peak Performance a book by Dr. Nick Begich, published by Earthpulse Press, 2006

"Extremely Low Frequency Transmitter Site Clam Lake, Wisconsin" a report by the United States Navy, 2001

"Sending Signals to Submarines" an article by David Llanwyn Jones, published in New Scientist, July 4, 1985

"Signaling Subs" an article by T.A. Heppenheimer, published in Popular Science, April, 1987

"ELF Electric Field Coupling to Dielectric Spheroidal Models of Biological Objects" a paper by Yih Shiau and Anthony R. Valentino, published by *IEEE Transactions on Biomedical Engineering*, vol. BME-28, no. 6, June 1981

"Electrical Potentials of the Human Brain" a paper by Alfred L. Loomis, E. Newton Harvey, and Garret Hobart, published by the *Journal of Experimental Psychology*, vol. XIX, No. 3, June, 1936

"Research in Electrical Phenomena Associated with Aerosols" a report by Bernard Vonnegut, Arnold W. Doyle, and D. Read Moffett, produced and published by Arthur D. Little, Inc., 1961

"How to Wreck the Environment" a paper by Gordon J. F. MacDonald as it appeared in *Unless Peace Comes: A Scientific Forecast of New Weapons* a book edited by Nigel Calder, published by The Viking Press, 1968

"On the Possibility of Directly Accessing Every Human Brain by Electromagnetic Induction of Fundamental Algorithms" a paper by Professor Michael A. Persinger, published in *Perceptual and Motor Skills*, June 1995

The Weather Matrix and Human Behavior by Michael A. Persinger, published by Praeger, 1980

“Mind-reading Tech? How Private Companies Could Gain Access to Our Brains” an article by Oscar Schwartz, published by *The Guardian*, October 24, 2019

“Microwave Weapons Are Prime Suspect in Ills of U.S. Embassy Workers” an article by William J. Broad, published by *The New York Times*, September 1, 2018

Microwave Auditory Effects and Applications a book by James C. Lin, PhD, published by Charles C. Thomas, 1978

U.S. patent #4,877,027 "Hearing System" by Wayne B. Brunkan, 1989

"Auditory Response to Pulsed Radiofrequency Energy" a paper by J.A. Elder and C.K. Chou, published by *Bioelectromagnetics*, 2003

"Transmission of Microwave-Induced Intracranial Sound to the Inner Ear Is Most Likely Through Cranial Aqueducts" a paper by Ronald L. Seaman

“New Technology Uses Lasers to Transmit Audible Messages to Specific People” published by The Optical Society, January 23, 2019

Tesla Man Out of Time a book by Margaret Cheney, published by Simon & Schuster, 1981

Prodigal Genius: The Life of Nikola Tesla a book by John J. O'Neill, published by Adventures Unlimited Press, 2008

Tesla vs. Edison: The Life-long Feud that Electrified the World a book by Nigel Cawthorne, published by Chartwell Books, 2016

The Jasons: The Secret History of Science's Postwar Elite a book by Ann Finkbeiner, published by Penguin Books, 2007

Planet Earth: The Latest Weapon of War a book by Rosalie Bertell, published by Black Rose Books, 2001

"Lightning to the Ionosphere?" an article by Otha H. Vaughn, Jr. and Bernard Vonnegut, published by *Weatherwise*, 1982

U.S. patent #3,019,989 "Atmospheric Space Charge Modification" by Bernard Vonnegut and Arthur D. Little, 1962

"Weather as a Force Multiplier: Owning the Weather in 2025" a report by Col. Tamzy J. House, Lt. Col. James B. Near, Jr., LTC William B. Shields (USA), Maj. Ronald J. Celentano, Maj. David M. Husband, Maj. Ann E. Mercer, and Maj. James E. Pugh, published by the United States Air Force, 1996

"Space Operations: Through the Looking Glass (Global Area Strike System)" a paper by Lt Col Jamie G.G. Varni, Mr. Gregory M. Powers, Maj Dan S. Crawford, Maj Craig E. Jordan, and Maj Douglas L. Kendall, published by the United States Air Force, 1996

Where Did the Towers Go? Evidence of Directed Free- energy Technology on 9/11 a book by Judy Wood, published by The New Investigation, 2010

"Regional Climate Responses to Geoengineering with Tropical and Arctic SO2 Injections" a paper by Alan Robock, Luke Oman, and Georgiy L. Stenchikov, published in the *Journal of Geophysical Research*, August 16, 2008

U.S. patent #7,629,918 "Multifunctional Radio Frequency Directed Energy System" 2009

U.S. patent #7,582,809 "Sorghum Aluminum Tolerance Gene, SBMATE" by Leon Kochian, Jiping Liu, Jurandir Vieira de Magalhaes, Claudia Teixeira Guimaraes, Robert Eugene Schaffert, Vera Maria Carvalho Alves, and Patricia Klein, 2009

The World According to Monsanto: Pollution, Corruption, and the Control of the World's Food Supply a book by Marie-Monique Robin, published by The New Press, 2010

"Weather Modification Programs, Problems, Policy, and Potential" a report by the Congressional Research Service, published by the University Press of the Pacific, 2004, reprinted from the 1978 edition

Interdepartmental Committee for Atmospheric Sciences reports 1960-1978, published by the Federal Council for Science and Technology

Clouds of Secrecy a book by Dr. Leonard A. Cole, published by Rowman & Littlefield, 1988

In the Name of Science a book by Andrew Goliszek, published by St. Martin's Press, 2003

Interdepartmental Committee for Atmospheric Sciences report number 15, March 1971, p22

"Toxic Properties of Materials Used in Weather Modification" a paper by William J. Douglas as it appeared in the *Proceedings of the First National Conference on Weather Modification*, 1968

"Climate Intervention: Reflecting sunlight to cool Earth" a report by the National Academies of Science, published by the National Academies Press, 2015

Chemtrails Confirmed a book by William Thomas, published by Bridger House Publishers, 2004

"The Environmental Applications of Wireless Sensor Networks" a paper by Ima Ituen and Guno Sohn

U.S. patent #6,315,213 "Method of Modifying Weather" 2001

"Operational Defenses Through Weather Control in 2030" a report by Michael C. Boger and the Air Command and Staff College, Air University, 2009

"Global Environmental MEMS Sensors (GEMS): A Revolutionary Observing System for the 21st Century" a report by John Manobianco, PhD, prepared for NASA Institute for Advanced Concepts, published by ENSCO, Inc., 2002

"Microelectromechanical systems (MEMS): fabrication, design and applications" a paper by Jack W. Judy, published in *Institute of Physics Publishing*, November 26, 2001

"GlobalSense System Stirs Winds of Change for Atmospheric Monitoring" an article published by *PRNewswire*, August 1, 2016

The MITRE Corporation: Fifty Years of Service in the Public Interest a book by the MITRE corporation, published by the MITRE Corporation, 2008

Morgellons: The Legitimization of a Disease a book by Dr. Ginger Savely, self-published, 2016

Jeb! And the Bush Crime Family: The Inside Story of an American Dynasty a book by Roger Stone and Saint John Hunt, published by Skyhorse Publishing, 2016

"Weather Modification as a Weapon" an article by Gordon J.F. MacDonald, published by *Technology Review*, October/November, 1975

"Atmosphere-Ionosphere Response to the M9 Tohoku Earthquake Revealed by Joined Satellite and Ground Observations. Preliminary results." A paper by Dimitar Ouzounov, Sergey Pulinets, Alexey Romanov, Alexander Romanov, Konstantin Tsybulya, Dimitri Davidenko, Menas Kafatos, and Patrick Taylor, published on *arXiv.org*, May 13, 2011

Ecoscience a book by Paul R. Ehrlich, Anne H. Ehrlich, and John P. Holdren, published by W.H. Freeman, 1977

"Bill Gates backs climate scientists lobbying for large-scale geoengineering" an article by John Vidal, published by *The Guardian*, Feb. 6, 2012

"Sir David Attenborough: If we do not control population, the natural world will" an article by Hannah Furness, published by *The Telegraph*, September 18, 2013

"Prince William warns that there are too many people in the world" an article by Victoria Ward, published by *The Telegraph*, November 2, 2017

The Deliberate Corruption of Climate Science a book by Timothy Ball, PhD, published by Stairway Press, 2014

"Implications of Worldwide Population Growth For U.S. Security and Overseas Interests" National Security Study Memorandum 200, a report by the National Security Council, 1974

"National Security Decision Memorandum 314" a memorandum by the National Security Council, November 26, 1975

The Georgia Guidestones a book by Bill Bridges, published by Elberton Granite Finishing, 1981

Chapter 11
The INFORMATION WAR

"The conscious and intelligent manipulation of the organized habits and opinions of the masses is an important element in democratic society. Those who manipulate this unseen mechanism of society constitute an invisible government which is the true ruling power of our country."
-Edward Bernays, from his book *Propaganda*

"The greatest challenge facing mankind is the challenge of distinguishing reality from fantasy, truth from propaganda."
-Michael Crichton, 2003

Because the people responsible for today's New Manhattan Project are mass murdering global populations and wrecking the environment, they feel a need to cover it up. Firstly they go about trying to make the public believe that this is not happening at all and secondly, if it was, that it would be justified or even beneficial. Over the years, in order to maintain their façades and accomplish their goals, they have produced and maintained an incredibly huge and well oiled disinformation apparatus. The group most probably responsible for this apparatus is the Central Intelligence Agency (CIA). This chapter examines the historical and ongoing psychological warfare campaigns lodged against the American people designed to simultaneously occlude and legitimize the New Manhattan Project. This chapter examines the New Manhattan Project's information war.

Efforts to occlude the New Manhattan Project involve a dismissal of the outward manifestations of the project as normal and natural. For example, the chemtrails so often seen in our skies are dismissed as unadulterated water vapor from jet airplanes. This is an effort to keep the project hidden by dismissing the notion that anything is going on whatsoever. Alternately, efforts to legitimize the New Manhattan Project revolve around the theory of man-made global warming. Geoengineers claim that spraying the Earth with tens of thousands of megatons of toxic materials annually would be justified because, you know, global warming. This is an effort to make it seem that, if planes *were* actually spraying us, then it would be legitimate and beneficial.

Today, the New Manhattan Project is in full swing, so we are simultaneously assaulted with both of these psychological warfare tactics. But before the lines in the sky began so regularly occurring, there was little need to occlude their existence, so the main focus was upon legitimizing the basis for any future actions. Early on, the main focus was upon legitimizing the theory of man-made global warming and promoting potential efforts ostensibly designed to head-off such a supposed threat.

Early geoengineering proposals

Glenn Seaborg

Appropriately enough, a most fundamental concept of what was synonymously called 'planetary engineering' goes back to a Manhattan Project scientist. Nobel Prize winning chemist and Atomic Energy Commission chairman Glenn Seaborg (1912-1999) championed the use of atomic energy as a panacea which could move mountains, divert rivers, and make the deserts bloom. As seen in Seaborg's proposals, Geoengineering is broadly defined as the large-scale manipulation of Earth's natural systems.

Not only was Seaborg the original promoter of modern geoengineering, he was also a population control advocate. The pertinence of this was outlined in the previous chapter. On March 5, 1970, at an event commemorating the career of Monsanto’s Charles Allen Thomas (ch 6), Seaborg spoke of a ‘limited population’ and using science and technology to attain and maintain ‘a reasonable number of human beings.’ Another speaker at the same event, the Chancellor of Washington University, Thomas H. Eliot said, “we seem to have just passed a critical milestone in population growth, and there are just too many people and too many machines for the environment to swallow without help." Through his use of particle accelerators, which were pioneered by his UC Berkeley colleague Ernest Lawrence, Seaborg added 9 elements to the periodic table. Seaborg first met Monsanto's Charles Thomas during the original Manhattan Project.

A 1945 document titled "Outline of Weather Proposal" by Vladimir K. Zworykin (a.k.a. 'The Father of Television') was also influential in geoengineering's early development. As geoengineering today has come to mean, more specifically, an effort to modify the Earth's climate, Zworykin's paper has been proven prescient. Along with some exogenous ideas such as the use of flamethrowers to modify climate, "Outline of Weather Proposal" originated other, more enduring geoengineering concepts such as the use of oil as an artificial surface coating for bodies of water. This document, more pertinently, also called for a global weather control program.

Vladimir Zworykin

Can We Survive Technology?

Ten years after the detonations of the world's first atomic bombs and the subsequent Japanese surrender, and two years before his death, in 1955 *Fortune* magazine published John von Neumann's article "Can We Survive Technology?" It is undoubtedly the first significant mention of the solar radiation management (SRM) geoengineering thesis and arguably the first mention of the modern theory of man-made global warming. Here we look into this seminal document in depth.

It is interesting to note that the first words written at the top of the first page read, "The New Goals - VI." If one glances at this phrase, one might read 'The New Gods' as the eye will have a tendency to amalgamate the 'a' and the 'l' in the word 'Goals,' turning these two letters into a 'd' and therefore forming the word 'Gods.' Maybe this is a subliminal message meant to deify von Neumann and his colleagues. And make no mistake, at the time, the world *was* standing in awe. After Manhattan Project scientists such as von Neumann had unleashed the terrible destructive power of the atom bomb and developed so many other technologies which enabled the Allied powers to win the war, top American scientists had a prominent role in drastically reshaping world power and were understandably respected. In fact, von Neumann in particular was a most highly revered scientist. Indeed, a strong argument can be made for von Neumann being *the most* revered American scientist of all time.

The opening sentence rings the alarm, stating, "'The great globe itself' is in a rapidly maturing crisis." Von Neumann then goes over how man's technological capabilities are stretching the Earth's resources and how this, if unchecked, might lead to global catastrophe. Von Neumann then rolls out both the theory of man-made global warming and the SRM geoengineering thesis. Here these two theses were fused together to form a giant, flaming sword of evil to be used against domestic populations worldwide. He begins these sections with quite a bit about weather modification. It reads:

"***Controlled climate***

"Let us now consider a thoroughly 'abnormal' industry and its potentialities - that is, an industry as yet without a place in any list of major activities: the control of weather or, to use a more ambitious but justified term, climate. One phase of this activity that has received a

good deal of public attention is 'rain making.' The present technique assumes extensive rain clouds, and forces precipitation by applying small amounts of chemical agents. While it is not easy to evaluate the significance of the efforts made thus far, the evidence seems to indicate that the aim is an attainable one.

"But weather control and climate control are really much broader than rain making. All major weather phenomena, as well as climate as such, are ultimately controlled by the solar energy that falls on the earth. To modify the amount of solar energy, is, of course, beyond human power. But what really matters is not the amount that hits the earth, but the fraction retained by the earth, since that reflected back into space is no more useful than if it had never arrived. Now, the amount absorbed by the solid earth, the sea, or the atmosphere seems to be subject to delicate influences. True, none of these has so far been substantially controlled by human will, but there are strong indications of control possibilities."

His reference to reflecting sunlight back into space here is the first ever significant mention of the SRM geoengineering thesis - probably the first of any kind. Also note that von Neumann writes of a wholly different type of weather modification program - not the conventional cloud seeding industry, which was already well established by the time of this article's publication. The New Manhattan Project is a wholly different kind of weather modification program. He makes two later mentions of such a program. He follows with the theory of man-made global warming. It continues:

"The carbon dioxide released into the atmosphere by industry's burning of coal and oil - more than half of it during the last generation - may have changed the atmosphere's composition sufficiently to account for a general warming of the world by about one degree Fahrenheit. The volcano Krakatoa erupted in 1883 and released an amount of energy by no means exorbitant. Had the dust of the eruption stayed in the stratosphere for fifteen years, reflecting sunlight away from the earth, it might have sufficed to lower the world's temperature by six degrees (in fact, it stayed for about three years, and five such eruptions would probably have achieved the result mentioned). This would have been a substantial cooling; the last Ice Age, when half of North America and all of northern and western Europe were under an ice cap like that of Greenland or Antarctica, was only fifteen degrees

colder than the present age. On the other hand, another fifteen degrees of warming would probably melt the ice of Greenland and Antarctica and produce worldwide tropical to semi-tropical climate."

Notice how in this passage von Neumann asserts the idea that ash spewing from an erupting volcano (in this case Krakatoa) will reflect sunlight back into space and therefore cool the Earth. Today's geoengineers claim this point *ad nauseum*. Although they claim that this idea originated in the late 1970s, we know now where they really got it from. It's just another deception in a long litany.

Among many other assertions, von Neumann then goes on to suggest that, "Probably intervention in atmospheric and climatic matters will come in a few decades, and will unfold on a scale difficult to imagine at present." And, "The most constructive schemes for climate control would have to be based on insights and techniques that would also lend themselves to forms of climatic warfare as yet unimagined." Von Neumann's imagination has become reality.

It is important to note the use of the word 'modern' here when describing von Neumann's assertion of the theory of man-made global warming. The term 'modern' is used here to describe a version of the theory of man-made global warming which has entered into popular culture and is connected to the SRM geoengineering thesis. The historical, or 'old' version of the theory of man-made global warming, was largely relegated to academia and not connected to the SRM geoengineering thesis.

The introduction of the SRM geoengineering thesis, in Johnny von Neumann's 1955 *Fortune* piece is the demarcation point between the 'old' and 'modern' versions of the theory of man-made global warming. It is reasonable to consider the period before von Neuman's paper as the era of the 'old' theory of man-made global warming and the period after von Neumann's paper as the era of the 'new' or 'modern' theory of man-made global warming. These assertions are being made for two reasons: "Can We Survive Technology?" was the first insertion of the theory of man-made global warming into the popular culture and "Can We Survive Technology?" was the first major publication of the SRM geoengineering thesis.

"Can We Survive Technology?" was a weaponization of the theory of man-made global warming and therefore a green light to the military-industrial-academic-intelligence complex. Up until the publication of "Can We Survive Technology?," the theory of man-made global warming was only something that existed as just that - a theory. When the theory of man-made global warming was linked to the SRM geoengineering thesis, the military-industrial-academic-intelligence complex had a new enemy to fight: the climate. A new enemy to fight means more academic study, more basic science, new weapons systems, and intelligence agency cover. In other words, it means a mountain of power and cash raining down upon them. In 1955 von Neumann handed the military-industrial-academic-intelligence complex an argument for tremendous expansion and they have taken him up on it ever since - spending vast sums on weather modification and the atmospheric sciences. In fact, it wasn't long after von Neumann's 1955 *Fortune* piece that the American military-industrial-academic-intelligence complex was the beneficiary of tremendous across-the-board spending increases catalyzed by Russia's Sputnik launch of 1957.

The modern theory of man-made global warming and the SRM geoengineering thesis have a symbiotic relationship. The SRM geoengineering thesis supercharged the old theory of man-made global warming. At the same time, the SRM geoengineering thesis has no reason to exist without the theory of man-made global warming. It only makes sense that both would be introduced and/or upgraded at the same time. It also makes dollars and sense that von Neumann's piece was published in *Fortune* magazine. As we saw in chapter 8, the evidence shows plenty of financial motives behind today's New Manhattan Project.

In 1955, when "Can We Survive Technology" was published, von Neumann was a chief consultant to the CIA as well as the recipient of Rockefeller Foundation funding.

The war rolls on

In the early months of 1958, the Solar Radiation Management geoengineering thesis was again mentioned in a paper co-authored by Bernard Vonnegut (ch 3) which appeared in the *Final Report of the Advisory Committee on Weather Control*. The paper reads, "The radiation properties of the atmosphere can be altered by the introduction of gases or aerosols and by cloud seeding." The *Final*

Report of the Advisory Committee on Weather Control is the most cited early weather modification document ever.

In June of 1958, the aforementioned (ch 1) weather modifier Howard T. Orville wrote an article for *Popular Science* titled "Weather as a Weapon" which makes another early mention of the SRM geoengineering thesis as well as an early mention of the modern theory of man-made global warming. The article's header describes weather control as, "a power more menacing than the H-bomb."

Orville was a Navy captain and a weather advisor for the WWII North African invasion as well as for Naval operations in the WWII Pacific Theatre. He was appointed by President Eisenhower to serve as chairman of Eisenhower's Advisory Committee on Weather Control. His article asserts the SRM geoengineering thesis when it states that 'Rocket-spread gas clouds' can be used to regulate the amount of solar energy striking the Earth. Whether it comes from rockets or jet airplanes, the principle is the same. The article also goes on to mention the SRM geoengineering thesis a second time when Orville notes that, "Air Force scientists are already experimenting with sodium vapor, ejected from jet planes, to intercept solar radiation." The article also states some other geoengineering proposals such as the spreading of surfactants across the world's oceans and states that regulating the distribution of heat in different parts of the Earth's atmosphere is the basis of global weather control.

Orville's article notes the importance of the ionosphere and atmospheric electricity to any potential global weather control - harkening forward to the ionospheric heaters of today's New Manhattan Project which use electromagnetic energy to manipulate the ionosphere. The article states, "We may find other ways to manipulate the charges of earth and sky and so affect the weather. One means might be an electronic beam to ionize or de-ionize the atmosphere over a given area."

Along with the two mentions of the SRM geoengineering thesis noted here, the article also states the catastrophic theory of man-made global warming. The article asserts:

"The last half-century, during which we have burned huge amounts of fossil fuels, has shown what an increase in atmospheric carbon dioxide can do. The amount spewed from chimneys and

automobiles in five decades has created the so-called 'greenhouse effect,' which has raised the earth's temperature by an estimated two degrees Fahrenheit - a significant rise.

"This so-far accidental result is already serious, according to Dr. Kaplan, and we must find means to counteract it. 'Melting polar ice will make ocean levels rise at least 40 feet, and inundate vast areas in the next 50 or 60 years,' he warns, 'unless atmospheric temperatures are controlled.'"

The article goes on to assert that global warming can cause, "the sinking of our East Coast cities beneath the Atlantic, and the transformation of the Mississippi Valley into a huge inland sea." If one gives Orville and Kaplan the maximum of 60 years which they predicted, then they, like so many other climate alarmists, have obviously missed their mark.

The first top-level government document to assert the modern theory of man-made climate change and the SRM geoengineering thesis was the aforementioned (ch 4) report "Restoring the Quality of Our Environment." This Orwellian 1965 report, written by the Environmental Pollution Panel of the President's Science Advisory Committee, explains how CO2 emitted by man is causing climate change and therefore we may need to inject small particles into the atmosphere in order to change it back. When a Presidential report is issued, all government departments under the executive branch are obligated to review it and to be in general compliance with it. That's what makes this document particularly relevant. The document reads:

"The climatic changes that may be produced by the increased CO2 content could be deleterious from the point of view of human beings. The possibilities of deliberately bringing about countervailing climatic changes therefore need to be thoroughly explored. A change in the radiation balance in the opposite direction to that which might result from the increase of atmospheric CO2 could be produced by raising the albedo, or reflectivity, of the earth. Such a change in albedo could be brought about, for example by spreading very small reflecting particles over large oceanic areas. The particles should be sufficiently buoyant so that they will remain close to the sea surface and they should have a high reflectivity, so that even a partial covering of the surface would be adequate to produce a marked change in the amount

of reflected sunlight. Rough estimates indicate that enough particles partially to cover a square mile could be produced for perhaps one hundred dollars. Thus a 1% change in reflectivity might be brought about for about 500 million dollars a year, particularly if the reflecting particles were spread in low latitudes, where the incoming radiation is concentrated. Considering the extraordinary economic and human importance of climate, costs of this magnitude do not seem excessive. An early development of the needed technology might have other uses, for example in inhibiting the formation of hurricanes in tropical oceanic areas.

"According to Manabe and Strickler (1964) the absorption and re-radiation of infrared by high cirrus clouds (above five miles) tends to heat the atmosphere near the earth's surface. Under some circumstances, injection of condensation or freezing nuclei will cause cirrus clouds to form at high altitudes. This potential method of bringing about climatic changes needs to be investigated as a possible tool for modifying atmospheric circulation in ways which might counteract the effects of increasing atmospheric carbon dioxide."

Note how at the end of the first paragraph, the authors state that injecting small particles into the atmosphere can not only save us from climate change, but also achieve weather modification. The truth is softly spoken.

Two years later, in the pages of the 1967 National Science Foundation's ninth annual weather modification report, we find another early example. It reads:

"ESSA [Environmental Science Services Administration] is also investigating the effect of cirrus clouds on the radiation budget of the atmosphere by studying aircraft-produced contrails which often spread into cirrus layers covering considerable fractions of the sky. One technique proposed for modifying lower cloud development has been the generation of a high level cirrus deck with jet aircraft. By intercepting solar radiation at high altitude it may be possible to influence larger scale cloud development elsewhere by reducing solar input and reducing convective cloud generation in areas where they are not needed."

Note how the authors of this report misrepresent chemtrails as contrails. Before 1965, the National Science Foundation (NSF) was not saying this. This is a psychological warfare tactic. They are trying to make us believe that an apple is an orange. If a lie is repeated often enough and with enough conviction, people will believe it. This is what the authors of this passage are counting on. This lie has been repeated *ad nauseam*. After the release of "Restoring the Quality of Our Environment," the focus became one of legitimizing the lines in the sky that were sure to proliferate with the New Manhattan Project and this 1967 NSF document is an early example from this phase of the NMP's information war. A little later in the same NSF report, it reads that their computer atmospheric simulations, or 'models' as they call them, might simulate, "...producing high-level cirrus cloud cover over an area by means of jet aircraft, inserting particulate matter into the upper atmosphere to alter the solar radiation balance and the like."

A 1970 paper, again originating from the Environmental Science Services Administration (ESSA), provided another early example. The author, P.M. Kuhn writes:

"*The spreading out of jet contrails into extensive cirrus sheets is a familiar sight. In the most common case, when persistent conditions exist from twenty-five to forty thousand feet, a few long contrails increase in number and gradually merge into an almost solid white sheet.*"

Kuhn continues:

"*From a descriptive point of view contrail development and spreading commonly begins in the morning hours as heavy jet traffic commences and finally extends from horizon to horizon as the air traffic peaks.*"

Conveniently for Mr. Kuhn, this paper was not peer reviewed or refereed.

In 1977 the Weather Modification Advisory Board expounded upon a familiar line of bull. Their authors wrote, "Introducing contrails in selective regions of the atmosphere can produce cirrus clouds, to keep heat in or out of an area." As mentioned in chapter 1, 1977 was also the year that the term 'geoengineering' was introduced in the premiere edition of the journal *Climatic Change*. The late 1970s was when today's SRM geoengineering thesis began taking a firm foothold.

In 1987, the World Meteorological Organization and the British Royal Air Force cast an abomination out upon the Earth. The caption reads:

"The sky is crossed in every direction by contrails of different ages. As they expand progressively they form fluffy or fibrous clouds and it is impossible to say with certainty whether there are also clouds of natural origin in the sky. Note the pendant swellings, like inverted toadstools, at 1-2, typical of recently formed contrails."

Royal Air Force, Farnborough (Hampshire, U.K.). 1 January 1945

Condensation trails (contrails)

The sky is crossed in every direction by contrails of different ages. As they expand progressively they form fluffy or fibrous clouds and it is impossible to say with certainty whether there are also clouds of natural origin in the sky. Note the pendant swellings, like inverted toadstools, at 1-2, typical of recently formed contrails.
From an anticyclone in the south-west a strong ridge of high pressure extended over the area. In the higher layers, advection of warm and rather humid air was in progress from the north-north-east. The associated warm front was at about 250 km west-north-west of Farnborough; it caused only light precipitation when it passed.

Chemtrails presented as contrails by the WMO and the Royal Air Force

A 1992 report by the National Academy of Sciences titled "Policy Implications of Greenhouse Warming" advocates the SRM geoengineering thesis. In this case, they suggest that, in order to save us from a slightly increased average global temperature, aircraft spray the Earth with megatons of sulfuric acid. Sound good?

As noted in chapter 2, in the mid-nineties, Lawrence Livermore National Laboratories scientists Edward Teller, Lowell Wood, Roderick Hyde, and others wrote multiple papers calling for small aluminum particles to be 'injected' into our atmosphere in order to 'scatter' sunlight. They wrote that this will save us from global warming.

The United Nations backed Intergovernmental Panel on Climate Change (IPCC) issued a 1999 report with a cover photo of three jet planes flying in formation and emitting trails entitled "Aviation and the Global Atmosphere." This report contained chapters titled, "Aviation-Produced Aerosols and Cloudiness" and "Modeling the Chemical Composition of the Future Atmosphere." The report bolstered lies about contrails by stating that they can form cirrus clouds.

In 2000, the Environmental Protection Agency (EPA) along with the National Aeronautics and Space Administration (NASA), NOAA, and the Federal Aviation Administration (FAA) compromised themselves by releasing a report titled "EPA Aircraft Contrails Factsheet." In this report, they tell lies about how contrails, "...last for hours while growing to several kilometers in width and 200 to 400 meters in height." This report included an image depicting chemtrails which were, of course, described as consisting only of harmless water vapor.

Another 2000 paper appearing in the proceedings of the 15th International Conference on Nucleation and Atmospheric Aerosols of the American Institute of Physics conflates contrails with chemtrails. Writing of effluents produced from jet engine exhausts, the paper states, "These gases and particles alter the concentration of atmospheric greenhouse gases and trigger the formation of condensation trails (contrails) and may increase cirrus cloudiness."

A 2001 paper appearing in the *Journal of Geophysical Research* says that 'contrail clusters' cause radiative forcing (the SRM geoengineering thesis) and, "...mature into diffuse cloudiness." The authors write:

"Clusters of spreading, persistent contrails sometimes form in heavy air-traffic regions when the warm exhaust emissions from subsonic jet aircraft mix with cold and relatively moist ambient air in the upper troposphere. Time series of satellite imagery have shown that these clusters can develop into large regions of cirrus-like cloudiness [Minnis et al., 1998a], which may affect climate by modifying the atmospheric radiative budget. As contrails are ice clouds, they have radiative effects that are similar to thin cirrus clouds [Fu and Liou, 1993] and may have significant regional climatic impacts [Liou et al., 1990]."

The Trilateral Commission produced a 2007 document advocating for SRM geoengineering. It is titled "Energy Security and Climate Change." The lead author was former CIA director John Deutch.

Since at least 2008, an organization at the Federal Aviation Administration has been producing misinformation about chemtrails. The Aviation Climate Change Research Initiative has produced a multitude of papers and reports that go on and on about the science of contrails and how 'persistent contrails' and 'contrail cirrus' are perfectly natural and normal phenomena. It's really quite amazing how the authors of their scientific reports, in particular, blather on and on with all this highly technical discussion of contrails without recognizing the elephant in the room. What will they say when we put the New Manhattan Project out of business and the lines in the sky suddenly disappear?

During a 2009 presentation, referencing a picture of a chemtrail laden sky, European Organization for Nuclear Research (CERN) scientist Jasper Kirkby, PhD said, "These are not smoke trails. These are clouds which are seeded by jets dumping aerosols into the upper atmosphere." May we get a further clarification of that Dr. Kirkby, or have you said too much already?

The U.K. Royal Society published a 2009 report titled "Geoengineering the Climate: Science, Governance and Uncertainty" in which they propose SRM geoengineering and state, "A wide range of types of particles could be released into the stratosphere with the objective of scattering sunlight back to space." This, while the British government has released official statements telling us that chemtrails do not exist.

The British House of Commons collaborated with the United States House of Representatives Science and Technology Committee to release a report on March 10, 2010 entitled "The Regulation of Geoengineering" in which they advocate geoengineering regulation. They are advocating for the legitimization of chemtrails.

During 2009 and 2010's aforementioned (ch 1) Congressional geoengineering hearings, testifying geoengineers proposed spraying substances from aircraft many, many times. Most testifying geoengineers characterized the practice as a cheap and effective way to mitigate global warming. Geoengineer Lee Lane writes the best example:

"Several proposed delivery techniques may be feasible (NAS, 1992). The choice of the delivery system may depend on the intended purpose of the SRM [solar radiation management] program. In one concept, SRM could be deployed primarily to cool the Arctic. With an Arctic deployment, large cargo planes or aerial tankers would be an adequate delivery system (Caldeira and Wood, pers. comm., 2009). A global system would require particles to be injected at higher altitudes. Fighter aircraft, or planes resembling them, seem like plausible candidates. Another option entails combining fighter aircraft and aerial tankers, and some thought has been given to balloons (Robock et al., 2009)."

They describe a program using retrofitted drone aircraft to spray particulates. In his response to a follow-up question by Chairman Bart Gordon, geoengineer Alan Robock writes, "Certainly studies should be done of the feasibility of retrofitting existing U.S. Air Force planes to inject sulfur gases into the stratosphere, as described by Robock et al. [2009], as well as of developing new vehicles, probably remotely-piloted, for routine delivery of sulfur gases or production of aerosol particles."

During these Congressional hearings, the 'c' word was used twice. Former congressman Brian Baird (D - WA 3rd district) said the word first, "And so I applaud you all for suggesting that we are not going to have this — to rescue us by, you know, chemtrails or whatever people want to distribute into the air." Chairman Baird used the word again when he jumped right into the conspiracies and said this:

"I will share with you, though, this idea of placing particles in the upper atmosphere. Are any of you familiar with the conspiracy theory known as chemtrails? Have you heard of this? It is a rather interesting phenomenon. I was at a town hall and a person opined that the shape of contrails was looking different than it used to, and why was that? I gave my best understanding of atmospheric temperature and humidity and whatnot, but the theory which is apparently pretty prevalent on the Net is that the government is putting psychotropic drugs of some sort into the jet fuel and that is causing a difference in appearance of jet fuel and allowing them to secretly disseminate these foreign substances through the atmosphere via our commercial jet airline fleet."

He conveniently got all that very wrong. This next excerpt is from the hearing's charter:

"...negative public perceptions of geoengineering may also prove to be a powerful catalyst for emissions reductions. A study by the British Market Research Bureau found that while participants were cautious or hostile toward geoengineering, several agreed that they would actually be more motivated to undertake mitigation actions themselves after a large-scale geoengineering application was suggested."

They are saying that the suggestion of geoengineering may encourage people to voluntarily adopt carbon mitigation efforts. They're saying that geoengineering can be used as a threat to intimidate the public into accepting a lower quality of life. Who are the real terrorists?

In October of 2010, the United States House of Representatives Science and Technology Committee issued a report called "Engineering the Climate: Research Needs and Strategies for International Coordination." On the final page, this report advocates:

"It is the opinion of the Chair, in agreement with U.K. Committee, that further collaborative work between national legislatures on topics with international reach, such as climate engineering, should be pursued. The Chair also agrees that there are a range of measures that could be taken to streamline the process and enhance the effectiveness of collaboration."

"Advancing the Science of Climate Change" by the National Research Council of the National Academies of Science is another top-level 2010 scientific report advocating for SRM geoengineering research.

In a 2011 report, the Congressional Research Service produced a paper called "Geoengineering: Governance and Technology Policy" in which the authors suggest an 'aerosol injection' - the dispersal of aerosols into the atmosphere by military aircraft.

A 2011 report by the Defense Science Board titled "Trends and Implications of Climate Change for National and International Security" advocates for solar radiation management geoengineering research.

The United Nations Educational Scientific and Cultural Organization (UNESCO) issued a report in November of 2011 titled "Engineering the Climate: Research Questions and Policy Implications." It proposes geoengineering and features a full color illustration of a jumbo jet spraying stratospheric aerosols (chemtrails).

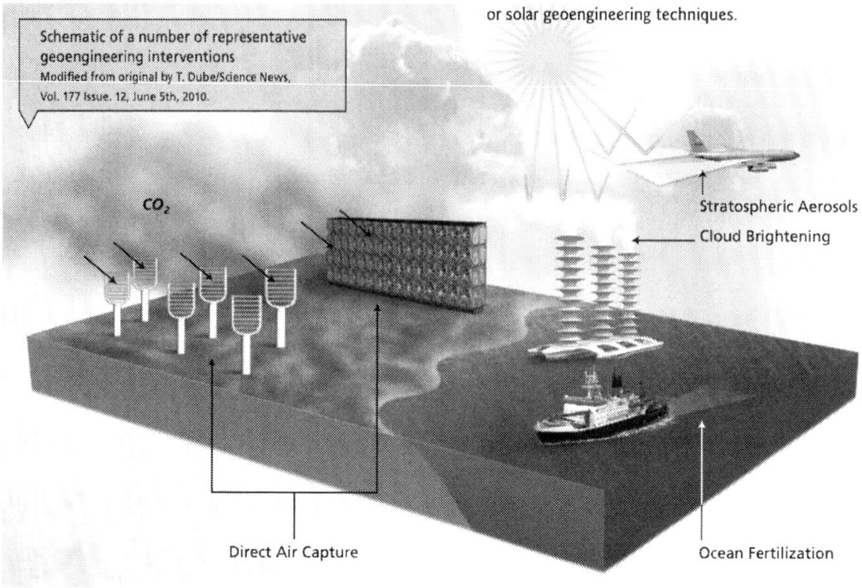

UNESCO's proposition

Another very boring read proposing the injection of stratospheric aerosols arose from a 2011 meeting of geoengineers in Lima, Peru. A report called "IPCC Expert Meeting on Geoengineering" was produced by the UN's Intergovernmental Panel on Climate Change (IPCC) and the World Meteorological Organization.

The RAND Corporation published a 2011 report titled "Governing Geoengineering Research: A Political and Technical Vulnerability Analysis of Potential Near-term Options" in which they advocate for SRM geoengineering and its regulation.

A 2011 report titled "America's Climate Choices" by the National Research Council of the National Academy of Sciences suggests SRM geoengineering.

A 2011 report by the Royal Society and the Environmental Defense Fund titled "Solar Radiation Management: The Governance of Research" prattles on about, "Introducing reflective aerosols into the stratosphere." The authors of the report suggest that, "A moratorium on all SRM-related research would be difficult if not impossible to enforce."

In 2013 the American Meteorological Society released a policy statement claiming that man's emissions of atmospheric carbon dioxide is a big problem which must be dealt with. They go on to suggest that putting reflective particles into the atmosphere (SRM geoengineering) might be a good idea and that more geoengineering research is needed.

A 2014 report by the IPCC titled "Climate Change 2013: The Physical Science Basis" states that SRM geoengineering might be a good way to reduce the impact of the dreaded global warming.

On November 8 of 2017, the United States House of Representatives Science, Space, and Technology Committee held another hearing on Geoengineering. This one was called *Geoengineering: Innovation, Research, and Technology*. All of the congresspeople present, both Republicans and Democrats, spoke about geoengineering as a promising new technology. If they can't agree upon anything else, both parties can apparently agree to look at spraying us with tens of thousands of megatons of toxic materials. At this hearing, geoengineers Phil Rasch, Joseph Majkut, Douglas MacMartin, and Kelly Wanser regurgitated old talking points about how funding for geoengineering research is needed to explore their superhero-like activities which could save us all from the dreaded climate change.

The Carnegie Foundation has been quite active here. The Carnegie Council for Ethics in International Affairs published a video of a geoengineering panel discussion held in early 2017 to commemorate the launch of their Carnegie Climate Geoengineering Governance Initiative which 'seeks to catalyze the creation of effective governance' for SRM geoengineering. The panelists blathered on for an hour and a half laden with CIA talking points and disinformation. It all amounted to a public relations event for SRM geoengineering.

On August 21, 2017 the Carnegie Endowment for International Peace published an article titled "Understanding Climate

Engineering." In this article, the author writes about spraying the Earth with particles from aircraft in order to save us from catastrophic global warming. On September 21 of 2017 the Carnegie Endowment for International Peace published an article titled "Could Geoengineering Save the Planet from Global Warming?" You can take one guess as to what they wrote about. On May 29, 2018 the Carnegie Endowment for International Peace published an article titled "Advancing Public Climate Engineering Disclosure." In this article, the three authors suggest that we might save ourselves from catastrophic global warming by spraying things out of planes.

As part of his bid for the 2020 Democratic nomination for president, on August 28, 2019 businessman Andrew Yang announced his support for SRM geoengineering.

On September 7 of 2019, *CNBC* released a short video titled "Why Bill Gates Is Funding Solar Geoengineering Research." In this high production-value video, SRM geoengineering is presented as a possible solution to man-made global warming.

NASA had a web page dedicated to brainwashing children. At the top of the page it read, "Contrail Watching for Kids." Alongside a picture of a chemtrail-laden sky, the text read, "Contrails are long clouds made by high-flying aircraft. Because kids are so good at watching clouds, they can easily be taught to identify contrails." Our friends at NASA have since expanded this information into a report titled simply "Contrails." It is to be used as a classroom teaching aide. The report expands their familiar line of bull and provides instructions for a classroom experiment that proves nothing. One might wonder why NASA is so concerned about making sure that children know about contrails.

The World Meteorological Organization has classified chemtrails as 'Cirrus homogenitus.' They classify chemtrails that have been spread out by wind as 'Cirrus floccus homomutatus' or 'Cirrus fibratus homomutatus.'

WMO Cirrus homogenitus

WMO Cirrus floccus homomutatus or Cirrus fibratus homomutatus

Our friends at NASA were providing disinformation once again when they presented the image below with a caption that described the many chemtrails as, "...contrails, cirrus clouds created by airplanes."

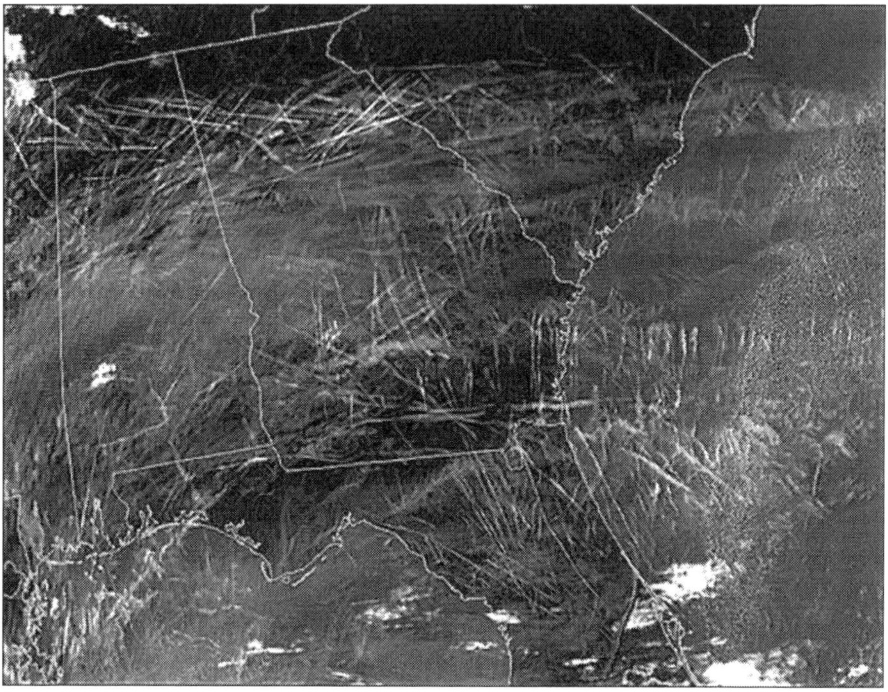

Chemtrail Clutter over Georgia

The CIA has been quite active in the areas of spreading disinformation about man-made global warming as well as SRM geoengineering. The CIA funded a study resulting in the 2015 National Academy of Sciences report titled "Climate Intervention: Reflecting Sunlight to Cool Earth." This study was also supported by some of the usual suspects: the National Oceanic and Atmospheric Administration, the National Aeronautics and Space Administration, and the Department of Energy. In "Climate Intervention" the authors cite the discredited Intergovernmental Panel on Climate Change (IPCC) data to assert that global warming is occurring. Then they write that the topic of SRM geoengineering needs more research and, without admitting that anything like it is currently taking place, they present SRM geoengineering as a potential solution. If one has read scores of these types of documents, one recognizes this standard line of bull immediately. To cement their position, on June 29 of 2016, the director of the CIA, John O. Brennan made a speech at the Council on Foreign Relations (CFR). He suggested that spraying the Earth with tiny particles can save us from global warming.

By the time John Brennan made his famous speech about SRM geoengineering at Pratt House (the CFR's headquarters), the CFR had already been heavily invested in SRM geoengineering for quite some time. Beginning in 1990, the Council started organizing study and discussion groups devoted to the topic of climate change and they have since gone on to issue many publications about SRM geoengineering.

In 2004, CFR member M. Granger Morgan teamed up with top geoengineer Ken Caldeira and six other authors to produce a paper titled "A Portfolio of Carbon Management Options." In it, they advocate for geoengineering Earth's climate with stratospheric aerosols.

In 2008 CFR member David G. Victor published a piece titled "On the Regulation of Geoengineering" in which he advocates for the establishment of an international framework of laws pertaining to geoengineering.

Also in 2008, the CFR released a report titled "Confronting Climate Change: A Strategy for U.S. Foreign Policy: Report of an Independent Task Force." In it, they discuss, "injecting reflective aerosols into the atmosphere."

2008 was a big year for geoengineering at the Council. In addition to the two documents noted above, in 2008 the CFR issued a paper titled "Unilateral Geoengineering: Non-technical Briefing Notes for a Workshop at the Council on Foreign Relations." In this document, the authors propose spraying aluminum oxide dust into the stratosphere.

Among many other subsequent articles about global warming and climate change, CFR members David G. Victor, M. Granger Morgan, Jay Apt, John Steinbruner (1942-2015), and Katharine Ricke co-wrote a piece published in 2009 by the title of "The Geoengineering Option." As one might guess, in this article the authors fear monger about climate change then suggest that we should 'launch reflective particles into the atmosphere' in order to save us from it. The authors are quite insistent, writing, "Governments should immediately begin to undertake serious research on geoengineering and help create international norms governing its use." Rare in such propaganda, the authors also note that geoengineering is *not* a new idea and harken back to the aforementioned 1965 presidential document "Restoring the Quality of Our Environment" among other early developments. The piece, of course, then goes back to fear mongering, then on to ridiculing the opposition, then, lastly to reiterating its psychotic thesis.

Also in March of 2009, *Foreign Affairs* published an article called "Q&A With David Victor About Climate Change: What governments, scientists, and big business can do about global warming." A woman named Lucy Berman interviewed David Victor and they chatted about spraying us with megatons of toxic chemicals. What a lovely topic of discussion!

Later in December of 2009, CFR member Granger Morgan wrote an article for the Massachusetts Institute of Technology's *MIT Technology Review* advocating for SRM geoengineering research. It was titled "Why Geoengineering?"

In January of 2010, the most prestigious journal *Nature* published an opinion piece by CFR member Granger Morgan and two other authors including top geoengineer David Keith. They called their piece "Research on Global Sun Block Needed Now." In it they advocate for SRM geoengineering research.

In July of 2010, CFR members Katherine Ricke and Granger Morgan had an article published in the prestigious journal *Nature Geoscience* titled "Regional Climate Response to Solar Radiation Management." In it they found that spraying stuff out of planes might 'stabilize' our climate.

CFR members Granger Morgan and John Steinbruner held a panel discussion at Pratt House in 2010 titled "Developing an International Framework for Geoengineering." During the discussion, Morgan stressed the urgency of the geoengineering situation as he raised the specter of a 'black program' currently conducting SRM geoengineering which everybody finds out about later. He said, "I think it would be truly disastrous if, you know, we discovered a few years from now that there was a black program that some government had stood up to sort of learn on the quiet how to do this." Yes, we've found out about it already, Mr. Morgan and we plan to make this as disastrous as possible for your public relations effort.

CFR member Granger Morgan contributed to the aforementioned report resulting from the Intergovernmental Panel on Climate Change's 2011 Expert Meeting on Geoengineering held in Lima, Peru. He suggests, "Adding small reflecting particles in the stratosphere."

CFR member Katharine Ricke (sounds like 'reich') published a doctoral thesis in August of 2011 titled "Characterizing Impacts and Implications of Proposals for Solar Radiation Management, a Form of Climate Engineering" in which she suggests that Solar Radiation Management geoengineering might be a good idea.

CFR members Frank Loy, Granger Morgan, and David G. Victor all contributed to the big 2011 report by the Bipartisan Policy Center's Task Force on Climate Remediation Research titled "Geoengineering: A National Strategic Plan for Research on the Potential Effectiveness, Feasibility, and Consequences of Climate Remediation Technologies." The authors note that, "it may be desirable or even necessary to enlist international fleets of aircraft, satellites, and hardware as well as international sources of funding and management capabilities."

CFR member Jay Apt wrote a 2012 paper with top geoengineer David Keith titled "Cost Analysis of Stratospheric Albedo Modification Delivery Systems" in which the authors write about spraying 1-5 million metric tons of materials from airplanes at an altitude of 18-30 km up in the sky every year. Their top choice for the job was the Boeing '747-400.' The Boeing 747 is a close relative of the Boeing KC-135 which the author has identified as the most probable make and model of the most prevalent type of dedicated chemtrail fleet aircraft (ch 5).

In the Spring of 2013, CFR member Granger Morgan and two co-authors published an article in *Issues in Science and Technology* titled "Needed: Research Guidelines for Solar Radiation Management." In it, they advocate for geoengineering research, call for help from the federal government, and write that, "a small fleet of specially designed aircraft could deliver enough mass to the stratosphere in the form of small reflecting particles to offset all of the warming anticipated by the end of this century for a cost of less than $10 billion per year."

Another article published by CFR members Morgan, Apt, *et al.* in March of 2013 was titled "The Truth About Geoengineering: Science Fiction and Science Fact." In it, they speculate that, "Flying a fleet of high-altitude aircraft that spray particles into the upper atmosphere would cost perhaps ten billion dollars per year" The authors also warn that, "Small-scale field trials in the upper atmosphere to test components of an SRM [Solar Radiation Management] system are particularly urgent."

In February of 2015, geoengineer Jane C.S. Long and CFR members Granger Morgan and Frank Loy had an article published in the prestigious journal *Nature* titled "Start Research on Climate Engineering." In it they advocate for small-scale SRM geoengineering experimentation.

On November 3 of 2015, CFR member and member of the Carnegie Institution Department of Global Ecology, Katharine Ricke published a paper, along with two other authors, titled "Climate Engineering Economics." In this paper the authors write, "The most likely approach to implementation is high-flying aircraft outfitted with aerosol precursor dispensing systems (McClellan et al., 2012)" As have many other authors, Ricke *et al.* found that using aircraft to achieve SRM is the cheapest way. In conclusion the authors urge, "Finally, we need to begin to explore specific mechanisms to ensure an efficient and equitable implementation of climate engineering technologies. While some early steps have been taken in this direction, we need to understand, from an economic perspective, how to create institutions that can accommodate these novel climate risk reduction strategies."

Chemtrails in popular media

Since the 1946 media explosion celebrating the pioneering weather modification efforts of three General Electric scientists (ch 3), airplanes spraying stuff have made regular mainstream media appearances. Today, the persistent lines in the sky left behind by airplanes are woven into the mainstream media's fabric. What is presented here is but a small sampling.

The 1959 Disney short educational film "Eyes in Outer Space: A Science-Factual Presentation" was made in cooperation with the Department of Defense. The famous conventional weather modifier Irving P. Krick served as a technical consultant. This movie depicts men in ~~Lawrence Livermore National Lab~~ a global weather center controlling the weather with drone aircraft. As a plane flies across the screen emitting a trail, the narrator says, "Robot planes seed the clouds from above." These 'robot planes' are then depicted saving us all from a nasty hurricane by spraying stuff into it. There is an edited version available on *YouTube*.

The 1986 video for the Level 42 song “Lessons in Love” depicts chemtrails. Here is a still from the video:

The 2006 movie *Cars* depicts chemtrails. Here is an enhanced still:

The 2008 feature film *Toxic Skies* starring Anne Heche is a fictional drama involving chemtrails.

In the 2012 Denzel Washington movie *Flight*, chemtrails were presented:

The Topps Company, famous for its baseball cards, produced a 2013 series of cards called *Curious Cases*. One of the cards in the set was about chemtrails. Here is the front of the card:

The back of the chemtrails card reads:

"CHEMTRAILS - THOSE BILLOWING WATER VAPOR STREAMS BEHIND AN AIRPLANE - ARE ACTUALLY BIOLOGICAL AGENTS BEING DISPERSED BY VARIOUS GOVERNMENTS FOR UNDISCLOSED REASONS (POSSIBLY MIND OR POPULATION CONTROL). AT LEAST THAT'S THE ALLEGATION BY CONSPIRACY THEORISTS."

The *BBC* opening credits for the Wimbledon tennis tournament of 2013 depict chemtrails:

Here is a billboard advertisement referring to chemtrails as 'decorative streamers:'

The February 22, 2016 episode of the *X-Files* titled "My Struggle II" featured a storyline involving an outbreak of a mass pandemic caused by chemtrails combined with microwave electromagnetic energy.

Chemtrails are consistently appearing in all types of advertisements. *GeoengineeringWatch.org* presented a 2012 article about it titled "Chemtrails in Advertisement and Media Everywhere - Why?"

Chemtrails have been drawn into video games. Here are a couple of examples; one from a Lego video game and another from Grand Theft Auto:

Our armed services are also getting in on the act. This is a screenshot of a webpage from the Office of Naval Research (ONR) depicting chemtrails:

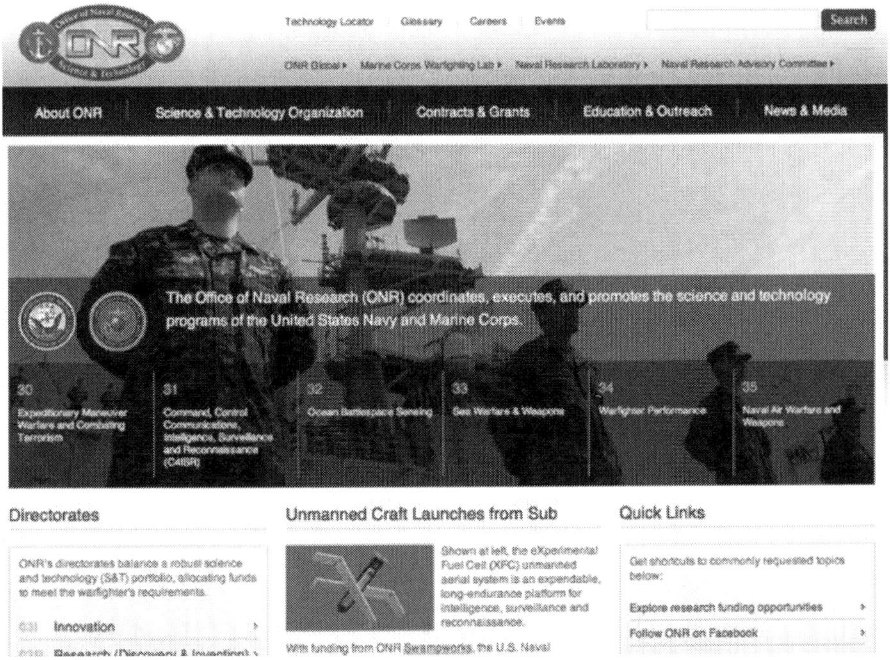

Office of Naval Research chemtrails

It figures. As we have seen throughout the pages of this book, Navy research, specifically the ONR, has been instrumental in the development of technologies used in today's New Manhattan Project and is probably to this day involved.

Controlling the narrative

In this chapter, we have seen many lies coming, in a one way conversation, from the establishment. The advent of the Internet has created a new dynamic, though. Now the average citizen gets to talk back. For when this talkback disagrees with the establishment position, new ways of squelching dissent have been developed.

President Obama's former administrator of the White House Office of Information and Regulatory Affairs, Cass R. Sunstein co-wrote a 2008 paper titled "Conspiracy Theories." Referring to theories about the 9/11 attacks, global warming, the Apollo moon missions, and many others, Sunstein and his co-author wrote that conspiracy theories can be very dangerous and must be dealt with. In stead of outright censorship, the co-authors opted for cognitive infiltration - a.k.a. psychological warfare. The authors write:

"*Government agents (and their allies) might enter chat rooms, online social networks, or even real-space groups and attempt to undermine percolating conspiracy theories by raising doubts about their factual premises, causal logic or implications for political action.*"

Let us not forget that our federal government's official position is that chemtrails do not exist and that the New Manhattan Project is a conspiracy theory.

Cass Sunstein

In his book, shortly before praising *Snopes.com* as a 'reliable and helpful reality check,' Sunstein suggests that, in order to stifle conspiracy theories online, the U.S. Government might: ban conspiracy theories, impose a tax on those that promulgate conspiracy theories, or engage in counterspeech contradicting conspiracy theories. Sunstein claims that he had no involvement in any implementation of his recommendations. That is not to say that implementation did not occur, but one can rest assured that, if such implementation did occur, our sweet little Sunstein did not participate. In his essay titled "Climate Change Justice," Sunstein also identifies himself as a devout adherent of the theory of man-made global warming.

A series of 2014 articles by Glenn Greenwald and others revealed documents, procured by Edward Snowden, showing how the British Government Communications Headquarters (GCHQ) has been running Internet psychological warfare operations against American citizens. The outward manifestation of these operations are myriad disinformation artists which almost always (sometimes immediately) pop up when anyone dares question the persistent trails in the sky or anything else the establishment doesn't want you to know about.

'Sock puppets' (also known as 'shills' and 'trolls') are the dastardly and diabolical creatures that almost invariably crawl out of the Internet's woodwork when one asserts that chemtrails are real. They have many different modes of operation and infest comments sections, forums, and disinformation websites. Although trolls employ many tactics, the most prevalent involve attempts to discredit sources and information. Incessant attacks upon an individual's or organization's character and credibility are used when the quality of the information coming from either cannot be impugned. The thinking goes that if one can discredit the source, then any other information produced by that source is also invalid - even if the other information produced by that source is unassailable. Trolls attempt to discredit information by building 'straw men.' A straw man is built by picking out a small aspect of a larger body of information and attempting to discredit that particular little bit. In this way, if one small part of a larger body of information can be discredited, then all of the larger body can also appear to be discredited.

Many, if not most trolls, are probably not Human. Way back in the early 1970s, people were communicating with computers. Not only that, but computer scientists at Stanford and UCLA were conducting wholly computerized correspondence between two computers over the ARPANET (the precursor to today's Internet). Today, this technology is commonplace. People ask their phones questions all the time. To have 'bots' automatically responding is probably simple and cost-effective; - therefore desirable and prevalent. The robotic manner in which computerized sock puppets respond often reveals their hand.

This is all quite chilling when considering that your author has been compared to domestic terrorists. In a March 14, 2015 article, the *Christian Science Monitor* (of all outlets) published a piece titled "EPA Debunks 'Chemtrails,' Further Fueling Conspiracy Theories." In this article, even though I have no criminal record, your author is compared to a married couple named Jerad and Amanda Miller who shot and killed three people including two police officers at a Las Vegas Wal-Mart before being killed themselves. This is a classic attempt to discredit the source and possibly set him up. That's not very Christian of them now, is it?

The CIA

It is most probably the Central Intelligence Agency (CIA) running the New Manhattan Project's media cover. Although their mandate states that they are not supposed to be operating domestically, the CIA, working with the CFR, has been commanding the American mainstream media for a long time.

It was David Rockefeller (1915-2017) who perhaps put it best when, apparently talking about the CIA's media operations, he said at the 1991 Bilderberg Group meeting:

"We are grateful to the Washington Post, the New York Times, Time Magazine and other great publications whose directors have attended our meetings and respected their promises of discretion for almost 40 years......It would have been impossible for us to develop our plan for the world if we had been subjected to the lights of publicity during those years. But, the world is more sophisticated and prepared to march towards a world government. The supranational sovereignty of an intellectual elite and world bankers is surely preferable to the national auto-determination practiced in past centuries."

It is well documented that the CIA has been directly controlling American mass media. In his 2015 masterpiece *Operation Gladio: The Unholy Alliance Between the Vatican, the CIA, and the Mafia*, author Paul L. Williams writes:

"The CIA, under Allen Dulles, initiated Operation Mockingbird in 1953. This operation involved recruiting leading journalists and editors to fabricate stories and create smoke screens in order to cast

the Agency's agenda in a positive light. Among the news executives taking part were William Paley of the Columbia Bradcasting System (CBS), Henry Luce of Time Inc., Arthur Hays Sulzberger of the New York Times, Barry Bingham Sr. of the Louisville Courier-Journal, and James Copely of Copely Press. Entire news organizations eventually became part of Mockingbird, including the American Broadcasting Company (ABC), the National Broadcasting Company (NBC), the Associated Press, United Press International, Reuters, Hearst Newspapers, Scripps Howard, Newsweek, the Mutual Broadcasting System, the Miami Herald, the Saturday Evening Post, and the New York Herald Tribune. With over four hundred journalists now involved, along with mainstream news outlets, the Agency could operate without fear of exposure."

According to David Talbot in his 2015 classic *The Devil's Chessboard*, noted Washington columnist Joseph Alsop (1910-1989), "once declared it was his patriotic duty to carry the agency's water." Joseph Alsop was a member of the aforementioned AVG Washington Squadron which supported the Flying Tigers (ch 5).

While vigorously suppressing any negative stories about themselves, the agency has been influencing the American news cycle in general. The CIA uses an old boy network to disseminate positive stories about the agency while enticing these same reporters with big, breaking exclusives gleaned from CIA intelligence activities. Early on, when the putative founder of the Agency, Allen Dulles (1893-1969) and Frank Wisner (1909-1965) were working in this area, the Agency's domestic media efforts were less obtrusive and enjoyed only varying degrees of success. One must remember that there used to be more journalistic integrity in America's mainstream media - as opposed, of course, to today's situation where we have people like CNN's Brian Stelter, with the intellectual heft and journalistic integrity of a kumquat running around repeating CIA talking points with extreme zeal. The American journalist of yesteryear was not so quick to sacrifice their rectitude. Nevertheless, early on, the Agency had success placing reporters under deep cover with reputable news outlets. The CIA's involvement in the domestic press was expanded after director Richard Helms (1913-2002) took office in 1966. Helms himself had been a reporter with United Press in Germany before WWII.

A good example of the CIA's tactics here can be found in their handling of the JFK assassination. In his meticulously researched book *Virtual Government*, Alex Constantine makes note of a declassified CIA dispatch outlining the best ways to conduct their disinformation operations. Constantine writes:

"One CIA dispatch, dated April 1, 1967, declassified nine years later under FOIA [Freedom of Information Act], advised planted 'assets' in the media on 'Countering Criticism of the Warren Report.' Features and book reviews 'are particularly appropriate for this purpose,' the CIA dispatch observed. Strategies to 'answer and refute' critics of the government's investigation of the JFK murder included accusations that they were 'financially interested,' 'hasty and inaccurate in their research,' 'infatuated with their own theories,' and 'wedded to theories adopted before the evidence was in.' The CIA's mouthpieces in the press were directed to emphasize: 'No significant new evidence has emerged,' 'there is no agreement among the critics,' or the ever-popular, 'Conspiracy on the large scale often suggested would be impossible to conceal.'"

As they attempt to throw us off the scent of today's New Manhattan Project, chemtrail shills follow these same guidelines.

Notable people & organizations speaking out

We are not alone. Not everyone works for the CIA. Millions of people worldwide are fully chemtrail aware and actively raising more awareness. Along with some governments and other organizations, many famous and notable people have been speaking out against chemtrails and geoengineering as well. What is presented here is undoubtedly a partial list.

Ted Gunderson (1928-2011), the former head of the Los Angeles, Dallas and Memphis FBI claims to have witnessed huge, unmarked chemtrail spraying planes on the ground at the Lincoln, Nebraska Air National Guard facilities. He also said that chemtrail spraying planes operate out of Fort Sill, Oklahoma.

U.S. congressman Dennis Kucinich used the word 'chemtrails' in proposed national legislation. The word appeared in the Space Preservation Act of 2001. The National legislation did not pass, but on

September 10, 2002 the City Council of Berkeley, California, in support of Congressman Kucinich's Space Preservation Act, enacted Resolution #61744 declaring the space 60 kilometers and above Berkeley as a space-based weapons-free zone. On September 14th, Congressman Kucinich was officially presented the resolution by Council member Dona Spring, who initiated the resolution before 700 people in Wheeler Auditorium. The audience gave Kucinich and Spring a standing ovation. On May 19, 2015 the City of Richmond, CA adopted a similar resolution in support of the Space Preservation Act to symbolically ban 'space-based weapons.'

In 2007, Dutch parliamentarian Erik Meijer raised the issue with the European Parliament. The European Parliament, of course, responded that contrails persist and that chemtrails do not exist.

There has been formal discussion of chemtrails in the New Zealand Parliament.

Multi-term Arizona state senator Karen Johnson has been an outspoken critic.

A former Canadian Minister of Defense named Paul Hellyer is chemtrail aware and has spoken out against them publicly.

Former CIA officer Kevin M. Shipp is chemtrail aware and in an interview with Dane Wigington has acknowledged as much publicly.

Former T.V. Weatherman Scott Stevens has spoken out against chemtrails consistently including in an appearance in the movie *Why in the World Are They Spraying?*

In 2012, former Georgia congresswoman Cynthia McKinney spoke out. She passionately cited a recurring pattern of government corruption as evidence for State involvement in today's chemtrail spraying operations. Her speech is presented in a *YouTube* video with the title of "Consciousness Beyond Chemtrails Conference - Cynthia McKinney."

A former mayor of Fairfax, CA by the name of Lew Tremaine is chemtrail aware. In a local 2012 article, he is quoted as saying, "You need only look up at the sky on any given day and watch what's happening." At the time, he was sponsoring the author's proposed, symbolic resolution declaring Fairfax, CA as a 'Chemtrail-free zone.' The article reads, "Tremaine went on to say that he believes the chemtrails are aluminum being used to seed clouds. He said he's seen an increase in aluminum in the water supply over the years, as well." Tremaine's colleague, former Fairfax mayor and town council member Pam Hartwell-Herrero went on to sponsor the proposed resolution after Tremaine's retirement.

As reported by KXLY, Spokane, Washington City Councilmember Mike Fagan is chemtrail aware. In an interview with the *Spokesman Review*, Fagan said, "What I would like to do is issue a challenge to the citizens out there, if they see an airplane in high atmosphere, watch the airplane's activities for a while," Fagan said. "You're going to see a trail come out of that airplane, that dissipates and enlarges itself, floating back down to the ground. Then what happens?"

On July 3, 2013, Arizona state senator Sylvia Allen posted a note on her *Facebook* page indicating that she thinks for herself. She is a very dangerous woman. She wrote:

"The planes usely[sic], three or four, fly a grid across the sky and leave long white trails streaming behind them. I have watched the chem-trails move out until the entire sky is covered with flimsy, thin cloud cover. It is not the regular exhaust coming from the plane it is something they are spraying. It is there in plain sight."

In October of 2013, a Canadian MP by the name of Alex Atamanenko presented to the Canadian Parliament a petition signed by over 1,000 people outlining concerns about chemtrails.

Former editor of the *Wall Street Journal* and U.S. Assistant Secretary of the Treasury Paul Craig Roberts wrote an Oct. 28, 2013 blog post titled "Ignored Reality Is Going to Wipe Out the Human Race" in which he expresses chemtrail awareness. Here is the text:

"Now we come to chemtrails, branded another 'conspiracy theory.' http://en.wikipedia.org/wiki/Chemtrail_conspiracy_theory However, the US government's efforts to geo-engineer weather as a military weapon and as a preventative of global warming appear to be real. The DARPA and HAARP programs are well known and are discussed publicly by scientists. See, for example, http://news.sciencemag.org/2009/03/darpa-explore-geoengineering Search Chemtrails, and you will find much information that is kept from you. See, for example, http://www.globalresearch.ca/chemtrails-a-planetary-catastrophe-created-by-geo-engineering/5355299 and http://www.geoengineeringwatch.org

"Some describe chemtrails as a plot by the New World Order, the Rothchilds, the Bilderbergers, or the Masons, to wipe out the 'useless eaters.' Given the amount of evil that exists in the world, these conspiracy theories might not be as farfetched as they sound.

"However, I do not know that. What does seem to be possibly true is that the scientific experiments to modify and control weather are having adverse real world consequences. The claim that aluminum is being sprayed into the atmosphere and when it comes to earth is destroying the ability of soil to be productive might not be imaginary. Those concerned about chemtrails say that weather control experiments have deprived the western United States of rainfall, while sending the rain to the east where there have been hurricane level deluges and floods.

"In the West, sparse rainfall and lightning storms without rain are resulting in forests drying out and burning down. Deforestation adversely affects the environment in many ways, including the process of photosynthesis by which trees convert carbon dioxide into oxygen. The massive loss of forests means more carbon dioxide and less oxygen. Watershed and species habitat are lost, and spreading aridity further depletes ground and surface water. If these results are the consequences of weather modification experiments, the experiments should be stopped.

"In North Georgia where I spend some summers, during 2013 it rained for 60 consecutive days, not all day, but every day, and some days the rainfall was 12 inches – hurricane level – and roads were washed out. I received last summer 4 automated telephone warnings

from local counties not to drive and not to attempt to drive through accumulations of water on the highways.

"One consequence of the excess of water in the East is that this year there are no acorns in North Georgia. Zilch, zero, nada. Nothing. There is no food for the deer, the turkeys, the bear, the rodents. Starving deer will strip bark from the trees. Bears will be unable to hibernate or will be able only to partially hibernate, forced to seek food from garbage. Black bears are already invading homes in search of food.

"Unusual drought in the West and unusual flood in the East could be coincidental or they could be consequences of weather modification experiments.

"The US, along with most of the world, already had a water problem prior to possible disruptions of rainfall by geo-engineering. In his book, Elixir, *Brian Fagan tells the story of humankind's mostly unsuccessful struggle with water. Both groundwater and surface water are vanishing. The water needs of large cities, such as Los Angeles and Phoenix, and the irrigation farming that depends on the Ogallala aquifer are unsustainable. Fagan reminds us that 'the world's supply of freshwater is finite,' just like the rest of nature's resources. Avoiding cataclysm requires long-range thinking, but humanity is focused on immediate needs. Long-range thinking is limited to finding another water source to deplete. Cities and agriculture have turned eyes to the Great Lakes.*

"Los Angeles exists because the city was able to steal water from hundreds of miles away. The city drained Owens Lake, leaving a huge salt flat in its place, drained the Owens Valley aquifer, and diverted the Owens River to LA via aqueduct. Farming and ranching in the Owens Valley collapsed. Today LA takes water from the Colorado River, which originates in Wyoming and Colorado, and from Lake Perris 440 miles away.

"Water depletion is not just an American problem. Fagan reports that 'underground aquifers in many places are shrinking so rapidly that NASA satellites are detecting changes in the earth's gravity.'

"If the government is experimenting with weather engineering, scientists are playing God when they have no idea of the consequences. It is a tendency of scientists to become absorbed by the ability to experiment and to ignore unintended consequences.

"Readers have asked me to write about Fukushima and chemtrails because they trust me to tell them the truth. The problem is that I am not qualified to write about these matters with anything approaching the same confidence that I bring to economic, war and police state matters.

"The only advice I can give is that when you hear the presstitute media smear a concern or explanation as 'conspiracy theory,' have a closer look. The divergence between what is happening and what you are told is so vast that it pays to be suspicious, cynical even, of what 'your' government and 'your' presstitute media tell you. The chances are high that it is a lie."

Former British Columbia Premiere (1986-1991) Bill Vander Zalm has repeatedly spoken out against chemtrails. In Dane Wigington's 2017 book *Geoengineering: Chronicles of Indictment*, Vander Zalm is quoted as saying:

"For many years, people have been asking me about Climate Control, Geo-Engineering, Global Warming, Climate Change, Greenhouse Gases and Chemtrails. My stock response is 'follow the money.' The most obvious one is Chemtrails (geoengineering aerosol dispersions), we can see them everywhere in the sky above us. What people don't seem to realize is that we are not just seeing contrails, which dissipate quickly, but instead a spraying into the atmosphere of sulfates such as aluminum, barium, and strontium, to reflect back the sun, a poisonous cocktail that eventually rains down on all living things be they animal or plant life.

"In the summer of 2013, after much pressure from fellow British Columbians, I requested, from the government of Canada, 'Freedom of Information,' their involvement in climate control programs. After a rather lengthy wait, I received a 47 page report of which 10 pages of relevant information were completely blanked out and 6 pages blank with only the statement 'information withheld pursuant to sections of the Access of Information Act.' So much for Freedom of Information.

"The report did state that 'Solar Radiation Management' was underway and dealt extensively with a number of approaches to combat greenhouse gases, Global Warming and Climate Change and identified two types of Geo Engineering those being 'Carbon Dioxide Removal' and 'Solar Radiation Management.' It goes on to say that

science is converging on the need to reduce global greenhouse gas emissions immediately in order to limit global warming to 2 degrees Celsius. The document refers to a program to increase the reflectivity of the atmosphere via sulfate injections, a process said to be by the anti-geoengineering activists, the spraying by airplanes, into the atmosphere, of aluminum sulfate, barium and strontium and evidenced by what appears to be 'contrails' - jet condensation trails - except that instead of disappearing in seconds, they spread and stretch from horizon to horizon, linger criss-cross and spread for thousand[s] of feet in the sky. The report also identifies 'stratospheric aerosol injection' (SAI) has a relatively cost effective process and that the spray evidence remains in the sky for only a few days."

On June 25, 2014 the Arizona Department of Environmental Quality held a public hearing about geoengineering and chemtrails. The meeting was sponsored by State Senator Kelli Ward. At the hearing, many local citizens testified to the dangers of ongoing geoengineering programs.

At a July 15, 2014 regular meeting of the Shasta County, CA Board of Supervisors, chemtrails were on the agenda. 300 people attended. Dozens testified. Among other speakers, we heard from a former American Airlines commercial airline pilot by the name of Jeff Nelson who testified to the fact that jet airplane exhaust does not persist. He said that contrails dissipate quickly. That means that these lines in the sky, often stretching from horizon to horizon, are not unadulterated contrails. We also heard similar testimony from another commercial pilot named Russ Lazuka. Additionally, we learned that a Shasta County, CA supervisor by the name of Bill Schappell recognizes geoengineering as an ongoing, secret operation. The supervisors voted to look into updating their local air monitoring equipment and to send a compilation of the information discussed at the meeting to appropriate government agencies.

Former Marin County Board of Supervisors, District 1 Representative Susan Adams has voiced interest in chemtrail legislation. She emailed the author stating, "I will continue to monitor your efforts and if there is federal legislation or regulation being considered which the county can send a letter of support for, please let me know."

Former Eleventh Senatorial District California State Senator Mark Leno recognizes chemtrails. He responded to the author's letter about chemtrails by stating:

"Thank you for writing to share the information on chemtrails. I appreciate your taking the time to voice your concerns regarding this important issue. As we begin to learn more about chemtrails and begin to have more in depth conversations about their use, I will keep the information you have forwarded in mind."

In August of 2015, geoengineering activists Patrick Roddie, Jim Lee, Amanda Bayes, and Max Bliss testified before the Environmental Protection Agency (EPA) in Washington D.C. They went on record about the hard evidence showing the mountains of toxins raining down upon us and this project's massive environmental fallout.

In late 2015, a former commercial passenger airline pilot named Willem Felderhof began speaking and writing publicly about his realization that those lines in the sky are not unadulterated contrails and that we are being sprayed.

In a December 2015 article appearing on *ActivistPost.com* titled "Saudi Princess Speaks Out Against Chemtrails and Geoengineeing," a member of the Saudi royal family, Princess Basmah Bint Saud went public with her concerns. She described the ongoing geoengineering program as, "a weapon of mass destruction; a method of slow poisoning which threatens all life on Earth." She expressed concern about the biological impacts. She expressed uneasiness about what kind of world we will leave for our children.

A North Atlantic Treaty Organization (NATO) commander by the name of Fabio Mini and a member of the Italian parliament by the name of Sandro Brandini have gone public. *Russia Today* and *Geoengineering Watch* covered all this in the *YouTube* video titled "NATO General Expresses Alarm Over Atmospheric Aircraft Spraying." Mini says that, "The pictures we can find online are really dreadful because the trails intersect regularly." He continues, "I cannot say the difference between the ones caused by the condensation and the ones left up there on purpose. I cannot speak on the ones left there on purpose, or what the purpose is. I mean, why are they left there?" He notes, "I think that, often exhaustive answers are not given because the knowledge of these phenomena is very limited." In closing he notes, "This situation increases the number of suspects. I understand the authorities approach much more not to give any explanation rather than totally deny."

Brandini says:

"What is sure is that it's not a mysterious phenomenon like some people may think. These trails are produced by the airplanes; especially the military ones. And in the course of time, they have changed their appearance. That's why I have said this phenomenon is real and it strongly concerns the citizens. Therefore the citizens need to be informed. Who has to do that? The government through the appropriate ministers; both the environment and the defense. If it was such a trivial issue, it wouldn't have been raised throughout the world. For instance, I know that in the U.S., the government made an investigation concerning this subject and there the conclusions weren't exhaustive either. Probably there's something related to military activities. That, I think, isn't shocking. The real problem is that when there are experiments and phenomena of this type, the government of the country is the first to be aware of them. In any case, the citizens should be reassured by saying they are experiments; or whatever they are. That they are not a concern for the health of the citizens."

The Cypriot government is taking action. Due to many complaints, Agriculture Minister Kouyialis has ordered an official investigation. The story was covered in a Feb. 17, 2016 article titled "Minister to Probe Mysterious Chemtrails."

Action star of movies and television Chuck Norris made a case for the reality of chemtrails in his April, 2016 piece written for *World Net Daily* titled "Sky Criminals."

In a May, 2016 piece appearing in the British publication *Sunday Express*, a former Essex County councillor by the name of John

Dornan expressed chemtrail awareness. He is quoted as saying, "There is a clear blue sky, then the planes come and the trails criss-cross over each other, until they spread out into long clouds. After that the sunlight dims." He continues, "Your eyes can't lie. It is happening everyday and the evidence is on this website [*GeoengineeringWatch.org*]."

In the fall of 2016, the Minneapolis Community College was offering classes about chemtrails.

Actress Hayden Panettiere is chemtrail aware. In a March 18, 2017 tweet, above a picture of a chemtrail laden sky she wrote, "What the heck r they spraying over us?! It's been happening for years but only this extreme for the past few months."

In late September of 2017, the Rhode Island general assembly passed historic anti-geoengineering legislation titled the Geoengineering Act of 2017. The act was sponsored by state Representative Justin Price (R). The act reads:

"The Rhode Island general assembly finds that geoengineering encompasses many technologies and methods involving hazardous activities that can harm human health and safety, the environment, and the economy of the state of Rhode Island. . .. It is therefore the intention of the Rhode Island general assembly to regulate all geoengineering activities."

The act requires that a person seeking to engage in geoengineering activities gets a license from the director of the Rhode Island Department of Environmental Management. Violators would be punished $500,000 per offense with not less than 190 days in jail.

~ ~ ~

Quite a few famous musicians have spoken out. Beck released a 2008 song titled "Chemtrails." The lyrics read like this:

So many people Where do they go?

You and me watching a sky full of chemtrails Watching the jet planes go by

Prince sang about chemtrails in his song “Dreamer,” from his 2003 album *Lotusflower*:
While the helicopter circles
And the theory’s getting deep
Think they’re spraying chemicals over the city
While we sleep?
From now on I’m staying awake
So you can call me a dreamer too
Wake up wake up

In 2010, Billy Corgan of the Smashing Pumpkins took time out from one of his live performances to point out the 'toxic death smoke' in the sky. The video is available on *YouTube*.

In a 2011 song titled "What I Hate," country music legend Merle Haggard sings, "What I hate is looking up and seeing chemtrails in a clear blue sky today." The song is available on *YouTube*.

The rock and roll band Cake is apparently chemtrail aware. Lyrics from multiple songs indicate awareness. Their 2001 song "Comfort Eagle" goes like this:
And the fluffy white lines that the airplane leaves behind are drifting right in front of the waning of the moon

Their 2004 ditty "Wheels" informs us:
In a wooden boat in the shipping lanes With the freighters towering over me I can hear the jets flying overhead Making lines across the darkening sky

Lastly, Cake's 2011 release "Easy to Crash" states:
Clouds hung hugely and oppressively Over our busy little cars
Clouds hung hugely and oppressively We didn't notice
We didn't care We didn't notice

~ ~ ~

A shill online once challenged your author to find any doctors or PhDs who have spoken out against chemtrails and geoengineering. At the time, I really wasn't aware of any. Since that fateful day, I have been compiling a list. To whomever that pathetic creature was… thank you for making us stronger. Here is the list:

Lenny Thyme, PhD, Dr. Russell L. Blaylock, Ann Fillmore, PhD, Leuren Moret, PhD, Dr. Bill Deagle, Dr. Joseph Puleo, Dr. Rosalie Bertell, Dr. Stephen D. McKay, Dr. Patrick Flanagan, Dr. Leonard G. Horowitz, Dr. Eric Karlstrom, Dr. Leonard A. Cole, Dr. Betty Martini, Dr. Thomas E. Bearden, Dr. Edward Group, Dr. Ilya Sandra Perlingieri, Dr. Sherry Tenpenny, Dr. Rima E. Laibow, Dr. Lorraine Hurley, J. Marvin Herndon, PhD, Dr. Steven Amato, Dr. Ted Broer, Dr. Stephen Davis, Dr. Barrie Trower, Dr. Hamid Rabiee, Cynthia McKinney, PhD, Dr. Frank Livolsi, Dr. Mark Whiteside, MD, Dr. Dietrich Klinghardt, Dr. Hans Kugler, Dr. Chuck Baldwin, Dr. Timothy Ball, PhD, & Dr. R. Michael Castle

Let history remember the people who had enough common sense to recognize the obvious and enough guts to tell the truth. When the New Manhattan Project is dead and gone, let the people who spoke out against it prosper. Also, let this book stand as a testament to what the weather was like and what the sky looked like under the tyranny of the New Manhattan Project. Let everyone see all the times that government agencies misrepresented chemtrails as contrails and told us that the lines in the sky were harmless water vapor that had always been. Let it be known about how the mainstream media ceaselessly excoriated 'tin-foil hat wearing conspiracy theorists' for stating the obvious. You see, they *knew for sure* that chemtrails did not exist. And they *knew for sure* that you were wrong. Well, as it turns out, *they* were wrong. But don't worry, they are sure to just act like they were always right and go on to their next line of B.S. Let this work stand as a testament to the establishment's lies.

References

The Making of the Atomic Bomb a book by Richard Rhodes, published by Simon and Schuster, 2012

Cult of the Atom: The Secret Papers of the Atomic Energy Commission a book by Daniel Ford, published by Simon and Schuster, 1982

"Science, Engineering and the Quality of Life" a report by the Monsanto Company, published by the Monsanto Company, 1970

"Outline of Weather Proposal" a paper by Vladimir K. Zworykin, published by RCA Laboratories, 1945

"Can We Survive Technology?" an article by John von Neumann, published in *Fortune* magazine, June, 1955

"John von Neumann: The Scientific Genius Who Pioneered the Modern Computer, Game Theory, Nuclear Deterrence, and Much More" a book by Norman Macrae, published by the American Mathematical Society, 1999

"A Study of Climatological Research as it Pertains to Intelligence Problems" a report by the Central Intelligence Agency, 1974

"The World as a Mathematical Game: John von Neumann and 20th century science" a book by Giorgio Israel and Ana Millán Gasca, published by Birkhäuser, 2009

Killing the Planet: How a Financial Cartel Doomed Mankind a book by Rodney Howard-Browne and Paul L. Williams, published by Republic, 2019

"The Future" a paper by Vincent Schaefer, Bernard Vonnegut, J.S. Barrows, and P.B. MacCready, Jr. as published in the *Final Report of the Advisory Committee on Weather Control*, a report by the Advisory Committee on Weather Control, 1958

"Weather as a Weapon" an article by Howard T. Orville as told to John Kord Lagemann, published by *Popular Science*, June 1958

"Restoring the Quality of Our Environment" a report by the Environmental Pollution Panel of the President's Science Advisory Committee, published by the U.S. Government Printing Office, 1965

National Science Foundation ninth annual weather modification report, 1967

"Contrail Effects on the Atmospheric Thermal Radiation Budget" a paper by P.M. Kuhn, produced by the Environmental Science Services Administration Research Laboratories, Atmospheric Physics and Chemistry Laboratory, Boulder, Colorado as it appeared in the Second National Conference on Weather Modification proceedings by the American Meteorological Society, 1970

"A U.S. Policy to Enhance the Atmospheric Environment" a paper by the Weather Modification Advisory Board as it appeared in a weather modification hearing before the Subcommittee on the Environment and the Atmosphere of the Committee on Science and Technology, U.S. House of Representatives, Ninety-fifth Congress, October 26, 1977

"On Geoengineering and the CO2 Problem" paper by Cesare Marchetti, published in *Climatic Change*, v1 n1, 1977

"Policy Implications of Greenhouse Warming: Mitigation, Adaption and the Science Base" a report by the National Academy of Sciences, published by the National Academy Press, 1992

"Global Warming and Ice Ages: Prospects for Physics-Based Modulation of Global Change" a paper by Edward Teller, Lowell Wood, and Roderick Hyde, published by Lawrence Livermore National Laboratory, 1997

"Long-Range Weather Prediction and Prevention of Climate Catastrophes: A Status Report" a paper by E. Teller, K. Caldeira, G. Canavan, B. Govindasamy, A. Grossman, R. Hyde, M. Ishikawa, A. Ledebuhr, C. Leith, C. Molenkamp, J. Nuckolls, and L. Wood, published by Lawrence Livermore National Laboratory, 1999

"Active Climate Stabilization: Practical Physics-Based Approaches to Prevention of Climate Change" a paper by Edward Teller, Roderick Hyde, and Lowell Wood, published by Lawrence Livermore National Laboratory, 2002

"Aviation and the Global Atmosphere" a report by the Intergovernmental Panel on Climate Change (IPCC), 1999

"EPA Aircraft Contrails Factsheet" a report by the Environmental Protection Agency, the National Aeronautics and Space Administration, National Oceanic and Atmospheric Administration, and the Federal Aviation Administration, 2000

"Plume Processing of Jet Engine Exhaust Aerosols Injected Into the Upper Troposphere and Lower Stratosphere" a paper by D.E. Hagen, P.D. Whitefield, J. Paladino, and O. Schmid as it appeared in the proceedings of the 15th International Conference on Nucleation and Atmospheric Aerosols of the American Institute of Physics, published by the American Institute of Physics, 2000

"Estimates of Cloud Radiative Forcing in Contrail Clusters Using GOES Imagery" a paper by David P. Duda, Patrick Minnis, and Louis Nguyen, published by the *Journal of Geophysical Research*, vol. 106, no. D5, p4927-4937, March 16, 2001

"Energy Security and Climate Change" a report by John Deutch, Anne Lauvergeon, and Widhyawan Prawiraatmadja, produced by the Trilateral Commission, 2007

"Chemtrails Confirmed: Climate Scientist Admits Jets Are 'Dumping Aerosols.'" Published by *Chemtrailsplanet.net*, January 9, 2015

"Engineering the Climate: Science, Governance and Uncertainty" a report by the U.K. Royal Society, 2009

"Contrails: Frequently Asked Questions" by the British Department for Transport
"The Government's View on Geo-engineering Research" by the British government

"The Regulation of Geoengineering" a report by the British House of Commons / U.S. House of Representatives Science and Technology Committee 2010

"*Geoengineering: Parts I, II, and III*" hearing before the Committee on Science and Technology House of Representatives 2009-2010

"Engineering the Climate: Research Needs and Strategies for International Coordination" a report by the United States House of Representatives Science and Technology Committee, 2010

"Advancing the Science of Climate Change" a report by the National Research Council of the National Academies of Science, 2010

"Geoengineering: Governance and Technology Policy" a report by the Congressional Research Service, 2011

"Report of the Defense Science Board Task Force on Trends and Implications of Climate Change for National and International Security" a report by the Office of the Undersecretary of Defense for Acquisition, Technology, and Logistics, the Defense Science Board, October, 2011

"Geoengineering the Climate: Research Questions and Policy Implications" a report by The United Nations Educational Scientific and Cultural Organization (UNESCO) November, 2011

"IPCC Expert Meeting on Geoengineering" a report by Intergovernmental Panel on Climate Change, 2011

"Governing Geoengineering Research: A Political and Technical Vulnerability Analysis of Potential Near-term Options" a report by Robert J. Lempert and Don Prosnitz, published by the Rand Corporation, 2011

"America's Climate Choices" a report by the National Research Council of the National Academy of Sciences, published by the National Academies Press, 2011

"Solar Radiation Management: The Governance of Research" a report by the Royal Society, the Environmental Defense Fund, and TWAS, published by the Royal Society, the Environmental Defense Fund, and TWAS, 2011

"Celebrating 50 Years of Success" a report by The Institute of Atmospheric Sciences at the South Dakota School of Mines and Technology, 2012

“Geoengineering the Climate System” a policy statement of the American Meteorological Society, published by the American Meteorological Society, 2013

“Climate Change 2013: The Physical Science Basis” a report by the Intergovernmental Panel on Climate Change, published by the Cambridge University Press, 2014

“Geoengineering: Innovation, Research, and Technology” hearings before the United States House of Representatives Science, Space, and Technology Committee, November 8, 2017

“Understanding Climate Engineering” an article by Deborah Gordon, published by the Carnegie Endowment for International Peace, August 21, 2017

“Could Geoengineering Save the Planet from Global Warming?” an article by Deborah Gordon and David Livingston, published by the Carnegie Endowment for International Peace, Sept. 21, 2017

“Advancing Public Climate Engineering Disclosure” an article by Deborah Gordon, Smriti Kumble, and David Livingston, published by the Carnegie Endowment for International Peace, May 29, 2018

“Why Bill Gates Is Funding Solar Geoengineering Research” a video by CNBC, published on *YouTube* September 7, 2019

"Climate Intervention: Reflecting Sunlight to Cool Earth" a report by the National Academy of Sciences, 2015

"A Portfolio of Carbon Management Options" a paper by Ken Caldeira, M. Granger Morgan, Dennis Baldocchi, Peter G. Brewer, Chen-Tung Arthur Chen, Gert-Jan Nabuurs, Nebojsa Nakicenovic, and G. Philip Robertson, published in *ResearchGate*, March, 2004

"On the Regulation of Geoengineering" a paper by David G. Victor, published in the *Oxford Review of Economic Policy*, 2008

"Confronting Climate Change: A Strategy for U.S. Foreign Policy: Report of an Independent Task Force" a report by the Council on Foreign Relations, published by the Council on Foreign Relations, 2008

"The Geoengineering Option: A Last Resort Against Global Warming?" an article by David G. Victor, M. Granger Morgan, Jay Apt, John Steinbruner, and Katherine Ricke, published by *Foreign Affairs*, March, 2009

"Q&A With David Victor About Climate Change: What governments, scientists, and big business can do about global warming" an article by Lucy Berman, published by *Foreign Affairs*, March 18, 2009

"Why Geoengineering?" an article by Granger Morgan, published by the *MIT Technology Review*, December 21, 2009

"Research on Global Sun Block Needed Now" an op-ed by David W. Keith, Edward Parson, and M. Granger Morgan, published in *Nature*, January 2010

"Regional Climate Response to Solar Radiation Management" a paper by Katharine L. Ricke, M. Granger Morgan, and Myles R. Allen, published in *Nature Geoscience*, August 2010

"Developing an International Framework for Geoengineering" a panel discussion conducted at the Council on Foreign Relations w/ Granger Morgan and John Steinbruner moderated by Ruth Greenspan Bell, Mar. 10, 2010

"IPCC Expert Meeting on Geoengineering: Meeting Report" a report by the Intergovernmental Panel on Climate Change, published by the Intergovernmental Panel on Climate Change, 2012

"Characterizing Impacts and Implications of Proposals for Solar Radiation Management, a Form of Climate Engineering" a doctoral thesis by Katharine L. Ricke, August, 2011

"Geoengineering: A National Strategic Plan for Research on the Potential Effectiveness, Feasibility, and Consequences of Climate Remediation Technologies" a report by the Bipartisan Policy Center's Task Force on Climate Remediation Research, published by The Bipartisan Policy Center, 2011

"Cost Analysis of Stratospheric Albedo Modification Delivery Systems" a paper by Justin McClellan, David Keith, and Jay Apt, published by *Environmental Research Letters*, Aug. 30, 2012

"Needed: Research Guidelines for Solar Radiation Management" an article by Granger Morgan, Paul Gottlieb, and Robert R. Nordhaus, published in *Issues in Science and Technology*, Spring 2013

"The Truth About Geoengineering: Science Fiction and Science Fact" an article by David G. Victor, M. Granger Morgan, Jay Apt, John Steinbruner, and Katherine Ricke, published by *Foreign Affairs*. March 27, 2013

"Start Research on Climate Engineering" a paper by Jane C.S. Long, Granger Morgan, and Frank Loy, published in *Nature*, Feb. 5, 2015

"Climate Engineering Economics" a paper by Garth Heutel, Juan Moreno-Cruz, and Katharine Ricke, published by the *Annual Review of Resource Economics*, Nov. 3, 2015

"Chemtrails in Advertisement and Media Everywhere - Why?" an article by Dane Wigington, published by *GeoengineeringWatch.org*, Nov. 10, 2012

"Conspiracy Theories" a paper by Cass R. Sunstein and Adrian Vermeule, produced by the Harvard University Law School and the University of Chicago Law School, 2008

Conspiracy Theories and Other Dangerous Ideas a book by Cass R. Sunstein, published by Simon & Schuster, 2014

"Snowden Docs Reveal British Spies Snooped on YouTube and Facebook" an article by Glenn Greenwald, published by *NBC News*, Jan., 27, 2014

"Snowden Docs Show UK Spies Attacked Anonymous, Hackers" an article by Mark Schone, Richard Esposito, Matthew Cole, and Glenn Greenwald, published by *NBC News*, Feb. 4, 2014

"Snowden Docs Show British Spies Used Sex and 'Dirty Tricks'" an article by Mark Schone, Richard Esposito, Matthew Cole, and Glenn Greenwald, published by *NBC News*, Feb. 7, 2014

"How Covert Agents Infiltrate the Internet to Manipulate, Deceive, and Destroy Reputations" an article by Glenn Greenwald, published by The Intercept, Feb. 24, 2014

"EPA Debunks 'Chemtrails,' Further Fueling Conspiracy Theories" an article by Brad Knickerbocker, published by the *Christian Science Monitor*, March 14, 2015

Operation Gladio: The Unholy Alliance Between the Vatican, the CIA, and the Mafia a book by Paul L. Williams, published by Prometheus Books, 2015

The Devil's Chessboard: Allen Dulles, the CIA, and the Rise of America's Secret Government a book by David Talbot, published by HarperCollins Publishers, 2015

The CIA and the Cult of Intelligence a book by Victor Marchetti and John D. Marks, published by Dell Publishing Co., 1980

Virtual Government: CIA Mind Control Operations in America a book by Alex Constantine, published by Feral House, 1997

"Space Preservation Act of 2001" H.R. 2977

"Resolution in Support of the Space Preservation Act and the Space Preservation Treaty to Permanently Ban the Weaponization of Space" City of Berkeley Resolution no. 61744, 2002

"People Fearing Space Weapons Target One California Town" an article published by *CBS News*, June 2, 2015

Why in the World Are They Spraying? a documentary by Michael J. Murphy, produced by Truth Media Productions, 2012

Geoengineering: Chronicles of Indictment: Exposing the Global Climate Engineering Cover-up a book by Dane Wigington, published by Geoengineering Watch Publishing and Media LLC, 2017

"Geoengineering Activist Patrick Roddie Testifies Before the EPA" an article by Peter A. Kirby, published by *Activist Post*, August 16, 2015

"Saudi Princess Speaks Out Against Chemtrails and Geoengineeing" an article by Patrick Roddie and Peter A. Kirby, published by *ActivistPost.com*, December 11, 2015

"Minister to Probe Mysterious Chemtrails" an article published by In-Cyprus.com, Feb. 17, 2016

"Sky Criminals" an article by Chuck Norris, published by *World Net Daily*, April 24, 2016

"CHEMTRAILS: Are jet planes REALLY secretly spraying chemicals to REVERSE climate change?" an article by Jon Austin, published in the *Sunday Express*, May 22, 2016

"Rhode Island Tackles 'Geoengineering'" an article by American Free Press published by *American Free Press*, October 27, 2017

Chapter 12
A CONVENIENT LIE

"Global warming is about politics and power rather than science. In science, there is an attempt to clarify; in global warming, language is misused in order to confuse and mislead the public."
-Richard S. Lindzen, former professor of meteorology at the Massachusetts Institute of Technology

"The whole aim of practical politics is to keep the populace alarmed (and hence clamorous to be led to safety) by menacing it with an endless series of hobgoblins, all of them imaginary."
-H.L. Mencken (1880-1956)

Today's geoengineers say that because of man-made global warming and/or climate change, it is a good idea that we might be sprayed with tens of thousands of megatons of toxic materials from aircraft annually. On its face that is a psychotic assertion. Their assertion presumes the death of probably hundreds of thousands of people, wholesale destruction of our environment, and untold losses of plants, animals, insects, and all other forms of life on this planet. But, not only is their suggestion horrible almost beyond comprehension, in fact, the *premise* of their assertion is incorrect as well.

The theory of man-made global warming and climate change is a lie. It is so incredibly false that many people promoting it must know better. The science supporting it is patently and provably false, but the establishment continues to use it as a battering ram in attempts to legitimize the New Manhattan Project and accomplish myriad socio-economic agendas. The global warming/climate change lie is so often repeated because there is a mountain of money and power behind it, not because of an organic social movement. The theory came from the top down, not the bottom up. As we will learn here, today's prevalence of the theory of man-made climate change is the product of a power-hungry elite, a political ideology, the military/industrial complex, and a corrupted scientific establishment. Calling the theory of man-made global warming 'an inconvenient truth' is an inversion of reality. It is a convenient lie. This chapter exposes the incorrect and unscientific nature of the theory of man-made climate change, its true history, and the corrupted scientific establishment behind it.

Incorrect

The theory of man-made global warming and climate change rests upon the notion that man's contributions of atmospheric carbon dioxide (CO2) are having a significant effect upon the Earth's climate. Man's burning of hydrocarbon fuels, which produces atmospheric CO2, is specified as the main contributing factor. So let's take a look at atmospheric carbon dioxide, shall we?

Atmospheric carbon dioxide comprises a minuscule .038% of Earth's atmosphere. Of that .038%, only 3% is from the burning of carbon-based fuels. 97% of atmospheric carbon dioxide comes from natural sources such as: the oceans, volcanoes, dead and dying vegetation, people and animals exhaling, forest fires, etc. 3% in any context is usually considered statistically insignificant. So how is it that a statistically insignificant 3% of atmospheric carbon dioxide, has a significant effect upon climate? The fact of the matter is that it doesn't.

In fact, as evidenced in ice core samples, current levels of atmospheric CO2 are the lowest in 600 million years. Although proponents of the theory assert that man's industrial activity is to blame, current levels of atmospheric CO2 are *lower* than before the Industrial Revolution and *nowhere near* all-time highs.

The best data (satellite data) shows that the Earth's average temperature is not rising. In fact, for about the last 15 years, the Earth's average temperature has gone down slightly. The much ballyhooed long-range atmospheric models predicting warming are based on faulty data and are the product of a collaboration of scientists and computer modelers working for lobbyists, bureaucrats, and politicians. Al Gore and Michael Mann's 'hockey stick' chart showing a huge upswing in average global temperature has been scientifically and systematically proven to be false. While levels of CO2 have risen over the last 20 years, Earth's average temperature has not.

In fact, the entire notion that increases in atmospheric CO2 cause an increase in average global temperature, which is what today's global warming alarmists assert, is completely backwards. In his book *The Deliberate Corruption of Climate Science*, climatologist Dr. Timothy Ball, PhD writes that ice core samples show that an increase in the Earth's average global temperature is ALWAYS FOLLOWED by an increase in atmospheric CO2 - not the other way around. Dr.

Ball writes, "Every record for any period shows that temperature increases *before* CO2. The only place a CO2 increase precedes a temperature increase is in IPCC [Intergovernmental Panel on Climate Change] computer models."

Although climate alarmists consistently assert otherwise, sea levels are currently not rising significantly and don't look to do so anytime soon. Sea levels have historically been rising for the past 700 years, but lately the rate of sea level rise has slowed to almost negligible amounts. The glaciers have been melting since long before man started burning large amounts of hydrocarbon-based fuels as well. This is because we are in what is known as an 'interglacial' period.

The history of Earth's climate is interspersed by interglacial periods and ice ages. It is much more probable that a new ice age is coming rather than some sort of anomalous warming period. In fact, in the case of global warming, periods of increased global temperatures are associated with more life on this planet, not less. More sun, increased temperatures, and more CO2 = more life. Atmospheric carbon dioxide is good. It makes plants grow. They breathe it like we breathe oxygen. With more CO2, foliage grows bigger and, in turn, provides us with what *we* need to grow: oxygen.

Being that Earth's average temperature has been rising since long before the Industrial Revolution, it is logical to assert that other factors, besides man's burning of hydrocarbon fuels, are responsible. An obvious choice is solar activity. Although, among many other deficiencies, the leading climate models assume a constant amount of energy coming from the sun (they assume that the Earth is a flattened sphere with the sun's energy distributed evenly across the surface), this is not the case. The sun goes through cycles where it produces more or less energy, thus having a warming or cooling effect upon the Earth.

The leading climate models also assume a constant amount of energy coming from the Earth. This too is not the case. Nuclear geophysicist Dr. Marvin Herndon, PhD asserted a long time ago that observed fluctuations in the heat energy produced by the Earth itself are a product of changes in the activity of a natural nuclear reactor which exists in the center of our planet. In fact, he asserts that most if not all planets have a similar reactor at their cores. For these observations and assertions, the good Dr. Herndon was excommunicated from the scientific establishment. You see, Dr.

Herndon's discoveries did not fit in with the establishment's theory of man-made global warming and so, they had no use for him.

Unscientific

Not only is the theory of man-made climate change incorrect, as it is promoted, it is unscientific. We the general public have heard from politicians, bureaucrats, businessmen, pundits, movie stars, athletes, and many other unknowledgeable people *ad nauseam* about how the science regarding man-made climate change is 'settled.' All this, while the most rudimentary understanding of the scientific method allows for the fact that true science is *never settled*. The notion that the science is settled is, in itself, unscientific. When someone asserts that the science regarding the theory of man-made climate change is settled, it only proves that they have been indoctrinated into a certain belief - and nothing else. The scientific thing to do is to attempt to disprove a thesis rather than prove it.

The popular talking point about 95% of scientists espousing the theory of man-made global warming is untrue. The Heidelberg Appeal of 1992 called the theory of man-made global warming into question and garnered the signatures of about 4,000 scientists, including 72 Nobel Prize winners. Secondly, 31,487 U.S. scientists (including 9,029 with PhDs), as part of something called the Global Warming Petition Project, have publicly signed a statement declaring that they consider the catastrophic man-made global warming hypothesis to be inconsistent with the evidence. These are just the scientists who have gone public. There are probably many more who don't want the public persecution associated with such viewpoints, and therefore keep their views private. We're talking here about a very significant portion of the scientific community bucking the theory of man-made global warming - possibly a majority. Even if 95% of scientists *did* agree with the theory of man-made global warming, that still proves nothing. As Dr. Herndon writes, "Popularity only measures popularity, not scientific correctness; science is a logical process, not a democratic process. In science, consensus is nonsense."

This example of media inaccuracies about the theory of man made global warming is the most prevalent, but there are many more. In fact, there is so much misreporting about man-made climate change that it is well beyond the scope of this book to cover it all. It suffices to say that just about *everything* reported in the media about man-made climate change is incorrect. But, since this myth about 95% of all scientists agreeing upon the theory of man-made global warming so strongly persists, let us continue to expose the truth. Author Mark Steyn produced a 2015 book titled *A Disgrace to the Profession* in which scores of reputable and/or famous scientists from around the world, in their own words, refute the work of leading climate scientist Michael Mann with many refuting the theory of man-made global warming entirely. The CO2 Coalition is comprised of many knowledgable people, including many prominent scientists, who refute the theory of man-made global warming. The 'About' page from their website says it all:

"The CO2 Coalition was established in 2015 as a 501(c)(3) for the purpose of educating thought leaders, policy makers, and the public about the important contribution made by carbon dioxide to our lives and the economy. The Coalition seeks to engage in an informed and dispassionate discussion of climate change, humans' role in the climate system, the limitations of climate models, and the consequences of mandated reductions in CO2 emissions.

"In carrying out our mission, we seek to strengthen the understanding of the role of science and the scientific process in addressing complex public policy issues like climate change. Science produces empirical, measurable, objective facts and provides a means for testing hypotheses that can be replicated and potentially disproven. Approaches to policy that do not adhere to the scientific process risk grave damage to the economy and to science."

Lastly, in a 2019 interview, Greenpeace co-founder and former president of Greenpeace Canada, Dr. Patrick Moore, PhD described the cynical and corrupt machinations fueling the narrative of anthropocentric global warming and climate change. Did you hear about that on CNN?

The true history

The theory of man-made global warming originated with a Swedish scientist by the name of Svante Arrhenius (1859-1927). During his time with something called the Stockholm Physics Society (which he founded), Arrhenius began his work on the subject on Christmas Eve of 1894, and his first paper on the subject was published exactly one year later. So the public had their first opportunity to know about his theory of man-made global warming around the beginning of 1896. Exactly one hundred years later, in 1996, large-scale, American domestic spraying operations began. Also in 1996, the most well-known English biography of Arrhenius was published. These are not the only uncanny connections between the theory of man-made global warming, its originator, and the New Manhattan Project.

Svante Arrhenius

Arrhenius worked on the theory of man-made global warming with his friend and colleague Vilhelm Bjerknes (1862-1951), the father of atmospheric modeling. In fact, many of Arrhenius' friends and colleagues worked in the atmospheric sciences, some even dabbled in the fledgling field of weather modification. Arrhenius himself worked extensively in other areas of the atmospheric sciences including atmospheric electricity (ch 3) and he produced at least one paper published in the local media which hypothesized about the means to artificially produce rainfall.

Arrhenius' man-made global warming investigation pertained to atmospheric carbon dioxide's influence upon the coming and passing of ice ages. Arrhenius warned that our present existence may be, "nothing but a short flourishing of civilization between two Ice Ages." So, according to Arrhenius, man-made global warming may have been happening, but it was a good thing. Although Arrhenius is not referenced, it is interesting to note that in 1997 Lawrence Livermore National Laboratory scientists Edward Teller, Lowell Wood, and Roderick Hyde released the aforementioned research paper titled "Global Warming and Ice Ages: Prospects for Physics-Based Modulation of Global Change" wherein the authors suggest that saturating the upper atmosphere with small aluminum particles could avert the next ice age.

In all honesty, there was one other scientist who had previously postulated that levels of atmospheric CO2 could have an effect upon climate. In 1861 England's John Tyndall (1820-1893) suggested as much. But Arrhenius is more notable here because he produced a model in attempts to prove it and he popularized the theory with multiple media publications and speeches. Also, Arrhenius' assertions about atmospheric CO2 having a warming effect upon Earth's climate were based upon the so-called 'greenhouse effect' of atmospheric vapors first postulated by Joseph Fourier (1768-1830) in 1824. The major difference here is that Fourier did not specify CO2. Although the term 'greenhouse effect' did not come into use until much later, it was probably Arrhenius who first used the hothouse metaphor. In America, Arrhenius' theories about atmospheric carbon dioxide were further investigated by geologist Thomas Chamberlin (1843-1928) at the Rockefeller founded and funded University of Chicago.

Following Arrhenius' early significant work on the subject, the next major development in the saga of the theory of man-made climate change really didn't come until 1955 when *Fortune* magazine published the aforementioned (ch 11) article written by John von Neumann titled "Can We Survive Technology?"

Although Arrhenius is the putative originator of the theory of man-made global warming, pertaining to this, he really originated nothing. Arrhenius' work here was unscientific from the beginning. He *first* had the idea that changes in amounts of atmospheric CO2 had an effect upon climate, *then* he created a model to prove it. The proper way to conduct science is to simply conduct experiments and see what happens. Arrhenius' work here was contrary to the scientific method and mathematical models are inherently of seriously questionable scientific validity - or as the British statistician George E.P. Box (1919-2013) stated, "Essentially, all models are wrong, but some are useful." For more about the philosophies behind proper scientific methods and practices, please refer to Marvin Herndon's excellent books *Maverick's Earth and Universe* as well as *Herndon's Earth and the Dark Side of Science*.

Arrhenius' models were based on what amounts to wild speculation. He did not have access to the vast amounts of accurate data needed to confirm his grand assertions. Networks for the collection of the necessary data did not exist at the time. His work in this area relied simply upon guesstimates and contributions from his peers - not any real empirical data. He was in way over his head and for these reasons, his work in this area holds no validity whatsoever.

It is interesting and vital to our investigation to note that members of Arrhenius' Stockholm Physics Society were, to a person, political progressives. Conversely, members of the progressive political parties of the time were also scientists. With the influence of these predominant political views surrounding him, Arrhenius became a progressive himself. Crawford, the author of *Arrhenius: From Ionic Theory to the Greenhouse Effect* writes:

"While the passion for science was the primary bond between members of the society, they also shared progressive political opinions. Several of them - Nils Ekholm, Hugo Hamberg, Arvid Högbom - had been radicalized as students at Uppsala in the 1880s and one of them, Arvid Högbom, could claim to have been the first dues-paying member of the radical student association Verdandi, founded in 1882. The political circles in the capital with which they were most likely to sympathize were liberal reformist ones. The Liberals favored extending voting rights, limited at the time to about six percent of the adult population (women did not have the right to vote), and improving social welfare. In these and other issues, they often made common front with the Social Democrats. There were scientists among the supporters of both parties; in fact, the Social democratic leader, Hjalmar Branting, had worked as an astronomer before becoming engaged full-time in politics and political journalism. It was a new

experience for Arrhenius to consort with colleagues whose political beliefs were translated into action. He had probably not given politics much thought before leaving Sweden. Although he never was much interested in politics, he nevertheless acquired the liberal reformist views that he would hold for the rest of his life."

Crawford continues:

"The members of the Physics Society and the broader scientific circles in which Arrhenius moved shared a distinctive conception of the role of science in society. Having first-hand knowledge of the many ways in which science had improved living conditions - electric lights at home and in the streets and better sanitary standards, for instance - they firmly believed that science fostered progress. In this they came closest to the 'optimistic evolutionism' that was an important part of late nineteenth century philosophies of progress."

This over-arching theme of political progressivism is of note to our discussion because, as previously stated, it was during this period of his work with the politically progressive scientists of the Stockholm Physics Society, that Arrhenius first developed the theory of man-made global warming. A theory which, to this day, is marked by political progressivism.

As previously noted, since the days of Arrhenius, the next major development for the theory of man-made global warming came in 1955 with the publication of John von Neumann's "Can We Survive Technology?" As detailed in previous chapters, von Neumann's article was followed by an avalanche of similar pieces expounding upon the problem, and the physics-based, engineering solutions. Tremendous expansions of military spending on weather modification and the atmospheric sciences subsequently occurred. Meanwhile, beginning in the late 1960s, the socio-economic solutions to the purported problem of man-made global warming were being formed at the Club of Rome and the United Nations (UN).

The Club of Rome was formed in 1968 and has been instrumental in promulgating the theory of man-made global warming and climate change. They describe themselves as, "an informal organization that has been aptly described as an 'invisible college.' Its purposes are to foster understanding of the varied but interdependent components - economic, political, natural, and social - that make up the global system in which we all live; to bring that new understanding to the attention of policy-makers and the public worldwide; and in this way to promote new policy initiatives and action." The Club of Rome has interlocking memberships with the Committee of 300 and the Bilderberg Group. The founding meeting of the Club of Rome occurred at David Rockefeller's estate in Bellagio, Italy.

Shortly after the formation of the Club of Rome, in 1969 the UN General Assembly decided to convene the first major inter-governmental conference on environmental issues called the UN Conference on the Human Environment. It was funded by the Rockefellers. UN Secretary-General, U Thant invited a trustee of the Rockefeller Foundation, Maurice Frederick Strong (1929-2015) to lead it as Secretary-General of the Conference and as Undersecretary General of the UN responsible for environmental affairs. The resulting 1972 meeting, known as the Stockholm Conference, adopted a declaration of principles and an action plan to deal with global environmental issues. The aforementioned Thomas Malone of The Travelers insurance company (ch 8) was in attendance. It was the kick-off of the establishment's phony environmental movement. Being that the theory of man-made global warming originated at the Stockholm Physics Society, it is interesting that this landmark conference was held in Stockholm, Sweden. Have you heard of Stockholm syndrome? Anyway, during this UN Conference on the Human Environment, the global management of natural resources was promoted. This was to be accomplished through a new global bureaucracy called the United Nations Environment Program (UNEP). Subsequently, in December of 1972, the UN General Assembly established the UNEP and elected Strong to head it. The UNEP

promptly began issuing a steady stream of news releases about, among other things, the pollution of the air by carbon emissions.

Also in 1972, in association with the Massachusetts Institute of Technology, the Club of Rome issued a report titled "The Limits to Growth." As one might guess from the title, this report outlines the scenario of an overpopulated world and suggests a reduction of the Earth's population. As noted in chapter 10, population reduction is an agenda inextricably linked to the theory of man-made global warming as well as to the New Manhattan Project. More pertinently to our discussion, "The Limits to Growth" characterizes man's emissions of atmospheric carbon dioxide as problematic because they say that such emissions could cause Earth's climate to change.

While all this was going on, the Rockefellers were cranking up the public relations machine. In order to better ~~brainwash~~ inform the public, the Rockefellers planned, promoted, and funded the first Earth Day on April 22, 1970. Among similar actions taken by other politicians, and in accordance with the event, Nelson Rockefeller, as the Governor of New York, established a state Environment Department. Earth Day participants were provided with a booklet called the *Environmental Handbook*. The *Environmental Handbook* advocated for population reduction. Paul Ehrlich, the famous population reduction advocate and author of *The Population Bomb* contributed to this booklet; writing that by 1979 all important animal life in the sea would be extinct. He wrote that the cause of the problem was, "too many cars, too many factories, too much detergent, too much pesticide, **multiplying contrails** [author's emphasis], inadequate sewage treatment plants, too little water, too much carbon dioxide - all can be traced easily to too many people." Less than three months after Earth Day, under the guidance of Henry Kissinger, President Nixon created the Environmental Protection Agency (EPA) with an executive order.

In 1976, Laurence Rockefeller, president of the American Conservation Association, warned in a *Reader's Digest* article of grave consequences to 'the climate change.' He wrote that if people kept on burning carbon-based fuels at will, then 'authoritative controls' would need to be imposed.

The large-scale global warming scam was launched in America at a 1988 Senate committee hearing arranged by U.S. Senator Tim Wirth

(D-CO) and featuring testimony by NASA's James Hansen who swore that he was certain that man's emissions of atmospheric CO2 were causing the Earth to warm. Years earlier, Hansen was warning of all the impending doom sure to happen from global *cooling*, but never mind that. Since 1971 the Earth had warmed - not cooled, so Hansen had changed his theories and now we were supposed to listen to him again, you see. In preparation for this hearing designed to alert the public to the dangers of global warming, Wirth and others called the Weather Bureau and found out what was most commonly the hottest day of the year for Washington and scheduled the hearing for that day. Not only that, but the night before, people went in and opened all the windows and made sure that the air conditioning was not working in order to make the hearing as hot and sticky as possible.

The senator who arranged these hearings, Tim Wirth was later interviewed by PBS and he said, "We knew there was this scientist at NASA, you know, who really identified the human impact before anybody else had done so and was very certain about it. So we called him up and asked him if he would testify." Former Senator Wirth has also been quoted as saying, "We've got to ride the global warming issue. Even if the theory of global warming is wrong, we will be doing the right thing..."

Tim Wirth

The NASA 'scientist' whose testimony was featured at these hearings, James Hansen was plucked from obscurity at NASA to be the director of their Goddard Institute of Space Studies (GISS) where he served in that position from 1981-2013. As part of his duties at the

GISS, instead of relying upon an already-established method of recording Earth's tropospheric temperatures using satellites and radiosondes, Hansen chose instead to use the wonky networks of land and ocean-based sensors. James Hansen has a long public record as a global warming activist.

James Hansen

Also in the same year these congressional hearings commenced, 1988, the Intergovernmental Panel on Climate Change (IPCC) was created. The IPCC was created as an ostensibly scientific organization used to promote the theory of man-made global warming and they have produced a series of reports that deceive and spin in favor of their chosen hypothesis. The aforementioned creator of the UNEP, Maurice Strong set up the IPCC through the United Nations' World Meteorological Organization (WMO). This allowed the IPCC to receive national funding. IPCC members were chosen from national weather organizations through the WMO with the caveat that they would identify human activities as the cause of global warming. In America, the national weather organization chosen to feed the WMO and the IPCC with both money and personnel was the National Oceanic and Atmospheric Administration (NOAA). The IPCC was to consider only man-made sources of atmospheric carbon emissions. Natural sources were to be ignored. As once again so succinctly put by Dr. Richard Lindzen, "The consensus was reached before the research even began." Insider meteorologist Roy Spencer wrote, "Politicians formed the IPCC over 20 years ago with an endgame in mind: to regulate CO2 emissions. I know, because I witnessed some of the behind-the-scenes planning. It is not a scientific organization. It was organized to use the government-funded scientific research establishment to achieve a policy goal."

Describing the experience of an honest scientist with the IPCC, German meteorologist and physicist Dr. Klaus-Eckert Puls writes:

"Ten years ago I simply parroted what the IPCC told us. One day I started checking the facts and data - first, I started with a sense of doubt, but then I became outraged when I discovered that much of what the IPCC and the media were telling us was sheer nonsense and was not even supported by any scientific facts and measurements. To this day I still feel shame that as a scientist I made presentations of their science without first checking it."

In anticipation of a big meeting planned for the next year, President of the Club of Rome, Alexander King (1909-2007) and Assistant Secretary Bertrand Schneider more specifically outlined the global warming socio-economic agenda in their 1991 book *The First Global Revolution*. In *The First Global Revolution*, the authors King and Schneider write:

"In searching for a common enemy against whom we can unite, we came up with the idea that pollution, ***the threat of global warming*** *[author's emphasis], water shortages, famine, and the like, would fit the bill. In their totality and their interaction these phenomena do constitute a common threat which must be confronted by everyone together. But in designating these dangers as the enemy, we fall into the trap, which we have already warned readers about, namely mistaking symptoms for causes. All these dangers are caused by human intervention in natural processes, and it is only through changed attitudes and behaviour that they can be overcome. The real enemy then is humanity itself."*

The founder of both the UNEP and the IPCC, Maurice Strong, was the mastermind behind the global warming deception, and in June of 1992, Strong rolled out the last pieces of the puzzle as Secretary General of the UN Conference on Environment and Development, known as the Earth Summit, in Rio de Janeiro, Brazil. Thomas Malone of The Travelers insurance company was in attendance here, too. The last pieces of the puzzle were The United Nations Framework Convention on Climate Change and Agenda 21. The United Nations Framework Convention on Climate Change defined the problem while Agenda 21 provided the solutions. The United Nations Framework Convention on Climate Change (UNFCC) provided the definition of man-made global warming which the IPCC used to guide their research. This allowed the IPCC to establish man's emission of atmospheric carbon dioxide as the demonic gas that was destroying the planet. The Agenda 21 documents finally provided directives and guidelines presented as socio-economic solutions to the purported problem of man-made global warming. This way, policymakers, the media, and misguided environmentalists had their marching orders. We see the legacy of Agenda 21 today in all the government policy and law that is geared towards the reduction of carbon emissions. The IPCC provided support for the science while Agenda 21 provided support for the policymakers. If the IPCC science was ever found to be inadequate, policymakers were encouraged to employ the precautionary principle and go ahead with carbon mitigation measures anyway. Something called the Conference of the Parties (COP) was also created as a forum for political leaders to make socio-economic decisions based on IPCC science. The first COP was held in Berlin in 1995 and the first meetings produced the Kyoto Protocol. The Earth Charter, touted by Maurice Strong at the 1992 Earth Summit, was written by Steven Rockefeller.

In 1992, the same year that he chaired the UNEP meeting in Rio de Janeiro, Maurice Strong became the chairman of Ontario Hydro, the public utility that controls all energy production for the province. He then implemented a green energy program for Ontario Province which quickly reduced the availability of energy and substantially increased the price. Other green energy initiatives throughout Europe have since been disasters as well. Following the 1992 Earth Summit, Strong participated in a slew of big environmental organizations, worked for the World Bank and the World Economic Forum, and began promoting slogans like 'think globally, act locally' and 'sustainable development.' From 1989 to 2009, the U.S. government spent $79B on climate research, while scientists who challenged it all were denied funding and marginalized.

Maurice Strong

The media operations regarding the theory of man-made global warming have thundered right along. Prominent IPCC member Stephen Schneider revealed the method when he said in *Discover* magazine, "Scientists need to get some broader based support, to capture the public's imagination … that, of course, entails getting loads of media coverage. So we have to offer up scary scenarios, make simplified dramatic statements, and make little mention of any doubts we may have … each of us has to decide what the right balance is between being effective and being honest." After 1998, when global temperatures began to move sideways rather than up, the rhetoric changed from 'global warming' to 'climate change.' Apparently in attempts to muddy the waters and move the goal posts, Obama's science czar John Holdren (ch 10) subsequently introduced terms like 'climate disruption' and the 'polar vortex.'

Although he had experience with the issue going back to at least the mid-1980s (ch 7), former vice president Albert Arnold Gore, Jr. really began making a lot of noise about global warming in the mid-2000s. His Oscar winning documentary *An Inconvenient Truth* premiered in 2006.

Al Gore

It was also in the 2000s that energy hog Gore along with a slew of investment bankers and other financial establishment-type people and organizations began creating a market for the trading of carbon credits. This effort was commonly known as 'cap-and-trade.' Cap-and-trade refers to the official limiting of industrial carbon dioxide emissions and the trading of carbon credits which can allow for the emissions of said carbon. In other words, under a cap-and-trade scenario, companies would be limited in the amount of carbon dioxide they would be allowed to emit and any carbon dioxide they do emit would only be allowed with their acquisition and spending of carbon credits. It's effectively the 'carbon tax' that Gore's cohort Gordon J.F. 'How to Wreck the Environment' MacDonald testified about in 1987 before the Senate Energy Committee (ch 7). MacDonald's employer, the MITRE Corporation has also been deeply involved in promoting carbon taxes. The aforementioned Enron (ch 8) was also a key proponent of cap-and-trade. In August of 1997, Enron CEO Ken Lay met with Gore and President Clinton to develop positions for the upcoming Kyoto Protocol negotiations. That same year, the U.S. Senate voted 95 to 0 against ratification of the Kyoto Protocol. But that didn't stop Gore. Al Gore continued on with the promotion of his business activities pertaining to the firm he founded and chaired called Generation Investment Management (GIM). The London-based firm invested in green businesses - the type of businesses that would thrive under a cap-and-trade scenario. David Blood, the CEO of Goldman Sachs Asset Management, was Gore's business partner. As of 1997, the only firm to trade carbon credits in the United States was the Chicago Climate Exchange (CCX). The CCX was set up so that when any of its corporate members such as Ford, DuPont, Dow, or others bought carbon credits, they were effectively investing in green energy firms. Gore's close associate Maurice Strong was a CCX board member. Enron (ch 8) was the largest trader of carbon credits. Al Gore was actively using his office as vice president in attempts to create a situation whereby American businesses would be forced to buy carbon credits on Maurice Strong's exchange. These purchases would, in turn, most likely fund businesses in which Gore was invested. Genius. Never let a good crisis go to waste. There are many more details to this story involving organizations such as Goldman Sachs, the Ford Foundation, and the Rockefeller Brothers Fund. For a more in-depth discussion, please refer to a 2007 *Human Events* article by Corey Barnes titled "The Money and Connections Behind Al Gore's Carbon Crusade."

The Rockefellers have provided massive funding for just about all of the biggest NGOs (non-government organizations) involved in the environmental movement. The Energy Foundation was created by the Rockefellers as a way to funnel funds to organizations that fund political campaigns and lobby for public policy pertaining to issues of climate. At universities and colleges throughout the country that rely on Rockefeller funding, environmental courses have become mandatory. George Soros' Open Societies Foundations have also funneled billions of dollars to environmental activist foundations pushing the climate change agenda such as: the Aspen Institute, the Tides Foundation, Earthjustice, the Presidential Climate Action Project, the ClimateWorks Foundation, and the Natural Resources Defense Council.

Climategate

In late 2009 a treasure trove of emails, leaked from the Climate Research Unit (CRU) of East Anglia University, made their way onto the Internet - in full view of everybody. This was not a good look for the climate alarmists at the CRU because the emails showed a pattern of highly unscientific behavior designed to support the theory of man-made climate change and squash any opposition. The New Media covered it immediately, but it took the frauds in the old, mainstream media *a week* to get around to mentioning it. When CNN, NBC, CBS and the rest finally got around to covering it, they all in unison decreed that, yes the emails were authentic, but the general public is too stupid and ignorant to understand what these super-intelligent scientists are talking about. The story was then promptly swept under the rug and has not been heard about in the Old Media since. But we in the New Media remember what happened very well.

First of all, it is important to understand what the Climate Research Unit at the University of East Anglia is. The CRU was founded by Hubert Horace Lamb (1913-1997) in 1972 - the same year of the Stockholm Conference. Lamb is generally considered the father of modern climate studies.

Hubert Lamb

In the 1980s, Lamb turned management of the CRU over to his colleague Tom Wigley. During this time frame, the Rockefeller Foundation provided the CRU with funding.

The CRU was the main source of the scientific climate data used in the IPCC reports and that data was controlled by a small group of scientists within the CRU. At the time of the Climategate scandal, most, if not all of the CRU's management were members of the IPCC. The modern version of the CRU evolved in symbiosis with the IPCC. In fact, the IPCC and today's CRU are essentially the same organization.

When the Climategate emails were leaked, the director of the CRU, Phil Jones believed that the emails were stolen and reported the theft to the police. This is how we originally knew that the emails were authentic. Jones later tried to downplay the significance of the emails - claiming that they portrayed nothing but the usual, banal banter between scientists. If that was the case, then why was he so concerned about their dissemination that he called the police? If one reads the emails, it quickly becomes apparent as to why Jones was so concerned.

The Climategate emails from the CRU clearly show time and time again concerted efforts to: fudge the data, squash and sabotage the opposition, propagate lies, hide data, strategically delete data, corrupt the peer-review process, threaten journal editors, and more. All of this, of course, was done in efforts to promote the theory of man-made global warming and climate change. These activities were unscientific to say the least.

Of particular interest to our discussion, a 2009 Climategate email from Kevin Trenberth to Tom Wigley mentions geoengineering. His message suggests that activities at the CRU are designed to promote geoengineering. The email reads:

"How come you do not agree with a statement that says we are nowhere close to knowing where energy is going or whether clouds are changing to make the planet brighter? We are not close to balancing the energy budget. The fact that we cannot account for what is happening in the climate system makes any consideration of geoengineering quite hopeless as we will never be able to tell if it is successful or not! It is a travesty!"

The CIA

Most people are completely unaware of the fact that the Central Intelligence Agency (CIA) has anything to do with the theory of man-made global warming. That's probably by design. Evidence shows that they have been behind it this whole time. Although they apparently don't want you to know about it, here we will examine the CIA's pervasive involvement with and their promotion of the theory of man-made global warming and climate change.

John von Neumann, the original promoter of the modern theory of man-made global warming (ch 11), was a CIA agent. Our good friend Gordon J.F. 'How to Wreck the Environment' MacDonald was a CIA agent, worked extensively with the Agency, and was a proponent of the theory of man-made global warming. Al Gore worked with the Agency on the Measurements of Earth Data for Environmental Analysis (MEDEA) project (ch 7).

In the previous chapter, we learned how the CIA controls much of our modern mass media. We also have direct evidence of their involvement in the creation of propaganda regarding the Earth's average temperature. The 1974, CIA-produced report titled "A Study of Climatological Research as it Pertains to Intelligence Problems" fear mongers incessantly about climate fluctuations and weather. They report on recent examples of extreme weather and the resultant death and destruction as evidence for a long-term shift in Earth's climate which they call 'the climate change.' Maybe they should have made a horror movie by that title. The authors of "A Study of Climatological Research" go on to state that the combination of 'the climate change' and Earth's growing Human population is a recipe for disaster. They write that the crop failures due to 'the climate change' are destabilizing governments all over the world. As they touch on climate modification, they write that this destabilization of governments can cause war. The authors write:

"Timely forecasting of climate and its impact on any nation is vital to the planning and execution of U.S. policy on social, economic, and political issues. The new climatic era brings a promise of famine and starvation to many areas of the world. The resultant unrest caused by the mass movement of peoples across borders as well as the attendant intelligence questions cannot be met with existing analytical tools. In addition, the Agency will be faced with tracing and anticipating

climate modification undertaken by a country to relieve its own situation at the detriment of the United States. The implication of such a modification must be carefully assessed."

You guessed it. They needed more funding. Considering the CIA's commanding presence over our mass media, this report appears to be a big part of the wildly reaching assumptions pervasive among global warming alarmists today claiming famine, disease, mass-migrations, war, and Humanity's extinction all due to the climate change. A small problem for today's global warming cult members, though, is the fact that this early example of climate change delusion concerned itself with *global cooling*, not global warming. If one tries hard enough, anything can be made to appear as a problem; especially when lots of money and power hang in the balance.

Later, in 2008, the CIA contributed to a standard global warming scare-fest titled "National Intelligence Assessment on the National Security Implications of Global Climate Change to 2030." At least this time the authors of the report didn't foresee state failure. We have something to be thankful for. Lastly, in 2015, *CNS News* and *Mother Jones* reported that CIA director John Brennan was regurgitating nonsense about all the impending doom sure to come from Earth's temperature fluctuations and how he is so deeply concerned about it.

The overt agenda

It is undeniable that gigantic socio-economic agendas are attached to the theory of man-made global climate change. This agenda is exemplified by the Green New Deal as well as the aforementioned Agenda 21. From the proposed legislation presented on congresswoman Alexandria Ocasio-Cortez's (D-NY) website, we see that, because of catastrophic man-made global warming, we are supposed to: transition to alternative forms of energy, upgrade all existing buildings, eliminate greenhouse gas emissions, overhaul transportation systems, reconstruct labor laws, and oh so much more. James Corbett of *The Corbett Report* says that we're looking at: $90 trillion in energy infrastructure investments, a $1 trillion green bond market, a multi-trillion dollar carbon trading market, a $391 billion climate finance industry, and more.

As we have seen from the evidence presented in this chapter, the theory of man-made global warming was concocted and promoted by the elites who own big businesses and big governments. So it's not about environmentalism or shutting down big business. In fact, only well-funded big businesses could easily make the transition to a world dominated by the Green New Deal and Agenda 21. You see, the overt, socio-economic agenda of the theory of man-made global climate change is about shutting down the smaller competition to big business, consolidating political and financial power, and firmly placing a boot on the neck of Humanity. That's what the Green New Deal does and they're doing it openly.

The covert agenda

What this book details is the *covert* man-made climate change agenda - the New Manhattan Project. As we learned in the previous chapter, the modern theory of man-made global warming has been perpetuated not only as a way to propel the theory's overt socio-economic objectives. It has also been used as a way to propel the covert New Manhattan Project. In 1965, when traitorous and illegitimate elements of our federal government released the document titled "Restoring the Quality of Our Environment," simultaneously establishing both the theory of man-made climate change and the SRM geoengineering theses, that was the official beginning of this aspect of the big lie. This lie about global warming served and serves as the justification for spraying tens of thousands of megatons of toxic garbage over Humanity and the rest of Earth's biota. This lie was, in part, needed as justification for the visible aspects of the planned New Manhattan Project. Let us not forget that "Restoring the Quality of Our Environment" came out of the executive branch of the federal government and that the CIA is designed to serve the executive branch. Throughout the pages of this book, we have seen the CIA's pervasive involvement - including their promotion of the theory of man-made global warming.

Consider the implications of this. Consider how almost all of the environmental movements and organizations in the Western world have been folded into this grand deception. Consider the countless millions of people who have marched in the street, in a satanic orgy, demanding the Earth's (and their own) destruction. Consider all the money that has been donated to perpetuate this treachery. Consider all the fools who have forsaken and attacked their fellow man because he did not want to be destroyed. The establishment has had all these people thinking that they are doing a good turn, all the while these people have been supporting everybody's and everything's utter destruction. In the church of global warming, there is an altar beneath the pulpit.

If you still espouse the theory of man-made global warming, and are not completely nihilistic and selfish, please relinquish your dogma. If you've read this chapter, then you now know the true facts of the matter. We're all being murdered and most of us really don't want to be killed. Heck, most of us don't even want a reduction in our quality of life. It's not necessary. This is what happens when we let a vampire into our house. Death is what happens when enough people believe the big global warming lie.

The corruption of science

What we have seen so far in this chapter and in this book demonstrates a fundamental corruption of the global scientific establishment. All this begs the question: How did things get so screwed up? As suggested in chapter 2, it all began during World War II. Here in America, it began with Vannevar Bush and his National Science Foundation and that's where it continues.

Before WWII, there was very little government funding of science. The National Science Foundation (NSF) was established in 1951 to provide support for post-World War II civilian scientific research. The process for funding civilian scientific activities developed by the NSF has since been adopted by virtually all other U.S. government science-funding agencies. The problem is that the science-funding process developed by the NSF is fundamentally flawed. Dr. J. Marvin Herndon, who we will soon recognize as a victim of this scientific establishment, enumerates four flaws in the NSF process of government science-funding.

The first flaw enumerated by Dr. Herndon pertains to today's peer-review process. Believe it or not, the peer-review process of today allows for *anonymous* peer-reviewers. The NSF came up with the idea that peer-reviewers should be anonymous. Before this, the concept of scientific peer-review did not exist. This concept of anonymous peer-review was later adopted by virtually all government science-funding agencies and almost universally by scientific journal editors. Scientists hidden by the anonymous peer-review process often make untrue and/or pejorative statements about their competition's work. Sometimes, in order to reject a paper, the editors of the journals themselves will make negative and untrue comments anonymously. Scientists have come to understand that the anonymous peer-review process makes it unlikely that work challenging the consensus viewpoint will be published, so they don't bother to engage in groundbreaking work. They stick to work that has a better chance of being successfully funded. In this way, the anonymous peer-review process stifles the introduction of new ideas.

Flaw number two enumerated by Dr. Herndon is that the NSF requires pre-planned results. This is akin to putting the cart before the horse because it is impossible to say what one will discover before one discovers it. This aspect of the funding process, too, was invented by the NSF and it has led today's scientists away from true scientific discovery and towards trivial, heavily bureaucratic work.

Flaw number three is unaccountability. There is no direct legal responsibility or liability for NSF-funded individual scientists' conduct. These responsibilities are usually deferred to a university or other large institution and this encourages unethical conduct amongst individuals.

Flaw number four pertains to how the publishers of scientific journals are paid by the government to publish journal articles and then those same publishers turn around and charge the public for access to these same journal articles. This amounts to double-dipping the taxpaying American citizen and stifles the free exchange of information in that, if it costs money to view an article, then less people will view it. Stifling the free exchange of this information results in less vital criticism of the scientific work in question - thereby rendering an overall lower quality of product. Not only all that, but the journal publishers get ownership of copyrights to the articles when the underlying scientific research was originally funded by the taxpayer! That's more like triple-dipping!

Dr. Herndon makes note of a website called *arXiv.org* as being particularly problematic. He writes that this website, "has become the preeminent means of scientific communication in the areas of science and mathematics it hosts." In a 2011 paper Dr. Herndon writes that the NSF has permitted *arXiv.org*, "to become an instrument for science-suppression, and for blacklisting and discrimination against competent, well-trained scientists worldwide."

Dr. Herndon should know. He has experienced the active suppression of scientific discovery first-hand. In the late 1970s Dr. Herndon made major discoveries pertaining to the Earth's inner core. As noted earlier in this chapter, Dr. Herndon found that there is a natural nuclear reactor at the center of our planet. This discovery overturned the apple carts of many an established man of science and surely did not fit with the emerging theories of man-made global climate change. But, if we had a highly functioning scientific establishment, Dr. Herndon's discoveries would have been thoroughly examined and discussed in attempts to either prove or, more importantly, disprove his findings. Dr. Herndon was more than prepared to weather these storms. These types of discoveries have brought people directly into the world of the scientific elite. Instead... crickets. His discoveries were published in *Naturwissenschaften* as well as in the *Proceedings of the Royal Society of London*, but that was where it ended. Dr. Herndon writes:

"In 1979, I published an entirely different idea of the inner core's composition. The scientific paper was communicated by Nobel Laureate Harold C. Urey to the Proceedings of the Royal Society of London and I received a complimentary letter from Inge Lehman [1888-1993]. But instead of debate, discussion, and experimental and/or theoretical verification/refutation, I received silence from the geophysics community, not only on that discovery, but on a host of discoveries that followed as a consequence."

Since then, Dr. Herndon's work has been stifled by the NSF and their peer-review process many times. Dr. Herndon's thoughts on all of this are as follows:

"There is at work, I regret to say, an ongoing malevolent political agenda, potentially devastating to humanity and to our planet, and it is being driven by a scientific community that cannot be trusted to tell the truth, a scientific community comprised of fund-recipients of DOE, NASA, and NSF grants/contracts, and U.S. Government scientists and administrators, who have proven themselves inept, irresponsible, and incapable of rendering valid and truthful scientific knowledge."

The New Manhattan Project and the theory of man-made global climate change are front and center in this malevolent agenda.

The NSF's incestuous relationship with large universities and their anonymous peer-review process amounts to fraud upon the American taxpayer. By statute, the NSF is supposed to fund individuals outside of universities, but that hardly ever happens. The National Science Board is supposed to keep the NSF in line, instead, the NSF has been allowed to create a situation where the process has been corrupted and the 'science-barbarians,' as Dr. Herndon calls them, have been allowed to run wild. Again Dr. Herndon writes:

"Shielded, unaccountable, and protected by decades of institutionally-sanctioned secrecy, the barbarians, invited in by the National Science Foundation, have progressively changed science from an arena of new ideas and open debate into an intolerant religious practice, replete with the promulgation of singular points of view, where non-doctrinaire new ideas and debate are to be suppressed or buried, a real-life parallel to Ray Bradbury's Fahrenheit 451 *fire department whose mission had changed to one of burning books."*

Many people apparently think that somehow scientists (and therefore science) are incorruptible. They say that they 'believe in science.' But, believe it or not, just like so many countless others have done, corrupt scientists will compromise themselves and their work for personal gain - be it financial, political, or social. Saying that one 'believes in science' is essentially the same as saying that one 'believes in big business' or that one 'believes in government.' It is much healthier to be critical of all of these things and to take them all on a case-by-case basis.

Conclusions

Your author has good news. As this chapter demonstrates, we don't have to worry about rising sea levels. The frequency of extreme weather anomalies is not increasing. The polar bears are fine. Regardless of what the Old Media says, the world is not warmer than it ever has been. Current fluctuations in average global temperature are not unprecedented. The world has been warmer than it is today for the vast majority of the last 10,000 years. Rejoice! They're just trying to browbeat us into submission like they always do.

The provably false theory of man-made global climate change is being propagated so furiously because the elites of this world want the absolute, centralized power that only a one world government can afford. It is being promoted as the global problem that requires global solutions which can only be most effectively administered by a global government. Of course, the New Manhattan Project is being promoted as one of the great solutions to this problem and the United Nations is the presumptive one world government just waiting for enough of us to submit.

References

Human Caused Global Warming: The Biggest Deception in History: The Why, What, Where, When, and How It Was Achieved a book by Dr. Tim Ball, published by Tellwell, 2016

The Deliberate Corruption of Climate Science a book by Dr. Timothy Ball, published by Stairway Press, 2014

"Scientific Misrepresentations and the Climate-Science Cartel" a paper by J. Marvin Herndon, PhD, published in the *Journal of Geography, Environment and Earth Science International*, December 2018

Maverick's Earth and Universe: Understanding Science without Establishment Blunders a book by J. Marvin Herndon, Ph.D., published by Trafford Publishing, 2008

"A Disgrace to the Profession" The World's Scientists in Their Own Words on Michael E. Mann, His Hockey Stick, and Their Damage to Science a book by Mark Steyn, published by Stockade Books, 2015

"Greenpeace Founder: Global Warming Hoax Pushed by Corrupt Scientists 'Hooked on Government Grants'" an article by Robert Kraychik, published by *Breitbart*, March 7, 2019

NASA: Politics Above Science a book by Marvin J. Herndon, self-published, 2018

Arrhenius: From Ionic Theory to the Greenhouse Effect a book by Elisabeth Crawford, published by Science History Publications, 1996

Fixing the Sky: The Checkered History of Weather and Climate Control a book by James Rodger Fleming, published by Columbia University Press, 2010

"Global Warming and Ice Ages: Prospects for Physics-Based Modulation of Global Change" a paper by Edward Teller, Lowell Wood, and Roderick Hyde, published by Lawrence Livermore National Laboratory, 1997

Herndon's Earth and the Dark Side of Science a book by J. Marvin Herndon, Ph.D., self-published, 2014

Killing the Planet: How a Financial Cartel Doomed Mankind a book by Rodney Howard-Browne and Paul L. Williams, published by Republic, 2019

MauriceStrong.net

"MALONE, Thomas F." An obituary published in the *Hartford Courant*, July 10, 2013

"The Limits to Growth" a report by Donella H. Meadows, Dennis L. Meadows, Jørgen Randers, and William W. Behrens III, published by the Club of Rome, 1972

"The Money and Connections Behind Al Gore's Carbon Crusade" an article by Corey Barnes, published by *Human Events*, October 3, 2007

"Energy Efficiency and Climate Change Mitigation" a report by the Energy Modeling Forum of Stanford University, published by the Energy Modeling Forum of Stanford University, 2011

"A Study of Climatological Research as it Pertains to Intelligence Problems" a report by the Central Intelligence Agency, 1974

"National Intelligence Assessment on the National Security Implications of Global Climate Change to 2030" a report by the House Permanent Select Committee on Intelligence, House Select Committee on Energy Independence and Global Warming, statement for the record of Dr. Thomas Fingar, Deputy Director of National Intelligence for Analysis and Chairman of the National Intelligence Council, 2008

"CIA Director Cites 'Impact of Climate Change' as Deeper Cause of Global Instability" an article by CNS News staff, published by *CNS News*, Nov. 16, 2015

"CIA Director Delivers Some Blunt Talk About....Climate Change" an article by Kevin Drum, published by *Mother Jones*, Nov. 17, 2015

James Corbett, "And Now For The 100 Trillion Dollar Bankster Climate Swindle …" *The Corbett Report,* February 24, 2016.

Climate Change: The Facts a book edited by Alan Moran, published by Stockade Books, 2015

"Forecasting Global Climate Change" a paper by Kesten C. Green and J. Scott Armstrong as it appeared in the book *Climate Change: The Facts* edited by Alan Moran, published by Stockade Books, 2015

Climategate: A Veteran Meteorologist Exposes the Global Warming Scam a book by Brian Sussman, published by WND Books, 2010

"Corruption of Science in America" a paper by J. Marvin Herndon, published in *The Dot Connector Magazine*, 2011

Chapter 13
The NAZIS and the CIA

Being that the Nazi regime of World War II had little regard for Human life, lots of cutting edge-technology, and an authoritarian power structure, one can see how their former scientists would fit neatly into something like the New Manhattan Project. Evidence suggests that this is what has taken place. In the years immediately following WWII, Operation Paperclip was conducted in order to make over 1,600 former Nazi scientists America's own. These scientists went to work for our military, the Central Intelligence Agency (CIA), large American corporations, prestigious universities, and other organizations. Operation Paperclip was originally supported by an investigation into Nazi science and scientists called Operation Alsos. Operation Alsos was an offshoot of the original Manhattan Project. In the waning days of the war in Europe, the members of Operation Alsos quickly filled in behind Allied soldiers reclaiming European territories. They discovered and claimed the riches of Nazi WWII science. At the same time, Allen Dulles of the American Office of Strategic Services (OSS), Nazi General Reinhard Gehlen and others were making a deal that would bring Nazis to America. This chapter tells the story of Operation Paperclip and the former Nazi scientists who came to America to work on technologies germane to the New Manhattan Project.

Allen Welsh Dulles (1893-1969), stationed in Bern, Switzerland at the time and working for the CIA's predecessor, the Office of Strategic Services, was among the most influential people in laying the groundwork for the later formation of the CIA. Although there is no officially recognized founder of the CIA, a strong argument can be made for Dulles being that person. He and a handful of others went on to fill the CIA's nascent ranks with Nazis. This process began before the conclusion of WWII with a secret mission called Operation Sunrise. Among the few Germans most instrumental in beginning this process were Nazi Generals Reinhard Gehlen (1902-1979) and Karl Wolff (1900-1984). After WWII, the bulk of the former Nazis came to America as part of Operation Paperclip. Allen Dulles ended up being one of the early directors of the Agency and to this day he is the longest serving director of the CIA. He ran the CIA for 8 years with virtually no congressional oversight and the impact of his legacy continues today. For these reasons, Allen Dulles and his more famous brother, Secretary of State John Foster Dulles (1888-1959) are persons of interest here and are referred to throughout this chapter.

Allen Dulles

Journalist Linda Hunt writes in her groundbreaking and thorough 1991 book *Secret Agenda*, "At least sixteen hundred scientific and research specialists and thousands of their dependents were brought to the U.S. under Operation Paperclip. Hundreds of others arrived under two other Paperclip-related projects [Project National Interest and Project 63] and went to work for universities, defense contractors, and CIA fronts." The CIA and their predecessor organizations were instrumental throughout this entire process. Many of the former Nazis who became CIA agents were drawn from something called Amt VI (Department 6) of the SS RHSA (*Reichssicherheitshauptamt* or Reich Main Security Office). This was Nazi Germany's equivalent of the CIA. During the war, the German intelligence experts working for Amt VI of the SS RHSA were commonly involved in hunting down and exterminating Communists and Jews *and* securing their wealth. Most of the other portions of the CIA's early ranks were filled with Americans - mostly Ivy League and Wall Street types; many from Yale University in particular; the home of the notorious Skull and Bones secret society.

One may wonder how this happened. Wasn't Nazi Germany horrendously evil? Why were they not completely destroyed? Didn't the Nazis systematically exterminate millions of innocent men, women, and children? They did. But, they were also supported by the Western deep state. And in that case, the rules of the game are very different. Even though Hitler was an outrageous anti-Semite and infinitely aggressive war-monger from the beginning and through to the end, the Nazis enjoyed tremendous early moral and financial support amongst many of the richest and most powerful people in the Western world.

The Western deep state built the Nazi war machine. Many of the biggest American and European corporations did tremendous business with the Nazis after World War One - during the building and maintenance of the Nazi WWII war machine. Many executives from the German subsidiaries of General Electric, Standard Oil, Ford and other large American corporations were part of an inner circle of Heinrich Himmler's friends doing business in Weimar Republic and Nazi Germany. Of course, later, when it became socially unacceptable to support or to have anything to do with the Nazis, and as they were losing the war, the public sieg-heilling amongst the Western elites came to an end. The evidence now shows that it simply went underground. But the facts remain. International Business Machines (IBM) custom built early punch card computers specifically to manage the Holocaust which they then leased and regularly maintained. Henry Ford was famously an acolyte of Hitler whose company did lots of very serious business with the Nazis. Allen Dulles himself was a big wheel at a Wall Street law firm by the name of Sullivan and Cromwell where his brother John Foster Dulles was a partner. John Foster Dulles was an international attorney for dozens of Nazi enterprises. Sullivan and Cromwell financed the German arms manufacturer Krupp AG and managed the finances of I.G. Farben, the German chemical company that *was* the Nazi war effort and manufactured almost all of the Zyklon

B gas used to exterminate millions of Communists, Jews, and other 'undesirables.'

From his landmark book *Wall Street and the Rise of Hitler*, Ivy League professor Antony Sutton (1925-2002) provides further illustration here as he chimes in right on time, "It is important to note as we develop our story that General Motors, Ford, General Electric, DuPont and the handful of U.S. companies intimately involved with the development of Nazi Germany were - except for the Ford Motor Company - controlled by the Wall Street elite - the J.P. Morgan firm, the Rockefeller Chase Bank and to a lesser extent the Warburg Manhattan bank." Most, if not all of these firms also donated (through their German subsidiaries) significant funds to the Nazi party's political campaigns. Sutton's work is based on documents which surfaced during the post-war Nuremberg Trials. Information refuting Sutton's assertions pertaining to these matters is based on hearsay.

We are linking the Nazis and their fellow travelers to the New Manhattan Project here because: they share an intertwined history with the CIA, they had cutting edge technology such as that which is employed in the New Manhattan Project, they were very militaristic (the NMP is a military project), and many of them, especially at the top, had little to no regard for the sanctity of life - such as that which is exhibited by the people omniciding Humanity and the Earth with chemtrail spray today as part of the NMP. For these reasons it is logical to assert that the Third Reich and its documented continuing legacy play an important role in the New Manhattan Project. It is for these reasons also that this chapter now expounds upon Operation Paperclip's indiscretions and, more importantly, the possible roles of former Nazi scientists in the production of the New Manhattan Project.

The main reason given for importing Nazi scientists into America was that we needed to deny the Soviets attaining these individuals. But the man running Operation Paperclip, U.S. Army Lieutenant Colonel William Henry Whalen was later convicted of being a Soviet spy. There is evidence that the dreaded Soviets had penetrated Operation Paperclip almost from the beginning. In reality, Operation Paperclip was carried out in order to maintain and grow the deep state and the New Manhattan Project. Our good friend and founder of the New Manhattan Project, Vannevar Bush (ch 2) advocated strongly on behalf of Operation Paperclip Nazi scientists. Sputnik in 1957 greatly accelerated their importation.

German Paperclip scientists were inserted into many organizations associated with the development and production of the New Manhattan Project including (but not limited to): Radio Corporation of America, CBS Laboratories, the Naval Ordnance Testing Station at China Lake, the Desert Research Institute, Pennsylvania State University, The Massachusetts Institute of Technology, and General Electric.

Another reason given by the CIA for secretly importing former Nazis is the high quality of German science. The German scientist in particular was good at what he did because Germany is the birthplace of modern science. Let us refer to the eminent historian John Cornwell and his excellent book *Hitler's Scientists:*

"By the end of the first decade of the twentieth century, Germany had become the international Mecca of science. Researchers, basic and applied, flocked to German universities from all over the world; learned German to read the leading science journals and to participate in conferences and seminars. Germany was well placed to take a leading role in the development of a new physics that would transform the technology of the century, involving from the outset Max Planck, Albert Einstein, Max Born, Werner Heisenberg and Erwin Schrödinger, German-speakers all, alongside scientists from Denmark, the Netherlands, France and Britain. In turn, the new physics led to quantum mechanics and, ultimately, to nuclear physics, the science of the atom and the hydrogen bomb."

Relevant to our investigation, in Germany during the Nazi era, an interesting meteorological thesis gained some traction. Both Adolph Hitler and his lieutenant Heinrich Himmler were adherents of a theory known as the *Welteislehre*. Parts of this theory, known in English as 'glacial cosmogony,' asserted that Earth's weather was due to the movements of massive amounts of ice floating around in outer space. This ice supposedly perturbed the sun which would, in turn, cause weather here on Earth. This cosmic ice would also, according to the theory, fall as rain once it entered Earth's atmosphere.

Although the Operation Paperclip scientists and researchers were generally not the worst offenders here, a great number of Operation Paperclip Nazis were guilty of horrific war crimes. American intelligence officials regularly covered-up these facts by illegally expunging evidence from the records of hundreds of Operation Paperclip Nazis. Because of this, we don't know all of what the Operation Paperclip Nazis did over the course of the war. They probably expunged the worst stuff. What we do know is not good. Some of the more infamous benefactors of Operation Paperclip's amnesty were: Karl Wolff, Otto von Bolschwing, Robert Verbelen, Klaus Barbie, Alois Brunner, Eugen Dollmann, Herbert Wagner, Georg Rickhey, and Otto Ambros. We will now take a brief look at a few of these Operation Paperclip Nazis.

The aforementioned General Karl Wolff was third in command of the entire SS and responsible for arranging the transportation of people to the concentration camps. A German court would later find Wolff complicit in the murder of three hundred thousand men, women, and children. During the war, Otto von Bolschwing (1909-1982) instigated a massacre of innocent civilians in Bucharest and was a senior aide to the 'Architect of the Holocaust' Adolf Eichmann. Bolschwing later worked for the CIA. Robert Verbelen (1911-1990) was sentenced to death in absentia for war crimes including the torture of two U.S. Air Force pilots. He also worked as a contract spy for the U.S. Army, which knew about his background. Klaus Barbie (1913-1991), also known as 'The Butcher of Lyons,' was the head of the Nazi Gestapo. During the war, working out of occupied Lyons, France, Barbie deported Jews to death camps, tortured and murdered French resistance fighters, and served as the local political police. Barbie went on to work for U.S. intelligence in Germany. Barbie also went on in 1971 to recruit a mercenary army of neofascist terrorists which conducted a three-day coup in order to install a narco-friendly regime in Bolivia. The resulting large increase in coca production benefitted his shipping firm Transmaritania. Otto Ambros (1901-1990) was a director of I.G. Farben who took part in the decision to use Zyklon B gas to murder millions of people. Hunt explains that he, "personally selected Auschwitz as the site of an I.G. Farben factory, which he later managed, because Auschwitz concentration camp prisoners could be used as slaves in the factory." At this factory the slave laborers were worked to death and often murdered. Hunt tells us that, "Ambros was found guilty of slavery and mass murder at Nuremberg, but he was sentenced to a mere eight years' imprisonment." Even during his time in jail, the U.S. government kept Ambros listed as available for employment. At the order of the High Commissioner of Germany, John McCloy, Ambros was released from prison in 1951. He immediately went to work as a consultant to W.R. Grace, Dow Chemical, other American companies, and the U.S. Army Chemical Corps operating out of Edgewood Arsenal, Maryland. At the Edgewood Arsenal, Ambros conducted poison gas experiments upon over 7,000 American soldiers. Now do you think that the people running this show would hesitate to spray our civilian population with coal fly ash?

More pertinently, some of the benefactors of Operation Paperclip's amnesty are also known to have gone on to do work in the atmospheric sciences and other areas directly relevant to the New Manhattan Project. These men include: Wernher von Braun, Walter Dornberger, Martin Schilling, Kurt Debus, Arthur Rudolph, Ernst Czerlinsky, Hans Joachim Naake, Albert Pfeiffer, and Hans Dolezalek. We will now take a look at each.

Wernher von Braun (1912-1977) was an SS officer who developed rockets with his mentor and fellow Operation Paperclip benefactor Nazi General Walter Dornberger at the hellish Peenemünde slave labor camp. Von Braun applied to join the Nazi SS in 1933, was a Nazi party member since 1937, and later joined the SS at the personal behest of SS chief Himmler in 1940. Von Braun's story is told in the most detail here because he was the *de facto* political leader of the Operation Paperclip Nazis. In fact, von Braun hand-picked about 120 researchers for the Operation. Many former Nazis worked under and around him. Von Braun's story is the story of many of his fellow Operation Paperclip Nazi scientists.

Wernher von Braun

Von Braun's early work in America involved the further development of missiles based on the German V-2 rocket which the Nazis had used to terrorize London and Antwerp late in the war. This work was conducted at the U.S. Army's Fort Bliss in Texas, at the White Sands Proving Ground in New Mexico, and at the Redstone Arsenal in Alabama. Von Braun and many of his fellow Operation Paperclip Nazi scientists later worked for NASA.

It was only a matter of months after its founding in early February of 1945 that the White Sands Proving Ground began receiving captured German V-2 rocket parts and equipment. By the end of January, 1946, 115 Nazi rocket scientists, including von Braun, were working out of Fort Bliss, Texas, near El Paso. From this nearest of Army bases, small groups of these scientists were periodically sent to the White Sands Proving Ground in New Mexico where they lived in barracks alongside men from the General Electric Company. On the 16th of April, 1947 the combined German-GE team launched their first rocket. White Sands ended up launching at least 64 V-2 rockets. The lead GE scientist at White Sands working on the V-2 rockets was named Richard W. Porter (1913-1996) who was also the head of the panel of scientists representing America during the International Geophysical Year (1957-1958). Chapter 7 recounts the use of modified versions of von Braun's V-2 rocket for atmospheric sounding and satellite delivery. Most relevant to our discussion, many upper-atmospheric experiments were conducted at White Sands. These experiments pertained to cosmic radiation, wind patterns, and temperature as well as biological experiments. The Army also had an Atmospheric Sciences Laboratory at White Sands where they performed research involving the use of early lasers to probe the atmosphere. Before GE became involved, the army was working on its budding missile programs with the Jet Propulsion Laboratory of the California Institute of Technology in Pasadena. The White Sands Proving Ground in New Mexico is also adjacent to the area where the world's first atomic bomb blast known as Trinity occurred.

White Sands Proving Ground circa 1946

V-2 rocket launching at White Sands

The Redstone rocket, which was a direct descendant of the V-2, twice put US astronauts into orbit. Modified Redstone rockets also launched America's first satellite Explorer 1. At White Sands, the Naval Research Laboratory developed the Viking sounding rocket and the Johns Hopkins University Applied Physics Lab developed the Aerobee atmospheric sounding rocket. The Viking rocket's success led to the Vanguard satellite program. The Aerobee line of sounding rockets is the most prolific and successful ever. The first Aerobee rocket was fired at White Sands in 1947 and the last was fired in early 1985. The Aerobee rocket was the first research rocket to reach 200 miles in altitude. The Nike Hercules was a missile used in the SAGE program (ch 7) and it was developed at White Sands.

General Electric's participation in the V-2 program came to an end in June of 1951. At that point, all V-2 operations transferred to U.S. Army ordnance. In 1958 the name of the White Sands Proving Ground was officially changed to the White Sands Missile Range. By 1959 the White Sands Missile Range had become the principal rocket and guided missile testing center in the United States.

The move to Huntsville began in April of 1950. Eminent science historian Dr. Michael J. Neufeld writes, "Moving several hundred personnel, plus shops and test equipment, out of Fort Bliss and setting them up in converted buildings at Redstone Arsenal was a job that took six months." Neufeld continues, "For Wernher and Maria von Braun, and for almost all the other Germans who came with them, Huntsville quickly became home. For the first time they could live outside of a fenced-in base and integrate themselves into American daily life." At the Redstone Arsenal, von Braun's initial title was project director of the Ordnance Guided Missile Center. As a result of a big order from the auto manufacturer Chrysler, von Braun expanded his operations. Neufeld tells us:

"The construction of this in-house industrial capacity, along with cumulative army decisions to build substantial laboratories for guidance, computers, and other fields, gradually rebuilt von Braun's empire to dimensions not seen since Peenemünde. In mid-1952 he and the other Paperclip Germans were converted to regular civil service status; in January 1953 he became the civilian head of a division for the first time: the Guided Missile Development Division of the renamed Ordnance Missile Laboratories, now commanded by Holger Toftoy. As part of a decentralization of authority in Ordnance, Toftoy had been promoted to brigadier general and sent down from Washington, displacing, much to von Braun's relief, Hamill and his 'regime of junior officers.' By September 1954 von Braun had 950 employees; four years later that number had quadrupled to 3,925.

Once again he proved to be a virtuoso in building and managing huge, complex, technically demanding programs."

In 1955 von Braun and 102 other Germans were sworn in as American citizens in Huntsville.

Operation Paperclip Nazi scientists

The 1957 successful Russian launch of the Sputnik satellite prompted the U.S. government to, among many other things, create the National Aeronautics and Space Administration (NASA). NASA was created in order to enable American scientists to bridge the apparent technology gap between the U.S. and Russia. Von Braun had quite a bit to do with the formation of NASA. In a matter of hours after the launch of Sputnik, Speaker of the House of Representatives Sam Rayburn (1882-1961) named a 'blue ribbon committee' that came to be known as the House Select Committee on Astronautics and Space Exploration. Wernher von Braun spent hours testifying before this committee and headed its space research team. It was initially decided by this committee that the space program was to be placed under the Redstone Arsenal in Huntsville, Alabama - headquarters of von Braun's Peenemünde rocket group. Lyndon Johnson (1908-1973), as majority leader in the Senate, appointed a like committee there, which he headed. When President Eisenhower signed the National Aeronautics and Space Act in 1958, creating NASA, the House Select Committee on Astronautics and Space Exploration went out of existence. A permanent, standing House committee with the name of the Committee on Science and Astronautics, chaired by Representative Overton Brooks (1897-1961), was then established. The House Committee on Science and Astronautics lives on today as the House Committee on Science and Technology. The House Committee on Science and Technology has held numerous hearings on geoengineering.

Von Braun and approximately 120 of his fellow Operation Paperclip Nazi scientists were officially transferred to NASA on July 1, 1960. As the first director of NASA's sprawling Marshall Space Flight Center in Huntsville, Alabama, von Braun worked with his fellow former Nazis producing the rockets that would launch America's first satellite and, separately, the gigantic Saturn rockets designed to take Americans to the moon. But more important to this discussion are his contributions to the field of atmospheric science. At NASA von Braun helped to further develop modified versions of the German V-2 rocket for use in atmospheric soundings. As part of his work in the atmospheric sciences, von Braun also participated in the aforementioned (ch 1) operations Argus and Hardtack which involved detonating nuclear bombs in the high atmosphere (lower-ionosphere). These operations allowed our scientists to better map the auroral electrojet - a major aspect of our planet's space weather, which has a direct effect on the weather we experience daily down here in the troposphere. It is interesting to note that operations Argus, Hardtack, and other similar operations were funded and coordinated by the newly formed Advanced Research Projects Agency (ARPA). ARPA was founded in 1958 with the help of von Braun's boss at the Redstone Arsenal, Major General John Bruce Medaris (1902-1990). ARPA's first director was GE Vice President Roy Johnson.

Because weather satellites play an important role in the NMP, it is also very interesting to understand that von Braun apparently had something to do with the production of the first line of dedicated American weather satellites. Dr. Neufeld writes, "When the Defense Department made it clear in early 1958 that space reconnaissance was an air force mission, a typically inventive von Braun told an RCA engineer: 'Let's look at clouds!' The army/RCA proposal evolved into a weather satellite project that later went to NASA as Tiros." Tiros was the first line of American satellites dedicated to weather observation.

In his later years, von Braun became the vice president for engineering and development at an American aerospace firm called Fairchild Industries. It was during his time with Fairchild in September of 1974 that von Braun travelled to the North Slope of Alaska to visit the Prudhoe Bay oil field - not a very popular destination. This is of note to our discussion because one of the world's largest and most versatile electromagnetic energy generators known to be able to control the weather in the fashion of the New Manhattan Project, HAARP (ch 3) is powered by the natural gas coming out of the ground at Prudhoe Bay.

Also of note to our investigation, von Braun was briefed on early wireless power technologies. Pioneering expert in the field, Dr. William C. Brown (ch 5) writes:

"NASA's Marshall Space Flight Center became interested in a space station in low-Earth orbit and was investigating how power could be transmitted from a central space station to daughter satellites. After learning about our activity at Raytheon, we were asked by Dr. Ernst Stuhlinger [Operation Paperclip/1913-2008] to make a presentation to Dr. Wernher von Braun and his staff on the subject of using microwave power transmission for this purpose. As a demonstration to Dr. von Braun and his staff, we beamed power from a small parabolic reflector at one end of a long table in the board room to the little rectenna at the other end of the table. The rectenna was attached to a small motor which ran convincingly when I held the rectenna in the beam. Dr. von Braun was impressed and gave approval to a modest program of work to further develop the technology of microwave power transmission."

We will now take a look at some other Operation Paperclip scientists.

Von Braun's mentor Walter Dornberger (1895-1980) was the head of V-2 rocket development at the Peenemünde slave labor camp. Later in the war, Dornberger also convinced Hitler to build the Nordhausen rocket factory where at least 20,000 prisoners from the nearby Dora concentration camp were worked to death. After the war, he was initially interned in British POW camps, but in 1947, immediately after his release, he went to work on a classified rocketry program at the then Wright Field near Dayton, Ohio. Today this installation is known as Wright-Patterson Air Force Base and, as detailed in chapter

5, it is the most probable location used to develop the proprietary aircraft of today's chemtrail fleet. Dornberger went on to become a senior Vice President of the Bell Aerosystems Division of the Textron Corporation. He was never officially questioned about his role at the Peenemünde death camp.

Martin Schilling (1911-2000) was a developer of the German V-2 rocket during WWII as well. In 1958 he went to work for Raytheon where he went on to attain the rank of vice president for research and engineering. Evidence suggests that Raytheon manages much if not all of the directed electromagnetic energy portions of today's New Manhattan Project (ch 3).

Kurt Debus (1908-1983) was a member of the SS, the SA Brownshirts, and two other Nazi groups. He was also the V-2 flight test director at the Peenemünde slave labor camp. Debus went on to become the first director of NASA's Kennedy Space Center. Over the course of his work in America, Dr. Debus launched more than 150 missiles and space vehicles including the USA's first earth satellite Explorer I. Debus was part of a group of German Paperclip scientists working on rockets under von Braun known as the Army's Guided Missile Development Division. As we have seen, NASA has produced many technologies pertinent to the New Manhattan Project. In 1959 the U.S. Army gave Debus its highest civilian decoration, the Exceptional Civilian Service Medal.

Von Braun's Peenemünde rocket group

Arthur Rudolph (1906-1996) was the director of NASA's Saturn V rocket program designed to send Americans to the moon and the former head of production at the Mittelwerk slave labor death camp.

Ernst Czerlinsky, a former member of the Nazi SS, went to work at the Air Force Cambridge Research Center. The Air Force Cambridge Research Center did work on technologies later used in the New Manhattan Project such as advanced air traffic control systems and ionospheric heaters.

Hans Joachim Naake was a radar specialist. Radar, and specifically over-the-horizon radar, plays an important role in the development of the New Manhattan Project as these things are the predecessors of the ionospheric heaters which modify the weather as part of today's NMP (ch 3). Being that Naake was a radar specialist, he was probably working on over-the-horizon radar because that was, along with longitudinal radar, the most cutting edge radar technology of his time.

At the U.S. Army's Edgewood Arsenal, Albert Pfeiffer, a high-ranking Nazi scientist during the war, worked on new ways to disseminate airborne chemical warfare agents. Does that sound familiar?

Lastly, the aforementioned Hans Dolezalek (ch 10) is a distinguished German meteorologist who worked for the Wehrmacht Weather Service during WWII. He was also an early member of the Nazi SA Brownshirts. Dolezalek visited America in 1958 in order to attend a major meteorological conference and subsequently accepted a job with AVCO (Aviation Corporation) in Wilmington, Massachusetts. In 1985, AVCO was acquired by the aforementioned Textron for $2.9B. With this acquisition, Textron nearly doubled in size. Initially the acquired AVCO became a division of Textron known as Textron Defense Systems. Textron Defense Systems evolved into today's Textron Systems Weapons & Sensor Systems - a leader in intelligence-gathering capabilities and 'advanced protection systems.' This is the same Textron Corporation that employed von Braun's mentor Walter Dornberger. The Aviation Corporation was organized largely by Wall Street interests in 1929. Shortly after its founding, AVCO acquired 55 percent of Fairchild Aviation Corporation's stock, bringing with it Fairchild's Farmingdale factory and Sherman Fairchild (1896-1971) as an AVCO vice president. A man named E.L. Cord gained control of AVCO in 1933 and brought it under his Cord Corporation which ran

AVCO under a subsidiary known as the Aviation Manufacturing Corporation, headquartered in Chicago. Dolezalek has worked extensively in the vein of the New Manhattan Project. He has done lots of work in the area of atmospheric electricity and artificial ionization. He has also done work for the aforementioned Office of Naval Research - an organization with strong implications for the development of the NMP. Dolezalek continued his work in atmospheric electricity until at least 1999 when he was a member of the international commission to NASA's 11th International Conference on Atmospheric Electricity.

Apparently Operation Paperclip-like activities continue. In 1989 the Department of Defense announced that they were filling exempted positions under the Program for Utilization of Alien Scientists. This program was being run by the Defense Advanced Research Projects Agency's Research and Engineering Enterprise. This is the group that took over Operation Paperclip after it was disbanded in 1962. Thirteen of the new recruits were to work at NASA. The new hires included an unnamed 'world-renowned climatologist.' These NASA scientists worked at: the Langley Research Center (near the CIA's headquarters), the Ames Research Center at Moffet Field, and the Goddard Space Flight Center.

Operation Paperclip had the official support of the United States and its military. This means that many high-ranking U.S. military officers were fully aware of the situation and what was going on. Although most military officers aware of this program were undoubtedly not happy about it, at least one was enthusiastic. Enjoying secretly bringing in the former Nazis was Army Air Corps Colonel Donald L. Putt (1905-1988).

Colonel Donald Putt was among those who initially reviewed captured Nazi aircraft at the Hermann Göring Aeronautical Research Institute in Brunswick, Germany. According to journalist Linda Hunt, after seeing the facilities, swept-back wing aircraft, and other inventions there, "Putt gathered the Germans together and, without approval from higher authorities in the War Department, promised them jobs at Wright Field if they would go with him to a holding center for captured personnel in Bad Kissingen. He also promised to send their families to the United States, then instructed the scientists to sell all of their belongings and to travel light."

The then Wright Field was where all the information gathered by Operation Alsos and other similar operations was organized and archived. By the Fall of 1946, there were 140 former Nazi scientists working under Colonel Putt at Wright Field. They included: Theodor Zobel, Adolf Busemann, General Herhudt von Rohden, and Rudolf Hermann. Although many of Putt's Nazis were known to have committed heinous acts of violence during the war, Putt coddled, protected, and covered for his subordinates every step of the way.

Donald L. Putt

Putt went on to be promoted to the rank of lieutenant general and to become the military director, as well as the director of research and development of the USAF Scientific Advisory Board (SAB). At the SAB, Putt worked closely with the aforementioned Dr. Theodore von Kármán and T.F. Walkowicz (ch 3). Putt also served on Vannevar Bush's National Advisory Committee for Aeronautics (NACA) - the precursor to NASA. To top it all off, Putt was deeply involved in the creation of the MITRE Corporation (ch 7). Colonel Putt was also the man in charge of modifying the B-29 bomber that dropped its nuclear payloads on Hiroshima and Nagasaki. Doors needed to be made larger to accommodate the size of the bombs. Putt had earlier helped in the original development of the B-29.

A top-ranking CIA man named Frank Wisner (1909-1965) handled the CIA investigations of all Operation Paperclip Germans and helped obtain visas for Paperclip scientists. Wisner was the officer who formally accepted the surrender of the aforementioned Nazi General Reinhard Gehlen. Later, as director of the CIA's Office of Policy

Coordination, Wisner was also instrumental in the founding and later purchase of the aforementioned Civil Air Transport (CAT). As detailed in chapter 5, CAT is one of the logical predecessors of today's dedicated chemtrail fleet. Wisner may be most famous for being the CIA man who ended up going crazy and subsequently blowing his own brains out with his son's shotgun.

Frank Wisner

Frank Wisner's son Frank Wisner, Jr. went on to serve as a U.S. ambassador to: Zambia, Egypt, the Philippines, and India. He has also served as the U.S. Under Secretary of Defense for Policy and the Under Secretary of State for International Security Affairs. Most curiously though, Jr. was serving as the acting Secretary of State in 1996 when the large-scale domestic spraying operations began.

Frank Wisner, Jr.

The CIA, working with the help of a former Nazi, is on record as having sponsored domestic aerial spraying of biological agents. With the help of former Nazi Kurt Blome, a researcher by the name of Frank Olson and another by the name of Norman Cournoyer conducted CIA sponsored research out of Camp Detrick, Maryland. David Talbot in *The Devil's Chessboard* writes:

"After the war, they [Olson and Cournoyer] *had traveled around the United States, supervising the spraying of biological agents from aircraft and crop dusters. Some of the tests, which were conducted in cities like San Francisco as well as rural areas in the Midwest, involved harmless chemicals, but others featured more dangerous toxins. In Alaska - where the two men sought to stage their experiments in an environment that resembled wintertime Russia - 'We used a spore which is very similar [to] anthrax,' Cournoyer recalled. 'So to that extent we did something that was not kosher.' One of their research colleagues, a bacteriologist named Dr. Harold Batchelor, learned aerial spray techniques from the infamous Dr. Kurt Blome, director of the Nazis' biological warfare program."*

Frank Olson was the guy who, one week after being covertly dosed with Central Intelligence Agency LSD, died when he leapt (or was pushed) from a tenth floor window of the Statler Hotel in midtown Manhattan.

Frank Olson

Nazi biological warfare research, led by Dr. Kurt Blome (1894-1969), included experimentations on prisoners in concentration camps. Blome was later hired by the U.S. Army Chemical Corps to conduct new biological weapons research.

Kurt Blome

In 1962, the organization that oversaw Operation Paperclip was disbanded. What remained of the Paperclip program was taken over by the Pentagon's Research and Engineering Department; an Advanced Research Projects Agency (ARPA) division. Today, as noted a little earlier, this department is known as the Department of Defense Research and Engineering Enterprise; a Defense Advanced Research Projects Agency (DARPA) department. Operation Paperclip was folded into DARPA. The evidence suggests that DARPA has been overseeing the New Manhattan Project from the beginning and on through to today.

The aforementioned Herbert York (ch 7) once had an opportunity to press former president Eisenhower about his farewell address comments pertaining to the 'military/industrial complex.' Eisenhower had characterized such a complex as a threat to the American way of life. York writes, "I pressed this line of questions further by asking him whether he had any particular people in mind when he warned us about 'the danger that public policy could itself become the captive of a scientific-technological elite.' He answered without hesitation: '(Wernher) von Braun and (Edward) Teller.'" Eisenhower warned us about the New Manhattan Project.

References

Operation Paperclip: The secret intelligence program that brought Nazi scientists to America a book by Annie Jacobsen, published by Little, Brown, and Company, 2014

Operation Gladio: The Unholy Alliance Between the Vatican, the CIA, and the Mafia a book by Paul L. Williams, published by Prometheus Books, 2015

The Devil's Chessboard: Allen Dulles, the CIA, and the Rise of America's Secret Government a book by David Talbot, published by HarperCollins Publishers, 2015

Secret Agenda: The United States Government, Nazi Scientists, and Project Paperclip, 1945 to 1990 a book by Linda Hunt, published by St. Martin's Press, 1991

Blowback: The First Full Account of America's Recruitment of Nazis and its Disastrous Effect on Our Domestic and Foreign Policy a book by Christopher Simpson, published by Weidenfeld & Nicolson, 1988

IBM and the Holocaust: The strategic alliance between Nazi Germany and America's most powerful corporation a book by Edwin Black, published by Dialog Press, 2012

George Bush: The Unauthorized Biography a book by Webster Tarpley and Anton Chaitkin, published by Executive Intelligence Review, 1992

Wall Street and the Rise of Hitler a book by Antony Sutton, published by Buccaneer Books, 1976

Hitler's Scientists: Science, War, and the Devil's Pact a book by John Cornwell, published by Penguin Books, 2004

"Nazi Whitewash in 1940's Charged" an article by Ralph Blumenthal, published by the *New York Times*, March 11, 1985

Von Braun: Dreamer of Space Engineer of War a book by Michael J. Neufeld, published by Random House, 2007

"Richard W. Porter, 83, Led Early Space Effort" an article by Robert Mcg. Thomas Jr., published by *The New York Times*, October 10, 1996

The General Electric Story: A Heritage of Innovation 1876-1999 a book by the Hall of Electrical History, Schenectady Museum, published by Hall of Electrical History Publications, 1999

Images of America: White Sands Missile Range a book by Darren Court and the White Sands Missile Range Museum, published by Acradia Publishing, 2009

17th Interdepartmental Committee for Atmospheric Sciences Report, May 1973

"Electronics and Aerospace Industry in Cold War Arizona, 1945-1968: Motorola, Hughes Aircraft, Goodyear Aircraft a PhD dissertation by Jason Howard Gart, published by UMI Dissertation Services and ProQuest, August 2006

A Congressional Record: The Memoir of Bernie Sisk a book by Bernie Sisk and A. I. Dickman, published by the Regents of the University of California and Panorama West, 1980

"The History of Wireless Power Transmission" a paper by William C. Brown, published by *Solar Energy*, vol. 56, no. 1, p 3-21, 1996

"John F. Kennedy Space Center Souvenir Book" a book by the Dexter Press, 1969

The Secret History of the CIA a book by Joseph J. Trento, published by Carrol & Graf, 2005

The CIA and the Cult of Intelligence a book by Victor Marchetti and John D. Marks, published by Dell Publishing Co., 1980

A History in the Making: 80 Turbulent Years in the American General Aviation Industry a book by Donald M. Pattillo, published by McGraw-Hill, 1998)

"The USAF Scientific Advisory Board: Its First Twenty Years 1944-1964" a report by Thomas A. Sturm, published by the USAF Historical Division Liaison Office, 1967

Jerome C. Hunsaker and the Rise of American Aeronautics a book by William F. Trimble, published by the Smithsonian Institution Press, 2002

"Dr. Kurt H. Debus: Launching a Vision" a paper by C. McCleskey and D. Christensen, published by the International Academy of Astronautics, 2001

"11th International Conference on Atmospheric Electricity" a report compiled by H.L. Christian, published by the National Aeronautics and Space Administration, June 1999

"The Early History of the MITRE Corporation: Its Background, Inception, and First Five Years, vol I" a report by Howard R. Murphy, published by the State University College of Oneonta, New York, 1976

The Secret Team: The CIA and its Allies in Control of the United States and the World a book by L. Fletcher Prouty, published by Skyhorse Publishing, 2008

Arms and the Physicist a book by Herbert F. York, published by the American Institute of Physics, 1995

Chapter 14
The POWER BEHIND the CHEMTRAILS

"Some of the biggest men in the United States, in the field of commerce and manufacture, are afraid of something. They know that there is a power somewhere, so organized, so subtle, so watchful, so interlocked, so pervasive, that they had better not speak above their breath when they speak in condemnation of it."
-U.S. President Woodrow Wilson

"Behind the visible government there is an invisible government upon the throne that owes the people no loyalty and recognizes no responsibility."
-U.S. President Theodore Roosevelt

The history of the New Manhattan Project as recounted here in the pages of this book speaks to a huge, coordinated effort carried out over the course of 125 years. Understanding this, it begs the question: Who has coordinated it?

The New Manhattan Project ultimately requires coordination between many disciplines such as: science, politics, military art, intelligence, industry, finance, academia, and media. To match the global nature of the NMP, all of these disciplines must necessarily be applied to a global scale.

Although we have seen the pervasive involvement of many extremely powerful organizations such as the Department of Defense, the Central Intelligence Agency, and the Council on Foreign Relations, the vast scope and ultimately unified vision and execution of the New Manhattan Project speaks to an even higher authority. Evidence of such a higher authority exists and that evidence, with specific relevance to the New Manhattan Project is explored here. This chapter explores the power behind the chemtrails.

The Network of Global Corporate Control

A 2011 study produced by three mathematicians from the Federal Institute of Technology in Zurich, Switzerland titled "The Network of Global Corporate Control" offers strong evidentiary support for a group capable of pulling off the New Manhattan Project. This report scientifically proves the existence of a supra-national mega-conglomerate with the capability to exert a highly disproportionate influence over the Western world. Intuitively, we have always known it is there. Now it is proven. Vitali, Glattfelder, and Battiston write:

"We present the first investigation of the architecture of the international ownership network, along with the computation of the control held by each global player. We find that transnational corporations form a giant bow-tie structure and that a large portion of control flows to a small tightly-knit core of financial institutions. This core can be seen as an economic 'super-entity' that raises new important issues both for researchers and policy makers."

The authors state that the nucleus of the Network consists of: Franklin Resources, Prudential Financial, Commerzbank AG, Morgan Stanley, Credit Suisse, AXA, Citigroup, Deutsche Bank AG, Merrill Lynch, Barclays PLC, Bank of America Corp., UBS AG, JP Morgan Chase & Co., State Street Corp., T. Rowe Price, and Goldman Sachs. Barclays is noted as the most powerful overall. AXA Re is a known participant in the weather derivatives market (ch 8). After the bankruptcy, UBS bought Enron's weather derivatives trading business.

As we saw in chapter 8, lots of money and power can be consolidated through control of the weather. In fact, control of the weather is god-like power. Isn't it interesting to note, then, that the purpose of the world's largest corporations is to acquire as much power and wealth as possible? It is their legal, fiduciary duty to their shareholders to do so. If a global weather modification project such as the NMP can help large corporations consolidate their power and generate more wealth, then doesn't it stand to reason that said corporations would involve themselves in such an effort?

Doesn't it make sense that these corporations, apparently acting together, would take part in helping to bring about the New Manhattan Project? Most of the corporations exposed here by Vitali *et al*. as being a part of the Network of Global Corporate Control are old enough to have assisted since at least the end of WWII. Most, if not all of them, or their predecessors, are much older than the modern NMP. In light of the substantial nature of today's Network of Global Corporate Control, it is reasonable to assume that collaborations between large multinational corporations go back much further than 1945. The

Network of Global Corporate Control, or collaborations similar to it, has most probably been going on for centuries.

Other evidence for large, multinational corporations apparently acting in collusion to consolidate their power and wealth exists. These other examples offer more specific implications for the New Manhattan Project.

The Round Table

"The world is governed by very different personages from what is imagined by those who are not behind the scenes."
-English Prime Minister Benjamin Disraeli, 1884

A specific group known as 'The Round Table' is capable of pulling off the New Manhattan Project. Let us understand that the first discernible origins of the New Manhattan Project date back to the mid-1890s with Svante Arrhenius and the Stockholm Physics Society (ch 12). The Round Table's origins predate these events. Let us also understand that The New Manhattan Project requires expertise in and dominance over many disciplines such as: science, politics, military art, intelligence, industry, finance, academia, and media. The Round Table is capable of providing what has been required. In chapter 8, we learned about all the motivations for such a group as the Round Table to want to propel the New Manhattan Project. The Round Table has had means, motive, and opportunity.

Let us now examine the history of The Round Table. It began with Cecil Rhodes and something called 'The Society of the Elect.' Cecil Rhodes (1853-1902) made his fortune exploiting the natural resources of Africa. His activities were mainly funded by the Rothschilds - specifically Nathaniel Mayer Rothschild (1840-1915).

By November of 1888, Rhodes had drafted a will naming Nathaniel Rothschild as the sole beneficiary of his estate. In the codicil to the will, Rothschild was instructed to use the wealth accumulated by Rhodes to establish a British secret society. Rhodes called this new secret society "The Society of the Elect" and it was to be modeled after the Society of Jesus; also known as the Jesuits. Rothschild advanced this task by bringing a host of influential aristocrats to the new Society of the Elect.

Cecil Rhodes

Nathaniel Mayer Rothschild

The main focus of the Society of the Elect was to chart a course of events that would culminate in what they called a New World Order. As a means to achieve this, Rhodes understood very well the power of media. Rhodes wrote, "The Society should inspire and even own portions of the press for the press rules the mind of the people." Paul Williams and Rodney Howard-Browne, in their 2019 epic masterpiece *Killing the Planet* write:

"The members of the Society of the Elect were cut from the same cloth. They were rich, aristocratic, Oxford or Cambridge educated, and freemasons. They were disciples of John Ruskin [1819-1900] and ardent imperialists, who believed the British were the master race. They shared a deep interest in the occult, including theosophy, Satanism, and communication with the spirit world. In addition, many, including Lord Brett, Lord Roseberry, and Lord Balfour, were pederasts. Rhodes, throughout his life, kept a stable of 'angels and lambs' (young boys) to satisfy his sexual needs."

After Rhodes' passing in 1902, the Society of the Elect, under the leadership of Alfred Milner (1854-1925), was renamed The Round Table. Milner then recruited a group of young men from aristocratic families - mostly Oxford graduates - popularly known as 'Milner's Kindergarten' to firstly serve as his administrators, governors, and bureaucrats in the South African government. After some months, Milner's Kindergarteners were dispatched to: the United States, Canada, Australia, New Zealand, and British colonies around the world. These Kindergarteners became the next generation of Round Table members. As supervisors of the Round Table Movement, Milner and Nathan Rothschild went on to establish branches of the Round Table in Australia, New Zealand, and Canada.

As a way to advance the agendas of The Round Table, 'Pilgrim Societies' were created. Before his death, Rhodes had envisioned these Pilgrim Societies as a way to bring America back into the fold of the British Empire. He had argued that these Pilgrim Societies should be planted in New York and London with the purpose of strengthening economic trade between the two countries. These Pilgrim Societies were also to serve as a locus for international coordination towards the establishment of their New World Order. One of the ways the members of the Pilgrim Societies went about doing this was, and continues to be, by the selection of Oxford Rhodes Scholars for indoctrination into their plans. Let us not forget that the Rhodes Scholar, William Jefferson Clinton was president of the United States when large-scale domestic spraying operations began stateside in 1996.

Rhodes Scholarships are partially funded by the Rockefellers and something called the Carnegie United Kingdom Trust and are granted to 'young colonists' with the purpose of, "instilling into their minds the advantages of the colonies as well as to England for the retention of the Unity of the Empire." Most Rhodes Scholarships go to young American students. It has been noted by journalist William Fulton of the *Chicago Herald Tribune* that Rhodes Scholars dominate the U.S. Department of State.

As a way to advance the goals of the Round Table in America, J.P. Morgan (1837-1913), at the bidding of Rothschild and Milner, established the American chapter of the Pilgrim Society at the Waldorf Astoria hotel in Manhattan. A London headquarters of the Pilgrim Society had been previously established. Morgan's new Manhattan chapter of the Pilgrim Society included: Percy Rockefeller, Andrew Mellon, Nelson Aldrich, and Paul Warburg. Later, members of the Pilgrim Society of more specific note to the New Manhattan Project would come to include: John Foster Dulles, Allen Dulles, Henry Luce, David Rockefeller, Zbigniew Brzezinski, George H.W. Bush, and George W. Bush. In 1912 the Pilgrim Society spawned an American Round Table. The financial cartel consisting mainly of the houses of: Rockefeller, Carnegie, Morgan, Schiff, and Rothschild has been the driving force behind The Pilgrim Societies and The Round Table.

The American Federal Reserve Bank System was hatched by members of the Pilgrim Society. They believed that a new, American central bank could be joined with the Bank of England. A partial 1914 listing of U.S. Federal Reserve Bank shareholders includes: Brown Brothers Harriman, National City Bank, William Rockefeller, Percy Rockefeller, and New York Edison. The Rockefeller-controlled National City Bank was the largest shareholder with 30,000 shares.

The Council on Foreign Relations

CFR logo

The Council on Foreign Relations is an offshoot of The Round Table and it is the key that unlocks the door to the New Manhattan Project. As we saw in chapter 11, the Council on Foreign Relations (CFR) has been going on and on about the need to spray us with incredibly massive amounts of toxic material from aircraft. We are also interested in the CFR because in 2016, the former director of the CIA John 'Stratospheric Aerosol Injection' Brennan famously advocated for SRM geoengineering at their headquarters in Manhattan known as Pratt House.

The Council on Foreign Relations serves two purposes. It is at once a globalist think-tank and an establishment mouthpiece. Most other high-profile think-tanks are not nearly as vocal. The CFR enjoys an extremely cozy relationship with both our U.S. State Department and the CIA. With alarming frequency, the ideas emanating from CFR study groups become official United States government policy. CFR study groups not only produce suggested policy, but also tactics for confounding and discrediting any opposition to their suggestions. Financing for these CFR study groups comes from the Rockefeller, Carnegie, and Ford Foundations. Policy and strategy worked out by the CFR reaches the President by way of the State Department.

Pratt House

The CFR was formed in order to influence public opinion. In 1919 American and British members of the Pilgrim Society met at the Majestic Hotel in Paris to create the Council on Foreign Relations as a think tank for the U.S. State Department. The founders of the CFR felt compelled to create such an organization because of their recent failure to establish their post-WWI League of Nations. James Perloff, in his book *The Shadows of Power* recounts the story:

"Well before the Senate's vote on ratification, news of its resistance to the League of Nations reached Colonel [Edward] *House, members of the Inquiry, and other U.S. internationalists gathered in Paris. It was clear that America would not join the realm of world government unless something was done to shift its climate of opinion. Under House's direction, these men, along with some members of the British delegation to the Conference, held a series of meetings. On May 30, 1919, at a dinner at the Majestic Hotel, it was resolved that an 'Institute of International Affairs' would be formed. It would have two branches - one in the United States, one in England.*

"The American branch became incorporated in New York as the Council on Foreign Relations on July 29, 1921."

Carrol Quigley in his book *Tragedy and Hope* describes the purpose of the CFR:

"The CFR (Council on Foreign Relations), established six years after the Federal Reserve was created, worked to promote an internationalist agenda on behalf of the international banking elite. Where the Fed took control of money and debt, the CFR took control of the ideological foundations of such an empire - encompassing the corporate, banking, political, foreign policy, military, media, and academic elite of the nation into a generally cohesive overall world view."

Allen Dulles and his more famous brother, Secretary of State John Foster Dulles were among a small group of influential and wealthy individuals who founded the CFR. John Foster Dulles was a founding member of the CFR and, beginning with the first issue, regularly contributed articles to their most widely-read publication called *Foreign Affairs*. In 1926 Allen Dulles joined the CFR and in 1927 he was elected as the first president of the Council on Foreign Relations. From his soundproof room at Pratt House, Allen Dulles wielded tremendous power. Top secret CIA planning meetings chaired by Dulles were held at Pratt House. David Talbot, author of *The Devil's Chessboard* writes:

"When it came to undertaking secret missions, Allen Dulles was a bold and decisive actor. But he acted only after he felt that a consensus had been reached within his influential network. One of the principal arenas where this consensus took shape was the Council on Foreign Relations. The Dulles brothers and their Wall Street Circle had dominated this private bastion for shaping public policy ever since the 1920s. Over the years, CFR meetings, study groups, and publications provided forums in which the organization's leading members - including Wall Street bankers and lawyers, prominent politicians, media executives, and academic dignitaries - hammered out major US policy directions."

Talbot continues, "If CFR was the power elite's brain, the CIA was its black-gloved fist."

Williams and Howard-Browne write, "by 2017, twenty secretaries of state, nineteen secretaries of the treasury, fifteen secretaries of defense, and hundreds of other federal department heads have been CFR members, along with twenty-one of the twenty-four CIA directors, and every chairman of the Federal Reserve since 1951." The authors also note that, "Mike Pompeo … is the only Secretary of State within the past 85 years without any known ties to the CFR."

Many people previously implicated here in the New Manhattan Project have been CFR members such as: Lauchlin Currie, Gordon J.F. 'How to Wreck the Environment' MacDonald, Henry Loomis, Glenn Seaborg, Alfred Lee Loomis, James Killian, Harlan Cleveland, James Conant, John Deutch, and Frank Wisner. Many corporations implicated in the production of the NMP are (or have been) corporate members of the CFR such as Boeing, Raytheon, IBM, and General Electric. Five of GE's directors and three of Boeing's have been CFR directors as well.

The United Nations

United Nations logo

The United Nations (UN) has appeared over and over again throughout the pages of this book. In 1961 President Kennedy spoke in favor of weather control before the UN (ch 1). Civil Air Transport was formed at least in part to fly United Nations Relief and Rehabilitation Administration supplies into the Chinese mainland (ch 5). Evergreen Aviation did a lot of work for the UN (ch 5). Thomas F. Malone of the Travelers insurance company was a member of their United Nations Educational Scientific and Cultural Organization (UNESCO). The UN has been into monitoring climate with their EARTHWATCH program, their International Global Observing Strategy as well as with their

World Meteorological Organization. They have been deeply involved in the promulgation of the theory of man-made global warming with their Earth Day, their United Nations Environment Program, their World Meteorological Organization, their Framework Convention on Climate Change, their Intergovernmental Panel on Climate Change, and their Agenda 21. Their 1999 report titled "Aviation and the Global Atmosphere" told lies about contrails, stating that they can form cirrus clouds (ch 11). Their UNESCO pops up again and again in the historical weather modification literature and has proposed SRM geoengineering.

Considering these facts, it is not surprising to learn that the United Nations was first conceived in a CFR study group funded by the Rockefeller family. The plans for the UN were then drawn up in the U.S. State Department. Williams and Howard-Browne write:

"The United Nations, as conceived by the Council on Foreign Relations and funded by the House of Rockefeller, was never meant to be an academic debating society. It was conceived to become an international regime that would control the world's weapons, wars, courts, tax collectors, and economy. The orders to create this new international organization came from the Rockefellers. The plans were drawn up in 1943 by a 'secret steering committee' under Secretary of State Cordell Hull. The members of this committee - Leo Pasvolsky, Isaiah Bowman, Sumner Welles, Norman Davis, and Myron Taylor - were all prominent CFR members. The draft for the massive international agency, after passing Nelson Rockefeller's inspection, was presented on June 15, 1944, to President Roosevelt, who promptly gave it his approval."

The Dumbarton Oaks Conference was then held in 1944 which subsequently produced the proposal for the UN. In order to finalize this proposal, diplomats from around the world gathered in San Francisco in April of 1945. The secretary-general of the San Francisco conference was Alger Hiss (1904-1996), a CFR member and Soviet agent. The American delegates to the San Francisco conference included Nelson Rockefeller and John Foster Dulles and almost all of them were CFR members. After two months of negotiations, the San Francisco conference unanimously approved the new UN charter. Since its founding, the CFR has advocated on behalf of the UN.

After moving their headquarters around quite a bit early on, John D. Rockefeller, Jr. offered the UN $8.5M to buy a piece of real estate situated alongside the East River in central Manhattan previously known as Turtle Bay. The United Nations headquarters were mainly designed by the same architect who designed Rockefeller Center. The buildings were completed in less than 6 years and the new, world famous UN headquarters complex opened in 1952.

Kurt Waldheim (1918-2007) was elected the fourth secretary-general of the United Nations in December of 1971. He served at that post for 10 years. For all of 1972, while Waldheim was in office as secretary-general, George H.W. Bush was concurrently the U.S. Ambassador to the United Nations. Waldheim was later found out to have been a Nazi and to have been lying about it repeatedly. Stanley Meisler in his book *United Nations: A History* writes:

"As a second lieutenant in World War II, he [Waldheim] *had been assigned to a German army unit in the Balkans that had rounded up thousands of Jews in Greece for deportation to Auschwitz and had killed thousands of innocent villagers in Greece and Yugoslavia as reprisals for attacks by the 'Partisans,' as Resistance fighters were known during the occupation."*

President Kennedy's aforementioned November 1961 speech at the United Nations pertaining to weather control was followed that December by the American presentation of a draft UN resolution which included a recommendation to, "advance the state of atmospheric science and technology so as to provide greater knowledge of basic physical forces affecting climate and the possibility of large-scale weather modification." The resolution was passed on Dec. 20, 1961. The Soviet Union was a co-sponsor of the resolution. Out of this resolution grew the UN World Weather Watch.

There is a slew of other UN organizations pertaining to the atmospheric sciences and/or weather modification and therefore relevant to out investigation. Here is a list of some: the World Climate Research Program Joint Scientific Committee, the World Climate Research Program Observation and Assimilation Panel, the World Climate Research Program Working Group on Numerical Experimentation, the World Weather Research Program, the Expert Team on Weather Modification, and the Global Atmosphere Watch.

The Central Intelligence Agency

"[The CIA is] a secret government - a shadowy government with its own air force, its own navy, its own fund-raising mechanism, and the ability to pursue its own ideas of the national interest, free from all checks and balances and free from the law itself."

-Senator Daniel Inouye (D-HI) in his closing statement to the Iran-Contra hearings, Aug. 3, 1987

Central Intelligence Agency logo

The Central Intelligence Agency (CIA) appears to be the overall, day-to-day manager of all aspects of the New Manhattan Project. In chapter 4 we learned about how a former director of the CIA was most probably central to the NMP's development. In chapter 5 we learned how the CIA has owned and operated large fleets of covertly operating aircraft. In chapter 11 we learned about their promotions of SRM geoengineering and how they can provide media cover for the New Manhattan Project. In chapter 12 we learned about their promotions of the theory of man-made global warming. In the previous chapter we learned about the probable involvement of former Nazis and the CIA's Nazi roots. To list here every documented instance of the CIA being connected to the NMP would be truly tedious. It suffices to say that their presence throughout the pages of this book has been pervasive - more so than any other organization.

In light of all this information it is not surprising that the CFR was instrumental in the founding of the CIA and both organizations share an inextricably intertwined history. The CIA is the door to the New Manhattan Project which is unlocked by the CFR's key.

The history of the CIA really starts with its most significant predecessor, the Office of Strategic Services (OSS). In June of 1941, six months before America declared war on Germany, the OSS was set up in Rockefeller Center. The first director of the OSS was William J. 'Wild Bill' Donovan (1883-1959). Wild Bill Donovan's wife Ruth Ramsey was a director of the Rockefeller Foundation. Similar to some of the first CIA agents, the first agents of the OSS were bankers, lawyers, businessmen, and accountants deeply embedded with the CFR and the international money cartel. Nelson Rockefeller, Paul Mellon, Carnegies, Du Ponts, and Allen Dulles all served as directors of the OSS. The OSS was the dominant U.S. intelligence agency during WWII.

Shortly after the conclusion of the war, President Truman abolished the OSS and placed its operations under something called the Strategic Services Unit (SSU). Within months, the SSU morphed into the National Intelligence Authority, then into the Central Intelligence Group. General Hoyt Vandenberg (1899-1954) then recruited Allen Dulles to draft a proposal for the organization that would become the CIA. It was all financed by Rockefeller and Carnegie. Williams and Howard-Browne write:

"In 1946, General Hoyt S. Vandenberg, as director of Central Intelligence (DCI), recruited Allen Dulles, a lawyer for the House of Morgan and the president of the CFR, 'to draft proposals for the shape and organization of what was to become the Central Intelligence Agency in 1947.' The meeting took place within offices within Rockefeller Center. By the time the CIA was officially established by President Truman in 1947, David Rockefeller assumed control of the inner core of the CFR and the Rockefeller and Carnegie Foundations allocated $34 million for the new Agency to engage in covert activities."

Williams and Howard-Browne continue, "As soon as Dulles took charge of the CIA, the Rockefellers became the private bankers for the new intelligence empire, and David became an important source for off-the-books cash to the spy agency."

Once established, as noted in the previous chapter, the CIA's ranks were filled with white shoe boys. New CIA recruits included: John D.

Rockefeller, Jr., Tommy 'the Cork' Corcoran (ch 5), who was a former governor of the Federal Reserve Bank, Richard Bissell, a Ford Foundation executive, William Colby, and Richard Mellon Scaife, the principal heir to the Mellon family fortune. Rear Admiral Roscoe Hillenkoetter became the executive director and former OSS official and Wall Street lawyer Frank Wisner (ch 13) was appointed head of covert operations.

President Truman apparently later regretted his role in the creation of the CIA, saying:

"Now, as nearly as I can make out, those fellows in the CIA don't just report on wars and the like, they go out and make their own, and there's nobody to keep track of what they're up to. They spend billions of dollars on stirring up trouble so they'll have something to report on... It's become a government all of its own and all secret. They don't have to account to anybody."

A most probable place for the New Manhattan Project to reside inside of the CIA is in their Directorate of Science and Technology (DS&T). This Directorate has historically been at the cutting edge of two areas of technology relevant to the New Manhattan Project: satellite reconnaissance (ch 7) and over-the-horizon radar (ch 2). Satellites can monitor atmospheric conditions and track atmospheric particles while over-the-horizon (OTH) radar is the direct predecessor of today's ionospheric heaters which can modify the weather in the fashion of the New Manhattan Project. The DS&T was first mentioned in chapter 8.

Largely funded by the air force, the CIA's Directorate of Science and Technology was created in 1963 by CIA chief John McCone. With the exception of a few, more esoteric systems, the technologies perfected by the Directorate are commonly turned over to the Defense Department. Shortly after the Directorate's formation, Secretary of Defense Robert McNamara (1916-2009) and the Pentagon led an effort to put the military in control of all technical collection which eventually became one of the reasons that Director McCone resigned. William Raborn's directorship followed.

Richard Bissell as the head of Clandestine Services, was a driving force behind the development of space satellites for intelligence purposes. The first director of the directorate, Albert 'Bud' Wheelon (1929-2013) worked extensively on satellites. Wheelon left the DS&T

in 1966 and went to Hughes Aircraft where he eventually became CEO and chairman of the board. During his time at Hughes, they became the world's preeminent supplier of communications satellites.

Albert 'Bud' Wheelon

The DS&T was instrumental in the development of over-the-horizon radar. The DS&T ran OTH radar facilities in Pakistan and later in Taiwan under the project names EARTHLING and CHECKWROTE respectively. The EARTHLING system began operations in March of 1961 with operations ceasing in September of 1965. In May of 1965 work began on the installation of the CHECKWROTE system. As of the Spring of 1969, the CHECKWROTE system was still in operation and the most advanced OTH system in the world. The DS&T also shared the use of the CHAPEL BELL OTH radar facility with the Office of Naval Research.

The Directorate enjoys a very cozy relationship with academia. The authors of *The CIA and the Cult of Intelligence* write that, "The Directorate of Science and Technology employed individual professors, and at times entire university departments or research institutes, for its research and development projects." Marchetti and

Marks continue, "In many cases, the CIA's research involvement on the campuses went much deeper than simply serving as the patron of scholarly work." The authors go on to note that, "the Clandestine Services had their own research links with universities, for the purpose of developing better espionage tools (listening devices, advanced weapons, invisible inks, etc.)." The authors then go on to note that, "The Clandestine Services at times have used a university to provide cover or even assist in a covert operation overseas."

Rockefeller

The Rockefeller Foundation was founded in 1913 (the same year as the private U.S. Federal Reserve Bank) to promote nothing less than the welfare of all mankind. The Rockefellers intended to use science to solve the global problems of: sickness, poverty, underdevelopment, and ignorance. Spraying megatons of toxic waste from aircraft is apparently their idea of how to achieve these goals.

Throughout this book and, most acutely, throughout this chapter, we have seen compelling connections between members of the Rockefeller family and the New Manhattan Project. Before this chapter we have seen: John D. Rockefeller, Jr. funding the MIT Rad Lab (ch 2), Laurance Rockefeller and T.F. Walkowicz embedding themselves in military activities pertinent to the NMP (ch 3), intimate Rockefeller involvement with the Flying Tigers (ch 5), involvement with Frederick Cottrell (ch 6), connections to the MITRE Corporation (ch 7), connections to the Travelers Weather Center (ch 8), funding of John von Neumann, especially during the time he wrote "Can We Survive Technology?" (ch 11), extensive involvement with the promotion of the theory of man-made global warming (ch 12), and involvement with the Nazi regime (ch 13). So far in this chapter, we have seen the Rockefellers' fundamental involvement with The Round Table's Pilgrim Societies, the CFR, the UN, and the CIA.

Chicago University was founded with funds from John D. Rockefeller, Senior. This is significant because lots of weather modification work has been done at Chicago University. Also, it was at the University of Chicago where the the first known self-sustaining nuclear reaction was created - an important development in the history of the original Manhattan Project.

Considering this family's apparent pervasive involvement in the production of the New Manhattan Project, it behooves us to take a longer look at the history of the Rockefeller family with an emphasis on yet more potential connections to the NMP. It is common knowledge that the Rockefeller family made their original fortune with old J.D. Rockefeller, Sr.'s (1839-1937) Standard Oil and their business activities in the energy sector. The subsequent details, especially those details that they probably don't want you to know, are less commonly understood.

John D. Rockefeller, Sr.

After members of the Rockefeller family successfully funded the Russian Revolution, they became enamored with the centralized control provided by communism and have been pushing it on the West, most notably America, ever since. This is contrary to what most people believe, but nonetheless true. Most people believe that the Rockefellers are staunch capitalists. But when one takes into account, among other things, their pervasive activities promoting the centralized power needed to deal with the supposed threat of man-made climate change, one sees that their actions and their political philosophy are consistent. The Rockefellers have simply *used* capitalism to achieve their goal of communism - with them still on top, of course.

During the Great Depression, members of the Rockefeller family successfully turned America towards socialism - the precursor to communism. After all, they had their man Franklin Delano Roosevelt (1882-1945) in the White House. FDR had come from the Rockefeller-controlled CFR. FDR's family had been involved in New York banking with the Rockefellers for decades. FDR's uncle, Frederic Delano (1863-1927), had been the Vice Chairman of the original Federal Reserve Board. Nelson Rockefeller (1908-1979) was FDR's principal advisor.

With the help of Roosevelt, during the Great Depression the Rockefellers consolidated their wealth and power, buying up all sorts of assets for pennies on the dollar. Their most significant accomplishment during this era was their seizure of control over the lion's share of the holdings of the House of Morgan. The centerpiece of this accomplishment was the Rockefellers' takeover of the Chase National Bank of New York which subsequently became today's Chase Manhattan Bank.

But it was their previously mentioned successful funding of the Russian Revolution that allowed the Rockefeller family to capitalize upon the building of the Nazi war machine. After funding the Revolution, the Rockefellers had seized control of the Russian oil fields and began selling their oil to Europe. This gave them the foothold they needed to profit from Hitler's rise to power during the pre-WWII era. Another big way in which the Rockefellers participated was in providing the Germans with the equipment and information needed to produce synthetic gasoline from Germany's plentiful domestic coal supplies. Today, everybody knows how to do stuff like this. Heck, we're making gas from shale now cost-effectively. Making gas from coal is easy, right? Yes, but in the 1930s, these types of technologies were cutting edge and proprietary. The Nazi war machine ran on hydrocarbon fuel and had a seemingly insatiable need. The Rockefellers' Standard Oil sated it. Standard Oil also provided the Nazi war effort with the means to produce synthetic rubber and ethyl lead: two other substances needed for waging their reigns of terror. In fact, the production of a war effort such as that raised by the Nazis necessarily needed all sorts of chemicals and raw materials. In many of these cases, the Rockefellers' Standard Oil worked closely with the aforementioned I.G. Farben to produce them.

The Rockefellers were represented on the board of American I.G. not only by W.C. Teagle (a director of Standard Oil), but they were also represented by C.E. Mitchell, the chairman of the Rockefeller's National City Bank. After all, the German parent company of American I.G. had been created with a $30M loan from the Rockefellers' National City Bank and Standard Oil and I.G. Farben had been big business collaborators ever since. I.G. Farben was so successful and so powerful that by 1933, Farben had become the main funding engine for the Nazi war machine. Let us not forget that it was I.G. Farben that, as noted in the previous chapter, manufactured almost all of the Zyklon B gas used to exterminate millions of Communists, Jews, and other 'undesirables.'

Not only did Standard oil provide the Germans with the means for producing necessary raw materials, throughout WWII Standard Oil directly provided the Third Reich with massive amounts of raw materials for their war effort. We're talking about 48,000 tons of oil and 11,000 tons of tungsten per month. By 1939, Standard Oil was providing the German air force with high quality, tetraethyl-lead ('no-knock') aviation gasoline while *refusing to provide similar assistance to the U.S. War Department*. These activities did not go unnoticed. Williams and Howard-Browne write, "In 1944, Standard Oil was indicted by Assistant Attorney General Thurman Arnold [1891-1969] for deliberately blocking the production of synthetic rubber. The company pleaded guilty and agreed to release the patents and processes for use by the United States."

While all this was going on, the Rockefellers wanted to make sure that Americans didn't see the light. With the help of publicist Ivy Lee (1877-1934), during WWII the Rockefeller family was actively promoting Nazi Germany's public image in America. Throughout WWII, the head of the Rockefeller family was John D. Rockefeller, Jr. (1874-1960).

John D. Rockefeller, Jr.

Laurance Rockefeller (1910-2004) was one of John D. Rockefeller, Jr.'s sons. During the post-WWII era, Laurence Rockefeller worked closely with the aforementioned Dr. T.F. Walkowicz (ch 3). This revelation initially sparked the author's curiosity and so we will now take a good look at Laurence Rockefeller here.

Going back to before his service in WWII, Laurence Rockefeller (L.R.) became firmly embedded in the American military aerospace industry - an area germane to the New Manhattan Project. In the years leading up to America's entry into the war, L.R. had helped finance the beginnings of McDonnell Douglas and had saved Eddie Rickenbacker's Eastern Airlines from bankruptcy. In 1940 L.R. was persuaded by Assistant Secretary of the Navy James Forrestal (1892-1949) to assist the Navy in managing and financing certain aviation companies, and the seeds of Laurance Rockefeller's post-war venture capital activities were planted. During WWII, L.R. was a naval lieutenant overseeing Bureau of Aeronautics aircraft production. Jonathan Lewis, author of *Spy Capitalism: Itek and the CIA* writes:

"Rockefeller's knowledge of the aviation industry, combined with his friendship with Forrestal, led to naval assignments that sharpened his appreciation of the relation between industry, technology, and national security. For most of the war, Rockefeller's job was to monitor the production and development of patrol planes and fighter aircraft. His knowledge of the aviation industry grew, and so did his list of contacts. But most important, he witnessed firsthand the creation of new technologies that transformed the face of warfare, and he became captivated by them. He realized that many of these same technologies, converted to civilian purposes, could have an equally powerful impact 'on the way people live in the postwar world.' His insight was a powerful one."

Laurence Rockefeller

After the war, in 1946 Laurance Rockefeller founded Rockefeller Brothers, Inc. as a venture capital investment firm. This is where Dr. T.F. Walkowicz worked and Walkowicz served as their main military liaison. Their firm invested in companies producing: jets, helicopters, rockets, aviation electronics, and nuclear research. Among many other firms, Rockefeller Brothers helped fund the Glenn L. Martin corporation which eventually became today's Lockheed Martin - developer of the Polaris missile (ch 4) and producer of the C-141 supertanker aircraft (ch 5). Lewis writes:

"After the war, Rockefeller returned to work at the family's office at 30 Rockefeller Center with a new sense of purpose. He assembled a

small staff of colleagues to commercialize in the private sector the breakthroughs in aviation, radar, communications, and nuclear energy that had been developed in the war. He needed a seasoned financial professional to review business proposals, and he recruited Randolph Marston, who had been a banker at Chase Manhattan and had known David Rockefeller when he worked there. He also wanted someone familiar with the key technologies developed during the war. Harper Woodward, who had been Gen. Hap Arnold's assistant for procurement in the Pentagon, was soon hired."

Lewis Strauss, the head of the Atomic Energy Commission under Truman and Eisenhower, was a member of Laurance Rockefeller's staff during this period. Laurence Rockefeller's investment in something called the Nuclear Development Corporation was a direct result of Strauss' influence and interests.

In the weeks before the launch of the Russian satellite Sputnik, the Defense Department had announced a series of defense spending cutbacks. But the findings of the Rockefeller Brothers Fund Special Study Project came out not long after Sputnik and called for a higher level of defense spending, especially on advanced-technology weapons systems. Those Rockefellers always seem to have perfect timing!

The Rockefellers have been thick as thieves with the CIA. The Rockefeller brothers, Nelson and David, were very close associates of CIA founder Allen Dulles. David Talbot writes that, "Nobody occupied a more central position in the Dulles brothers' power circle than the Rockefeller brothers." Talbot also writes that, "The Rockefeller brothers served as private bankers for Dulles's intelligence empire." John Foster Dulles was a trustee of the Rockefeller Foundation from 1935 to 1952 and also served as chairman of the board. He was also a Rockefeller family in-law. Allen Dulles had extensive offices at Rockefeller Center. All throughout the post-WWII era, the CIA engaged in operations that protected Rockefeller family interests. The Rockefellers and the CIA were so close that the Agency allowed Nelson Rockefeller to store highly classified documents within a vault in a carriage house at Nelson Rockefeller's estate in Pocantico, New York.

Rockefellers have also historically dominated the CFR. Laurence Shoup in his book *Wall Street's Think Tank* writes that, "The Rockefeller Foundation is the foundation most closely tied to the

CFR." David Rockefeller (1916-2017) was the longest serving director of the CFR from 1949-1985 (36 years) and chairman from 1970 to 1985. During the 1920s, J.D. Rockefeller, Jr. became a fixture at Pratt House. Other CFR members have included: J.D. Rockefeller IV, David Rockefeller, Jr., Nicholas Rockefeller, and Steven Rockefeller. The Rockefellers' JP Morgan Chase is a corporate member of the CFR. The CFR is controlled by the Rockefellers, so it is no surprise that policies prepared at the CFR for implementation by the State Department have been designed to further the Rockefellers' interests. In 1948, the Rockefellers and the CFR were instrumental in the establishment and implementation of the post-WWII Marshall Plan. The first president of the CFR, John W. Davis, was a trustee of the Rockefeller Foundation. The first vice president, Paul Cravath, was the head of a law firm that protected Rockefeller family interests. Many CFR founders had strong ties to the Rockefeller Foundation. Being that the Rockefeller Foundation along with the Carnegie Foundation provided the funding for the organization's headquarters, all this should not be surprising.

Because of their extensive use of fictitiously-named shell companies, it is impossible at this time to exactly determine all the holdings of the Rockefeller empire. We do know that Exxon Mobil, Amoco, Chevron, ARCO, Sunoco, and Conoco all originally came from John D. Rockefeller-owned companies and that by 1976 the Rockefellers held a controlling interest in Atlantic Richfield, IBM, and Boeing. Also by 1976, DuPont and Monsanto maintained interlocking directorates with the Rockefeller Group. Citibank is part of the Rockefeller empire. As we can see, some of the companies listed here such as Atlantic Richfield, IBM, Boeing, and Monsanto have been implicated in the New Manhattan Project. Atlantic Richfield in particular is very interesting because their subsidiary APTI was the assignee of all of the HAARP patents (ch 3).

Believe it or not, it appears that the Rockefellers and their Standard Oil Company are the original founders of Lawrence Livermore National Lab. As your author wrote in chapter 7, Lawrence Livermore National Lab is my best educated guess as to where a NMP command base might be, so this information is very important. Ernest Lawrence's biographer Herbert Childs writes:

"As the AEC [Atomic Energy Commission] *was interested in developing private industrial cooperation on the West Coast, the*

California Research and Development Company was organized to undertake development of the project by the California Research Corporation, a subsidiary of Standard Oil Company of California. This was accomplished after Regent Neylan brought Ernest [Lawrence] *and Standard Oil Chairman R. Gwin Follis, and his predecessor, Harry Collier, together. It was an ambitious project for a new company, far larger than any ever undertaken by its parent, the California Research Corporation. The pilot plant would be constructed at a onetime naval air station near Livermore, just across the hills from Berkeley, near the edge of the San Joaquin Valley. A full-scale production plant was contemplated for Weldon Spring, Missouri. The project was announced on March 31, 1950, as an AEC classified research program."*

John D. Rockefeller IV as a senator representing West Virginia was a member of the Select Committee on Intelligence which oversaw John 'Stratospheric Aerosol Injection' Brennan's confirmation as Director of the Central Intelligence Agency in 2013.

Today the Rockefeller Brothers Fund advances so-called 'solutions' to the purported problem of man-made climate change. Are we really supposed to believe that they are doing this so selflessly? Yes we are. But now that we know the actual histories of the Rockefeller family and the theory of man-made climate change, as well as all the centralized power that can be afforded by a society's acceptance of such things, the author suggests that you don't. In fact, *please* don't believe the Rockefellers.

Carnegie

Before this chapter, throughout the pages of this book we have seen the name Carnegie come up many times. Top geoengineers Ken Caldeira and Katharine Ricke are Carnegie Institution members and Caldeira runs a Carnegie Foundation laboratory named after him. Vannevar Bush was President of the Carnegie Institution and their headquarters in Washington became the headquarters of Bush's National Defense Research Committee (NDRC) which produced the original Manhattan Project (ch 2). Alfred Loomis was a trustee of the Carnegie Institution of Washington (ch 2). Frederick Cottrell was made chairman of an organization funded by the Carnegie Foundation (ch 6). Monsanto's Charles Thomas was a trustee of the Carnegie

Corporation (ch 6). Gordon J.F. 'How to Wreck the Environment' MacDonald worked for the Carnegie Institution (ch 7). In chapter 11, we saw that Carnegie organizations have produced quite a bit of propaganda about SRM geoengineering.

In this chapter we have also seen that the Carnegie United Kingdom Trust funds the Rhodes Scholarships program. The House of Carnegie is a member of the financial cartel that has been the driving force behind The Round Table. The Carnegie Foundation funds CFR study groups and provided funding for the CFR's headquarters. A Carnegie served as the Director of the OSS. And the Carnegie Foundation provided initial funding for the CIA.

Andrew Carnegie (1835-1919) made his fortune in the railroad and steel industries. Although he believed in American supremacy, not British, Carnegie was indoctrinated into the Round Table Movement through William Stead (1849-1912) and Archibald Primrose (1847-1929). Carnegie also believed that government should supervise the distribution of wealth as they provide for the needs of the citizenry in the fashion of socialism and communism. Carnegie was funded by the Rothschild empire.

Andrew Carnegie

Stead persuaded Carnegie to provide the initial funding for an international court known as the Peace Palace at the Hague to house a permanent court of arbitration. This international court of arbitration was designed to at least partially dissolve America's sovereignty. In fact, Carnegie was persuaded to give much of his money to the cause of a one-world government.

In order to create the League of Nations, complete with an international police force, Carnegie first went about subversion and control of America's political and educational institutions. In order to do this, Carnegie created the Carnegie Endowment for International Peace. This Endowment was designed to make America's political and educational institutions dependent upon them and therefore, under their control. In 1910 the Carnegie Endowment for International Peace established its headquarters in Washington, D.C.. Like the CFR, their purpose was to 'educate' public opinion. The Carnegie Endowment for International Peace is a longtime promoter of globalism with offices around the world. It has received funding from the U.S. State Department, the Defense Department, the Energy Department as well as from foreign governments, individuals, corporations, and foundations. It is closely connected to the CFR. Many of its executives are CFR members. The man behind the IBM punch card computers used to manage the Holocaust, Thomas J. Watson (ch 13) was the chairman of the Carnegie Endowment for International Peace. In 1946 John Foster Dulles was the chairman of the board of the Carnegie Endowment for International Peace. His choice for chairman of that body was the aforementioned Alger Hiss. Dulles was also a trustee.

In 1939 the Carnegie Institution held the press conference which informed the world about the discovery of nuclear fission - a major discovery leading up to the original Manhattan Project.

Rothschild

Before this chapter, throughout the pages of this book there has been little mention of the Rothschilds. This is apparently because the House of Rothschild has had less direct exposure to the New Manhattan Project. Rather, the evidence indicates that members of the Rothschild family have *funded* people and organizations with more direct involvement in activities associated with the New Manhattan Project. They operate with at least one degree of separation.

Nonetheless, there is some evidence to be discussed and we will take a brief look at it here.

This chapter has provided the most compelling evidence for Rothschild involvement in the NMP with Nathan Rothschild's deep connections to Cecil Rhodes and The Round Table as well as associated groups. As we have also seen in this chapter, Andrew Carnegie was funded by the Rothschilds. Also of note to our discussion is the fact that Rothschild North America is a corporate member of the CFR.

As far as other evidence is concerned, the authors of *Killing Uncle Sam*, Williams and Howard-Browne describe Enron as "a shell company tied to the House of Rothschild." This is of note to our discussion because, if you will recall, in chapter 8 we went into detail about Enron and their connections to the weather derivatives market. Perhaps the most well known bit of evidence linking the House of Rothschild to the New Manhattan Project is their 2011 acquisition of a majority stake in Weather Central. *BusinessWire* describes Weather Central as, "the world's leading provider of interactive weather graphics and data services for television, web, and mobile." Today on their Rothschild & Co. website, one can read about how the Rothschilds are all so concerned about 'greenhouse gas emissions and climate change.' Uh-huh.

Bush

Throughout the pages of this book and in this chapter, we have already seen many compelling connections between members of the Bush family and the New Manhattan Project. In chapter 6 we learned how the Bush family has owned the business that is most probably the main American harvester of fresh coal fly ash: Dresser Industries. In chapter 8 we learned about the Bush family connections to Enron. In this chapter we have learned that both Bush Sr. and Bush Jr. were members of the Pilgrim Society and we will now learn much more. Your author will attempt to bring you everything they probably don't want you to know - and there's a lot of it!

The Bush family political dynasty is up to their eyeballs in this thing. Yes, the family we all love to hate has apparently been at it again. We all know how cool these guys look on TV in their $10,000 suits and how Bush Senior has recently been canonized by the lamestream media, but here are the facts. The Bush family bank, Brown Brothers Harriman funded the Nazis from their anti-Semitic, hate-filled, war-mongering beginnings to LONG AFTER they invaded Poland. The Bushes have no excuses and we shouldn't give them any. It doesn't end there. On Oct. 20, 1942, 10 months after the U.S. had entered the war, under the Trading with the Enemy Act, the U.S. government took over the Union Banking Corporation of which Prescott Bush was a director. Three of the shareholders who had their shares seized were Nazis. Two other businesses associated with the Union Banking Corporation were also seized. The following month, the Nazi interests invested in the Silesian-American Corporation, a company long managed by Senator Prescott Bush, were seized under the Trading with the Enemy Act as well. It gets darker. Roger Stone, in his book *Jeb! And the Bush Crime Family* suggests that old Prescott Bush (1895-1972) was getting his hands very dirty in Nazi Germany. Stone writes, "According to a Dutch intelligence agent, Prescott Bush managed a portion of the slave labor force from Auschwitz." The Bush family has all sorts of other connections to the Nazis involving people and organizations such as: William Stamps Farish, William Stamps Farish III, Fritz Thyssen, William H. Draper, Jr., William H. Draper III, the Rockefellers, I.G. Farben, Standard Oil, Hermann Schmitz, and the U.S. Agency for International Development.

The Bush family also has ties to the original Manhattan Project. Senator Prescott Bush (Bush Senior's daddy) sat on the board of directors of something called the Vanadium Corporation which supplied unrefined uranium to the Manhattan Project. Also, as noted in chapter 6, Prescott Bush sat on the board of directors of the aforementioned Dresser Industries which produced special pumps used in the production of the refined uranium needed for the world's first atomic bombs. Now that's called playing both sides and that's how the old boy network does it. Dresser Industries is a company that has worked frequently with the CIA. Prescott Bush was a member of the Dresser Industries board of directors for 22 years until 1952 when he entered the Senate. Once there, Senator Prescott Bush became Allen Dulles' day-to-day CIA contact man.

The Carlyle Group is a defense contractor private equity firm and one of the Bush family's main family businesses. The Carlyle Group is of import to our discussion here because in 1992 they bought a portion of the aforementioned Ling-Temco-Vought Corporation (ch 3, 5 & 6) and therefore may have been involved in the production and delivery of technological military hardware needed to produce the New Manhattan Project. Nazi collaborator George Soros is famously invested in the Carlyle Group. What more does one need to know? Here's more: the Carlyle Group is a corporate member of the CFR and Bush Sr. was a director of the CFR.

Lastly, George H.W. Bush (Bush Sr.) in particular has a few interesting things about him. As mentioned earlier, he was the United States ambassador to the UN. In 1989 former MITRE Corporation president Bob Everett received the National Medal of Technology from President George H.W. Bush. And he gave the aforementioned Laurance Rockefeller the Congressional Gold Medal for environmentalism.

It suffices to say that it truly goes on and on with these people, to the point where one just becomes numb to it all. I mean, the Bushes are into all of it: depopulation and eugenics, mind control, media manipulations, weapons trafficking, drug trafficking, money laundering, the CIA, the CFR, the Nazis and their death camps, Enron, secret societies, the New World Order, bioweapons, overthrowing governments, assassination plots, taxpayer rip-offs, *ad nauseam*. By all appearances and simply put, the Bush family is a very public face of the American deep state and probably would have no qualms about mass murdering the domestic population and omniciding the planet in the fashion of the New Manhattan Project. To truly understand the Bush family, one must understand the mind of a psychopath. That's not easy for people with good intentions.

Harlan Cleveland

One individual in particular from the highest levels of world power appears to have been instrumental in the development of the New Manhattan Project. Dr. Harlan Cleveland (1918-2008) was first mentioned in chapter 1. In the first chapter we learned that Dr. Cleveland, as chairman of the Weather Modification Advisory Board, testified before the U.S. congress, suggesting a new method of weather modification involving the use of electromagnetic energy. His group also produced a seminal weather modification report which mentioned that it would, "soon be possible to influence the weather more reliably and in a much greater variety of ways."

Cleveland was the U.S. ambassador to the North Atlantic Treaty Organization (NATO), a Rhodes Scholar and, along with his brother Harold van Buren Cleveland (1917-1993), a CFR member. He served as Assistant Secretary of State for International Organization Affairs, a United Nations relief administrator in China and Italy, and a foreign aid manager under the Marshall Plan. In his role as an Assistant Secretary of State during the Kennedy and Johnson administrations, Cleveland was apparently instrumental in the founding of the UN's World Weather Watch as well as outer space treaties and arrangements to divvy up the electromagnetic frequency spectrum.

Harlan Cleveland

In his book *The Conspirator's Hierarchy: The Committee of 300*, author John Coleman names Cleveland as a member of this Committee of 300. Williams and Howard-Browne describe the Committee of 300 as an offshoot of the British Pilgrim Society tasked with setting the stage for World War One. Just like WWII, World War One was a boon for large, multinational corporations and elite families. Coleman describes the Committee of 300 as being another extremely powerful supra-national organization.

Cleveland became LBJ's ambassador to NATO in 1965. The following year, French president Charles de Gaulle pulled France out of NATO and banished it from French soil. Of Mr. Cleveland, Coleman writes:

"Cleveland was a top official in the Club of Rome and what one intelligence officer described as a 'Socialist specialist on China.' Cleveland got his Socialist education at Oxford in 1938. He later became Assistant Secretary of State for International Affairs, and the UNRRA [United Nations Relief and Rehabilitation Administration] delegate to China and a director of the ECA [Bureau of Educational and Cultural Affairs] China program. Cleveland had a perception, that the way America was going was not the way the Socialists wanted it to go and methods had to be devised to make America change direction. The Institute for Social Relations and the Club of Rome were the vehicles he felt could be instrumental in bringing about the changes needed to socialize America."

In 1976, the MITRE Corporation sponsored the Aspen Institute's Symposium on Living with Climatic Change. One of the transcripts published in the proceedings of these talks was something titled "The Climate for a New World Order." This was Harlan Cleveland's talk. The gist of his speech was that the validity of the science of man-made climate change is not important because the theory is simply a good way to bring about a necessary, authoritarian New World Order. Apparently alluding to NBC weatherman Al Roker, he said:

"There are two options, each of them backed by weighty evidence, or at least by weighty meteorologists. One is that the weather is getting warmer. The other is that the weather is getting colder. You can pay your scientists and take your choice."

Cleveland served as the director of international affairs at the Aspen Institute from 1974 to 1980.

Cleveland was a long-time, staunch supporter of a program of global weather control. In his 1977 book *The Third Try at World Order*, Cleveland writes that, "the international control of weather modification had better be just around the corner." In this book Cleveland also writes that a concerned 'community of nations' needs to, among many other things, "control the modification of weather at human command." He writes that, "It is this impressive agenda, taken as a whole, that will amount to a third try at world order." Further still in this book, Cleveland writes that new or adapted international institutions are needed to, "collect and disseminate better information and guesses about the weather - and move toward international standards for weather modification." He classifies this proposed effort as, "among the most urgent systems waiting to be born." In a short 1978 piece titled "The Management of Weather Resources," Cleveland once again advocates for weather modification. During the 1980s, Cleveland served two three-year terms as a trustee at large of the University Corporation for Atmospheric Research (UCAR) in Boulder, Colorado. In Cleveland's 1990 book *The Global Commons: Policy for the Planet*, he writes, "As the sciences and technologies create the capacity to modify the weather at human command, new concepts of law and practice will have to be invented to match humankind's new power to change its physical surround." As it became a big political issue, Cleveland became a staunch supporter of the theory of man-made climate change. He stated the problem and supported the solution. In his 1993 book *Birth of a New World: An Open Moment for International Leadership*, Cleveland wrote, "Global warming already looms as the biggest environmental issue on the political horizon." In the same book, he noted that, "Some short-term weather can be modified at human command."

References

The Conspirator's Hierarchy: The Committee of 300 a book by Dr. John Coleman, published by World Intelligence Review, 4th edition, 2006

The Killing of Uncle Sam: The Demise of the United States of America a book by Rodney Howard-Browne and Paul L. Williams, published by River Publishing, 2018

"The Network of Global Corporate Control" a paper by Stefania Vitali, James B. Glattfelder, and Stefano Battiston, 2011

Killing the Planet: How a Financial Cartel Doomed Mankind a book by Rodney Howard-Browne and Paul L. Williams, published by Republic, 2019

Tragedy and Hope: The History of the World in Our Time a book by Carroll Quigley, published by Macmillan, 1966

The Shadows of Power: The Council on Foreign Relations and the American Decline a book by James Perloff, published by Western Islands, 1988

The Brothers: John Foster Dulles, Allen Dulles, and Their Secret World War a book by Stephen Kinzer, published by Times Books

Bohemian Grove and Other Retreats: A Study in Ruling-Class Cohesiveness a book by G. William Domhoff, published by Harper and Row, 1974

Wall Street's Think Tank: The Council on Foreign Relations and the Empire of Neoliberal Geopolitics, 1976-2014 a book by Laurence H. Shoup, published by Monthly Review Press, 2015

United Nations: A history a book by Stanley Meisler, published by Grove Press, 2011

"Weather Modification as a Weapon" an article by Gordon J.F. MacDonald, published by Technology Review, October/November, 1975

The Wizards of Langley: Inside the CIA's Directorate of Science and Technology a book by Jeffrey T. Richelson, published by Westview Press, 2001

"An Interview with Former DDS&T Albert Wheelon" an article by Ed Dietel, published by *Studies in Intelligence*

"Honoring the Founder of CIA's Directorate of Science and Technology" an article by Robert J. Kohler and R. James Woolsey, published in *Studies in Intelligence*

"Present and Future Capabilities of OTH Radars" a report by Nicholas R. Garofalo, published by *Studies in Intelligence*, Spring 1969

"Albert D. Wheelon, Architect of Aerial Spying, Dies at 84" an article by Douglas Martin, published in *The New York Times*, Oct. 2, 2013

Spy Capitalism: Itek and the CIA a book by Jonathan E. Lewis, published by the Yale University Press, 2002

Foundations of the American Century: The Ford, Carnegie, and Rockefeller Foundations in the Rise of American Power a book by Inderjeet Parmar, published by the Columbia University Press, 2012

"The Rockefeller File" a book by Gary Allen, published by 76 Press, 1991

"MITRE: The First 20 Years" a book by the MITRE Corporation, published by the MITRE Corporation, 1979

IBM and the Holocaust: The Strategic Alliance Between Nazi Germany and America's Most Powerful Corporation a book by Edwin Black, published by Dialog Press, 2011

The Devil's Chessboard: Allen Dulles, the CIA, and the Rise of America's Secret Government a book by David Talbot, published by HarperCollins Publishers, 2015

Area 51: An Uncensored History of America's Top Secret Military Base a book by Annie Jacobsen, published by Back Bay Books, 2011

An American Genius: The Life of Ernest Orlando Lawrence, Father of the Cyclotron a book by Herbert Childs, published by E.P. Dutton & Co., 1968

"E.L. Rothschild LLC Acquires a Majority Stake in Weather Central, LP" an article published by *BusinessWire*, Jan. 31, 2011

The MITRE Corporation: Fifty Years of Service in the Public Interest a book by the MITRE corporation, published by the MITRE Corporation, 2008

George Bush: the Unauthorized Biography a book by Webster Tarpley and Anton Chaitkin, published by Progressive Press, 2004

"None Dare Call It Conspiracy" a book by Gary Allen, published by Buccaneer Books, 1971

Jeb! And the Bush Crime Family: The Inside Story of an American Dynasty a book by Roger Stone and Saint John Hunt, published by Skyhorse Publishing, 2016

Initiative in Energy: The Story of Dresser Industries a book by Darwin Payne, published by Simon and Schuster, 1979

Prelude to Terror: The Rogue CIA and the Legacy of America's Private Intelligence Network a book by Joseph J. Trento, published by Carrol & Graf, 2005

The Iron Triangle: Inside the secret world of the Carlyle Group a book by Dan Briody, published by John Wiley and Sons, 2003

"The BCCI Affair" a report by Senators John Kerry and Hank Brown, 1992

Wall Street and the Rise of Hitler a book by Antony Sutton, published by Buccaneer Books, 1976

Birth of a New World: An Open Moment for International Leadership a book by Harland Cleveland, published by Jossey-Bass, 1993

"Harold Cleveland; U.S. Economist, 76, Aided Marshall Plan" an article by Wolfgang Saxon, published by *The New York Times*, March 13, 1993

Nobody in Charge: Essays on the Future of Leadership a book by Harlan Cleveland, published by Jossey-Bass, 2002

Proceedings of the Symposium on Living with Climatic Change Phase II a book published by the MITRE Corporation, 1977

"Harlan Cleveland; Dean, Author, Statesman and Lifelong Learner" an article by Joe Holley, published in *The Washington Post*, June 6, 2008

"Harlan Cleveland, Diplomat and Scholar, Dies at 90" an article by Dennis Hevesi, published in *The New York Times*, June 13, 2008

The Third Try at World Order: U.S. Policy for an Interdependent World a book by Harlan Cleveland, published by the Aspen Institute for Humanistic Studies, 1977)

"The Management of Weather Resources" by Harlan Cleveland, published by *Science*, Vol. 201, No. 4354, August 4, 1978

The Global Commons: Policy for the Planet a book by Harlan Cleveland, published by the Aspen Institute and University Press of America, 1990

Chapter 15
FUNDING

We have seen that since just after the end of WWII the New Manhattan Project has been in active development and, since 1996, it has been in active deployment. We have also now seen much compelling evidence as to who has developed and deployed it. Chapter 8 told us about the motives of the New Manhattan Project. Another logical question is to that of funding. In this chapter we will see that there is no shortage of ways to supply the trillions of dollars needed to carry this ultra-massive project forward over the course of 74 years. We will start with the infamous black budgets.

Black Budgets

It is not a secret that every year the United States taxpayer funds billion-dollar military budgets which include so-called 'black' funds. Although this type of spending is prohibited by our Constitution, these 'black budgets,' as they are called, have been disbursed at the discretion of the President, the Secretary of State, and the Director of Central Intelligence (the boss of the CIA) since WWII. Black budgets began, appropriately enough, with the original Manhattan Project. They became a permanent part of the government with the creation of the CIA in 1947. Not only that, but the CIA has the legal ability to send and receive money to and from any and all U.S. Government offices - providing them with a way to distribute and receive any funds they wish. For these reasons, and in consideration of the evidence presented here implicating the CIA, it is reasonable to assert that these black funds have been used to, at least in part, fund the New Manhattan Project.

The overt defense budget

Providing for even more possible funding, the annual *overt* defense budget is filled with code words, euphemisms, and other camouflage that allow our military to obfuscate many clandestine efforts. There is no shortage of ways for the United States government to directly fund the development of the New Manhattan Project through normal channels and we can assume that this is what has taken place. After all, elements of our United States military have repeatedly stated their position that climate change is a matter of national security and must be dealt with. In fact, the Department of Defense issued a 2014 report outlining their 'climate change adaption roadmap' detailing the horrors of global climate change and all the wonderful things they are doing to help stop it. There are many other similar documents.

Missing money

Many trillions of dollars are apparently 'missing' from the Pentagon. We all know how Rumsfeld held a press conference way back in 2001 on the day before the 9/11 attacks to let us know that $2.3T was somehow missing from the Pentagon. Well, as of 2018, a short article (with a collection of videos) appeared on Catherine Austin Fitts' website about how $21T of our taxpayer dollars has met the same fate. $21T is probably more than enough to have brought the New Manhattan Project forward over the decades.

The latest development in the Pentagon's saga of financial irresponsibility entails our friends at the DOD notifying us that in 2019 alone they made $35T in unsubstantiated accounting adjustments. No, that doesn't mean that they *found* $35T, of course it means that they *lost* $35T. The mainstream media made quick mention of it, then continued attacking the President.

The CIA's global narcotics racket

Even more interesting, though, is the notion that today the domestic spraying operations and everything else that the New Manhattan Project entails is funded by the Central Intelligence Agency's $400B global narcotics trade. For this assertion there is quite a bit of precedent.

Oh yeah, the Cocaine Importation Agency (CIA) has been running megatons of narcotics. There is no doubt about that. It would be quite tedious to go through all the credible books, movies, news reports, court cases, and other such solid evidence supporting this assertion. Further, eminent historians such as UC Berkeley's Peter Dale Scott can tell you that, at least since the time of China and Britain's Opium Wars in the mid-19th century, the deep state and their many nefarious activities have been funded by the narcotics trade. There is a certain chapter of this long, sordid history which is particularly relevant to the New Manhattan Project. It is something called Operation Gladio. Operation Gladio is explored here because the proceeds from these types of operations may have provided funding for the development of the New Manhattan Project and similar types of operations may be funding the NMP today.

Operation Gladio was a North Atlantic Treaty Organization (NATO) and CIA collaboration that involved staging terrorist attacks throughout Europe to be blamed on communist factions while, in actuality, the attacks were perpetrated by fascist (capitalist) ones. This was done in order to galvanize public opinion for capitalism and against communism. As we learn of their involvement here, we can see how the CIA participates in the most dastardly of deeds in order to grow their power while creating the perception of virtue. It's kind of like how geoengineers propose spraying the Earth and all of its biota with megatons of toxic materials in order to save us from global warming. Journalist Richard Cottrell, in his groundbreaking 2012 book *Gladio: NATO's Dagger at the Heart of Europe* writes:

"The sound of bullets, screams, the whumpf of great explosions, shattering glass, the scream of emergency vehicles, fell cross Europe. Bombs planted in a prominent city bank, in railway stations and trains, and in beer tents overflowing with jolly drinkers. Snipers fired at will into squares heaving with crowds of peaceful demonstrators, or supermarkets as weekend shoppers queued to pay for their groceries. A slaughter of the innocents. To meet its own political ends, the United States fostered the mirage of communist subversion of the European continent. In so doing it worked through the CIA and secret war departments of every NATO country, and ruthlessly milked and nourished latent forces of fascism left over from WW2."

As a direct result of Operation Gladio, 491 people were killed and 1,181 injured or maimed.

To understand how all this came about and to understand its relevance to the New Manhattan Project, we must, once again, go back to the beginning - to the devil's pact between Allen Dulles and the Nazis. As explained in chapter 13, towards the end of WWII, Allen Dulles of America's Office of Strategic Services (OSS) made a deal with high-ranking German officials. This deal was instrumental in the creation of the CIA, but it also resulted in the creation of the so-called 'stay behind' armies ostensibly designed to be a capitalist (fascist) failsafe against a post-WWII European invasion by communist Russia. If the Russians were to invade a weakened Europe, the theory went, then these stay behind armies would be activated to defeat the communist aggressors. It's not like the Russians were not weakened as well. They had just borne the brunt of WWII - suffering somewhere north of 20 million casualties and they had been our ally during the war. But nonetheless, this was the stated fear and this was the story. Creating new, secret armies consisting of Nazis and directed by Nazis and Nazi sympathizers apparently wasn't a concern. To the contrary - according to the OSS, it was good.

Shortly after the conclusion of the Yalta Conference in February of 1945, when it became crystal-clear that Germany was going to lose the war and preliminary arrangements for the post-war Europe had been made, Dulles arranged for the transportation of the aforementioned German General Reinhard Gehlen and his top people to Fort Hunt, Virginia where it was agreed that Gehlen would return to Germany under U.S. protection to establish the Gehlen Organization which was to be funded by United States taxpayer dollars. The Gehlen Organization subsequently set up shop in a former Waffen SS training facility near Pullach, Germany. The primary purpose of the Gehlen Organization was to create and expand the stay behind armies with men taken from the ranks of the soon to be defeated German military. The soldiers of these stay behind armies were known as 'gladiators' with the operation being named Gladio after the short swords the gladiators of ancient Rome used to finish their opponents.

Reinhard Gehlen

Also involved here was an Italian member of the Catholic Black Nobility by the name of Prince Junio Valerio Borghese (1906-1974) also known as The Black Prince. Borghese commanded troops for the Italian fascists under Mussolini during WWII and later organized many thousands of men from Italy, Germany, France, and Austria into the stay behind armies in Italy. By 1946, when the OSS was morphing into the CIA, hundreds of Gladio units were already in place throughout Western Europe in countries such as: Italy, Germany, Turkey, Belgium, Sweden, France, and Greece.

The $200M of original funding for the stay behind armies came from the Rockefeller and Mellon Foundations, but these Western European stay behind armies needed a more stable, recurring, and covert source of funding. Not only was a stable source of funding desirable just for Operation Gladio, it was desirable for the continued funding of all of the CIA's covert intelligence activities. Here's where it gets really interesting.

In the mid-to-late 1940s, around the end of WWII, it was Colonel Paul E. Helliwell's (1915-1976) job in the Far East as Chief of Special Intelligence for the OSS to support Chiang Kai-shek's Nationalist Chinese forces. He observed that a contingency force of Chiang Kai-shek's army operating in an opium producing region of Burma was selling the drug locally in order to fund their continuing efforts against Mao Zedong's communist forces. The sale of the opium at that time, though, was done in the ancient ways of many middlemen and slow ground transportation - a very inefficient process. Helliwell had the idea to begin transporting opium by aircraft to major market centers directly. This could allow not only for the more efficient movement and sale of the opium, but also for a more efficient movement and sale of the arms Chiang Kai-shek's soldiers needed to fight the communist aggressors. The arms could be flown in, thc opium could be flown out, and the Chinese Nationalist forces would be better supplied - a win win. Helliwell ran his idea by his boss, the aforementioned OSS chief General William 'Wild Bill' Donovan, and Wild Bill shared it with Dulles. Everybody loved it. They went about bringing Helliwell's idea into existence, and the airline that eventually began taking care of these missions was the aforementioned Civil Air Transport (ch 5). As alluded to earlier, Civil Air Transport (CAT) was the CIA's proprietary airline which came after the Flying Tigers, preceded Air America, and was (as were the Flying Tigers and Air America) most probably a direct ancestor of today's chemtrail fleet. In his book *Operation Gladio: The Unholy Alliance Between the Vatican, the CIA, and the Mafia*, well-respected researcher Paul Williams gives us more insight into the founding of CAT:

"By the close of World War II, Helliwell and a number of fellow Army intelligence officers - E. Howard Hunt, of Watergate fame; Lucien Conein, a former member of the French Foreign Legion with strong ties to the Corsican Mafia; Tommy 'the Cork' Corcoran, a lawyer serving the Strategic Service Unit; and Lt. General Claire L. Chennault, the military advisor to Chaing Kai-Shek and the founder of the Flying Tigers - had created the Civil Air Transport."

The original funding needed to get these guns-for-drugs airlift operations in the Far East off the ground came from Nazi gold that had been laundered and manipulated by Dulles and a man named William 'Little Bill' Stephenson (1897-1989), a British master spy. This was all done through something called the World Commerce Corporation. The World Commerce Corporation was formed by OSS chief Wild Bill Donovan with the help of a small group of very wealthy friends including: Nelson Rockefeller, Joseph C. Grew, Alfred du Pont, and Charles Jocelyn Hambro of the Hambros Bank. From its founding onward, the World Commerce Corporation's primary function was to trade U.S. surplus weapons and munitions to groups such as those producing the opium in Burma. CAT began handling these operations on October 10, 1949. Some CAT pilots were ex-Nazis.

Once the opium had been airlifted out of the wilds of Burma, it was brought to Bangkok, Thailand where General Phao Sriyanonda, the director of Thailand's national police, had his officers load it onto the freighters of the Sea Supply Company, a CIA front run by Helliwell. It was helpful that Donovan had recently been named the U.S. ambassador to Thailand. The opium was then shipped to Cuba where it was turned into heroin with the help of individuals from the most important families of the American Mafia as well as Schiaparelli, one of Italy's most respected pharmaceutical companies. The heroin was then distributed by the mob to major cities throughout America. The target customers were the African-Americans of our inner-city ghettoes. All this, again, was Helliwell's idea. Coupled with the subsequent burgeoning urban jazz scene of the 1950s, this venture proved to be a smashing success. An entire drug culture had been created and the CIA was in the money.

Opium and heroin production was later expanded to include opium production fields in Laos, heroin production laboratories located in Marseilles, France run by the Corsican mafia, as well as Turkish opium brokers. By 1971, there were more than 500,000 heroin addicts in the United States, producing a cash flow of $12 billion annually. Some of the heroin was smuggled into the United States in body bags containing dead soldiers. The CIA expanded their operations again later in the 1980s with street gangs and cocaine - once again targeting America's black populations living in urban ghettoes, but this time feeding off of the gangster rap phenomenon to create the surrounding culture. We can now stop wondering as to how the CIA has become so powerful. Their narcotics trafficking has been a resounding success for a very long time. The CIA and their cohorts have also organized similar weapons-for-drugs programs used to topple governments and install 'friendly' regimes all over the world, but it all started with Gladio.

The profits from the early CIA heroin production and distribution ventures of the late 1940s, 50s, and 60s detailed here were laundered through The Institute of Religious Works, also known as the Vatican Bank. The funds were then distributed to the administrators of Operation Gladio through Swiss banks. This is how Operation Gladio became sustainably funded.

The CIA used their global narcotics trade to finance extremely heinous acts of depravity during Operation Gladio and subsequent operations. They got away with it the first time. Why would they not do it again? By all appearances, the CIA still runs the global narcotics trade today. As we have seen, the CIA is also most probably responsible for the massively expensive and destructive New Manhattan Project. In consideration of all the evidence presented in this chapter and in this book, it is reasonable to assert that the CIA and their Nazis have been using the proceeds from their global narcotics racket to fund the New Manhattan Project.

It also should be noted that the global narcotics business is so profitable ($400B annually according to the UN) that a second, breakaway CIA running the global drug trade and the NMP may be totally independent. Who needs congress and all those bothersome, do-gooder bureaucrats, politicians, and taxpayers constantly peering over your shoulder? They don't even necessarily need to call themselves 'The CIA.' The organization running the global narcotics trade today apparently calls itself simply 'The Enterprise.' Providing evidence for these assertions, Joseph Trento in his book *The Secret History of the CIA* notes the coalescence of a second, secret CIA. This would be the group most probably mass murdering American citizens and wrecking our environment today. Trento writes:

"In 1976, George Bush was heartbroken at not being selected to continue as DCI [head of the CIA] when Jimmy Carter was elected. That non-hiring of George Bush gave momentum to the creation of a CIA in exile. This was a group of out-of-work agents that started to form with Colby and Kissinger's 1975 cutbacks in Clandestine Services. On All Hallows' Eve 1977, Bush's successor, Admiral Turner, speeded up the process by taking an even bigger axe to Clandestine Services. The unofficial leader of the exiled spooks was Berlin Base's Ted Shackley. His relationship to George Bush eventually resulted in the Iran-Contra mess during the Reagan presidency.

"By the time Reagan and Bush took office, they had a choice of two CIAs they could do business with - one that required oversight by Congress, and another off-the-books group made up of the old boys"

The Research Corporation for Science Advancement

The Research Corporation for Science Advancement (RCSA/ Research Corporation) has probably been funding basic scientific studies in support of the New Manhattan Project. To understand how this came about and to gain a deeper understanding of the RCSA, we will now take a closer look.

The story of the Research Corporation goes back to the early 1900s and has everything to do with the aforementioned (ch 6) Frederick Gardner Cottrell. Here we join up with our intrepid Cottrell just after he and his business associates had successfully installed machines known as electrical precipitators which were suddenly purifying the emissions of industrial factories. His successful business ventures were known as the aforementioned Western Precipitation and International Precipitation Companies.

Shortly after Cottrell's successes with industrial electrical precipitation, he resigned from his position as a professor at UC

Berkeley and took a job as a physical chemist with the Bureau of Mines. Cottrell took the job because the then director of the Bureau, a man by the name of Joseph A. Holmes, was cultivating an expanded role for the newly created Bureau. One of Holmes' initiatives was that of organizing efforts related to ameliorating the problem of pollution caused by the burning of coal. Holmes saw air pollution from the burning of coal as an issue for the Bureau of Mines because coal is a mined substance and the organizations burning said coal had intimate business relationships with the miners. Cottrell's experience in electrical precipitation made him a natural fit.

It was not long after his arrival in Washington, D.C. for his job with the Bureau of Mines in June of 1911, when Cottrell started taking steps towards forming the Research Corporation, as it was originally called. With help from Holmes as part of this process, Cottrell began negotiations with the Smithsonian Institution to let them take over ownership of his patents pertaining to electrical precipitation. It was suggested that the Smithsonian Institution could use funding from the royalties generated by the business development of International Precipitation Company's patents to fund scientific efforts which had the potential to improve Americans' daily quality of life. You see, unlike Bill Gates, George Soros, or so many others, Cottrell was a *real* philanthropist. He actually wanted to use his patents to foster development of the greater good. He may have been naïve, but his intentions were good.

Cottrell soon returned to the San Francisco Bay Area where he went about organizing a Bureau of Mines office and laboratory in San Francisco. Also at this time, he and his business associates (Miller, Heller, and O'Neill), as a preliminary step to handing their patents over to the Smithsonian, signed over both the Western Precipitation Company and the International Precipitation Company (without patents) to one of Cottrell's former pupils named Walter A. Schmidt, who became the manager of both the International Precipitation Company and the Western Precipitation Company. By this time, there was huge interest and activity associated with the Cottrell group's new technologies and the job offers were coming in from all over the world. But, with the signing over of the companies, Cottrell and his associates received only modest compensation. With the disposition of his businesses complete, Cottrell plunged into his work at the Bureau of Mines.

In December of 1911, the Regents of the Smithsonian decided to accept Cottrell and associates' offer with a caveat. The Regents of the Smithsonian advised that a stock corporation be organized which was to take title of the patents. The Regents advised that the Smithsonian Institution be directly represented in this new corporation by the Secretary of the Smithsonian Institution - at the time, a man by the name of Charles D. Walcott (1850-1927). From this arrangement and from the royalties generated by the International Precipitation Company patents, the Smithsonian Institution, they suggested, would stand to benefit financially. This suggested corporation would later be formed as the Research Corporation. At the time of the Smithsonian's decision, Cottrell happened to be back east again in Washington D.C. attending an important meeting. Being that the Smithsonian's Executive Committee was ready to act, Cottrell had extended his stay.

After attending the executive session of the Smithsonian's Board of Regents where they announced their decision pertaining to Cottrell's patents, Walcott (Secretary of the Smithsonian) and Cottrell adjourned to a restaurant across the street where they ran into Arthur Dehon Little (1863-1935). The presence of Arthur D. Little is important to our discussion because his eponymous corporation (a research organization, no less) was later to do lots of serious work in the vein of the New Manhattan Project (ch 3). From little acorns mighty oaks grow. Little was a staunch supporter of Cottrell's efforts to establish this new corporation - he even suggested the name 'Research Corporation.' Little offered lots of other advice and volunteered to steer Cottrell towards the 'right' people. It was Little who put Cottrell in touch with T. Coleman du Pont (1863-1930) who enthusiastically took a seat on the nascent Research Corporation's board of directors. Little took a seat on the first board as well. DuPont stayed on the Research Corporation board of directors from 1912 to 1930 while Arthur D. Little remained as a Research Corporation board member from 1912 to 1921.

The newly christened Research Corporation needed more members of the board to oversee their important work. For upwards of two months following the Smithsonian's decision, Cottrell and Bureau of Mines chief Holmes communicated with about one hundred men from many different vocations in order to find directors for the new Research Corporation. They eventually decided upon 14 well-known men from academia, government, and industry, including a man named Elihu Thomson (1853-1937) who was the founder of something called the Thomson-Houston Company. The Thomson-Houston Company was one of the precursors of the General Electric Company. Thomson was also the president of the Massachusetts Institute of Technology from 1920 to 1923. Another original board member was one Charles A. Stone (1867-1941) who was a trustee of the Massachusetts Institute of Technology. Smithsonian Secretary Walcott took his seat on the Research Corporation board as well. By early February of 1912, the Research Corporation's original board of directors had been chosen and its Articles of Incorporation were ready to be filed. Although years later the Research Corporation bought back all of its stock, members of its original board initially paid for founder's stock which filled the nascent Research Corporation's coffers and gave the corporation its seed money needed for starting operations.

Over the years, many other luminaries of the New Manhattan Project have served on the Research Corporation's board of directors including: Karl T. Compton (1887-1954), James R. Killian, Jr. (1904-1988), Alfred Lee Loomis, and Vannevar Bush. Alfred Lee Loomis was a Research Corporation board member from 1930 to 1933, then again from 1948 to 1959. In 1938 the Research Corporation awarded a grant to Vannevar Bush. Vannevar Bush was a Research Corporation board member from 1939 to 1946. Later, in 1952, Bush wrote Cottrell's National Academy of Sciences eulogy.

On February 16 of 1912 the Research Corporation was incorporated under the laws of the state of New York, a one-room office was leased at 63 Wall St., and a manager by the name of Linn Bradley was hired. Cottrell returned to his office work for the Bureau of Mines in San Francisco. A prospectus was printed up and, in order to increase revenue, Bradley went about finding new contract opportunities. Four years later, Cottrell and his wife moved to Washington D.C., where they then lived for the next 28 years.

In the first two years of its existence, the Research Corporation got its revenue almost entirely from the fees associated with their engineering consultancy work pertaining to the design, installation, and maintenance of electrostatic precipitators. In the first few years of its existence, the Research Corporation mostly went about building up cash reserves. During this time there were large precipitators being built around the country. By January 1915, about three years after its creation, the Research Corporation had $65K in cash and $100K in secured notes, subsequently, the previously issued founders' stock was bought back in that same year.

With the exception of an early grant to Cottrell as part of a company promotion, the Research Corporation didn't really start giving out money for the advancement of science until 1923. In 1923 the Research Corporation made their first real grant when they paid $5K toward the atmospheric sounding rocket experiments of the American rocketry pioneer Robert H. Goddard (1882-1945). This was an investment in the atmospheric sciences as sounding rockets are used to gather atmospheric data. From the beginning, the Research Corporation was funding projects highly relevant to the New Manhattan Project.

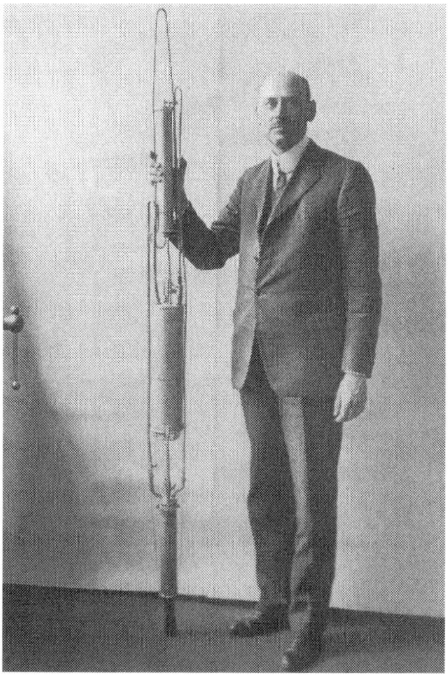

Robert Goddard

At the Research Corporation, the way it worked was that Howard Poillon ran day-to-day operations while Cottrell was largely responsible for developing new inventions and ideas, accepting new patents, and the granting of funds. Cottrell's official status was that of a paid consultant. Another early recipient of Research Corporation money was Ernest Orlando Lawrence (1901-1958): Manhattan Project scientist and the father of the cyclotron. In 1931 the Research Corporation made the first of many grants to Ernest Lawrence for the development of his cyclotron.

Ernest O. Lawrence

In late July of 1931, after Lawrence had successfully produced a million-volt proton with an eleven inch accelerator, Fred Cottrell encouraged his friend from UC Berkeley to ask the Research Corporation for funding. Before the end of the month, Lawrence was in New York, asking the Research Corporation for \$10-\$15K. Cottrell came along to personally introduce Lawrence and plead his case before the board of directors. The Research Corporation was impressed with Lawrence and they decided to make a \$5K grant, even though they had to go to the bank and borrow the funds because they were in the throes of the Great Depression. William Buffum of the Chemical Foundation followed up with a promise of another $2.5K. On different occasions between 1931 and 1940, the Research Corporation gave to Lawrence's work at the Berkeley Radiation Laboratory: $5K, $1.8K, $2K, $2K, $3K, $1K, $5K, $1.7K, $7.5K, $5K, and $50K. That's a grand total of $84K, or, conservatively, about $1.5M in today's dollars. The Research Corporation also applied for one of Lawrence's patents for him. Their collaborative patent, "Method and Apparatus for the Acceleration of Ions," with the Research Corporation as the assignee, covers methods of the cyclotronic production of radioactive substances discovered by Lawrence. Cottrell's biographer Cameron writes, "In the thirties, when the period of intensive study of atomic energy began, the earliest diversified support and encouragement of the development of the techniques of nuclear physics in America came from Research Corporation." He also writes, "The Lawrence patent had, by 1949, been licensed free of royalties to twenty-eight universities and scientific institutions to build cyclotrons. More than that, it was income from the precipitation patents that helped in several instances to build or operate these cyclotrons which the corporation licensed." The Research Corporation funded cyclotron work at: Columbia University, the University of Rochester, the University of Chicago, and Cornell University, among others.

The Research Corporation also funded Robert Van de Graaff's (1901-1967) electrostatic generator, also known as a Van de Graaff-Trump accelerator after the Donald's uncle, John G. Trump (ch 3). Van de Graaff's work was conducted at MIT. Cameron writes, "[The Van de Graaff group of patents on the electrostatic generator] were assigned to Research Corporation under a general agreement with Massachusetts Institute of Technology and in turn were ultimately licensed back to the High Voltage Engineering Corporation formed by Van de Graaff and his associates." Van de Graaff's associates included John Trump. As readers of this book already know, John Trump was the MIT professor who looked over Nikola Tesla's posthumously confiscated documents (ch 3).

As the years went on, the Research Corporation's electrostatic precipitator business grew and grew. In 1928 Research Corporation had 43 contracts in hand to build and install electrostatic precipitators. In 1941 Research Corporation had 95 contracts to build and install electrostatic precipitators. In 1942 Research Corporation had 130 contracts. At some point during this period, the Research Corporation moved out of its original offices on Wall St. and moved into offices in the steeple of the Chrysler Building in Manhattan.

During WWII something called the Research Construction Company, which was formed by the Research Corporation, served as a 'model shop' for MIT's Radiation Laboratory (ch 2). Under contract to MIT, the Research Construction Company produced over $12M worth of radar apparatus. In their book commemorating their centennial anniversary, the Research Corporation for Science Advancement tells us, "Research Construction Company made small production runs for immediate military needs. Successful prototypes were turned over to government contractors for mass production."

The Research Corporation for Science Advancement writes of significant developments in the post-war era, "In 1954, as a result of [a] change in the tax law, Research Corporation reorganized its precipitator business as Research-Cottrell, a wholly-owned but taxable subsidiary." The authors continue, "From 1957 to 1967 Research Corporation was supported mainly by the earnings of its commercial precipitation subsidiary, Research-Cottrell, and royalties from inventions in its patent portfolios." "By the mid-1960s, further federal efforts to limit nonprofit control over commercial activities prompted Research Corporation to lower the percentage of its ownership in the precipitator business, and Research-Cottrell became a publicly held company in 1967.

"The [Research Corporation] Foundation's endowment was established during the second half of the twentieth century, coming primarily from the Research-Cottrell stock offering. Research Corporation finally divested itself of all of the precipitator firm's stock in the 1980s." The authors continue, "The Foundation's precipitator business, Research-Cottrell, survives as Hamon-Research Cottrell, a public corporation headquartered in New Jersey that is a major provider of air-pollution control technology."

The Research Corporation has historically given many grants to scientific studies conducted in areas relevant to the New Manhattan Project. The Research Corporation has consistently funded work in the areas of: plasma physics, astrophysics, microwave spectroscopy, radioactive tracers, atmospheric physics, electromagnetic fields, particle physics, meteorology, ionospheric research, nucleation, biology, and more. Famous scientists at least tangentially implicated in the New Manhattan Project such as Merle Tuve (1901-1982), Isidor Rabi (1898-1988), and CalTech's Lee DuBridge (1901-1994) have all been Research Corporation grantees. In 1965 Alfred Y.F. Wong of the University of California at Los Angeles' Plasma Physics Lab received $6K for work in, "Experimental studies of fundamental plasma physics." Again in 1972 Wong received $10K from the Research Corporation. The following year Barrett H. Ripin, Reiner L. Stenzel, and the aforementioned Alfred Wong of UCLA got $10,200 to study ion beam-plasma wave interactions. This is the same Alfred Y.F. Wong who went on, in the 1980s, to found and direct the High Power Auroral Stimulation (HIPAS) ionospheric heater in Alaska (ch 3). The HIPAS ionospheric heater was a precursor to the infamous High-Frequency Active Auroral Research Program (HAARP) antenna. Further, my science advisor tells me that weather modification work prominently mentioned by Navy Vice Admiral William F. Raborn, Jr. (ch 4) was conducted at the HIPAS facility. HIPAS was run by the Navy.

This is also the same Alfred Y.F. Wong that got busted for fraudulently producing phony invoices pertaining to U.S. government contract work. He pled guilty, was sentenced to 5 days behind bars, as well as six months home detention, and was ordered to pay nearly $1.7M in restitution. Wong is currently listed on UCLA's website as a Professor Emeritus of 'Exp. Plasma & Environmental Physics.'

Oddly enough, the cover of the Research Corporation for Science Advancement's 2008 annual report features a question mark made of clouds high above the Earth, appearing as though it has been written in the sky by aircraft. In this same report they reveal that their so-called Scialog program, which they originally disclosed in the previous year's annual report, is all about tackling 'global climate change.' Oddly enough as well, so is the New Manhattan Project. In their 2009 report they refer to global climate change as a 'looming challenge.'

Cover of the RCSA's 2008 annual report

The Research Corporation has always funded what we call 'basic science.' These are the small, highly specific studies that produce results which are often in turn used as the building blocks of much larger scientific efforts. Over the years, the Research Corporation has funded thousands of basic science studies. Furthermore, we know that one of the 3 main ways that the Research Corporation has funded itself over the years is from corporate stock dividends - the other two being patent royalties and electrostatic precipitator engineering services fees. For many decades now, the Research Corporation has maintained an investment portfolio consisting mostly of the dividend-paying stocks of large, American companies. Doesn't it make sense, then, that the Research Corporation would fund basic science studies which might produce new discoveries helpful to the corporations that fund the Research Corporation? This is probably what has been taking place.

Given this, doesn't it also make sense, then that, over the years, the Research Corporation may have funded basic science studies which have produced new discoveries that were then used as building blocks of the New Manhattan Project? The Research Corporation has owned the dividend-paying stocks of companies like General Electric, Boeing, and Standard Oil which, as we have seen, all have strong implications for the NMP. As noted earlier, one of the Research Corporation's founding members, Elihu Thomson was also a founding member of General Electric and General Electric appears to be the most instrumental corporation in the NMP's history.

For the production of the New Manhattan Project, use of an operation like the Research Corporation would be very advantageous. The New Manhattan Project, being that it is the biggest scientific project in history, has necessarily required gargantuan boatloads of basic scientific studies. And here is a self-sustaining organization that, for over 100 years and to this day, due to the efforts of Frederick Cottrell, produces an endless amount of this stuff. It's even better than making the taxpayer pay for it!

References

Blank Check: The Pentagon's Black Budget a book by Tim Weiner, published by Warner Books, 1990

"2014 Climate Change Adaptation Roadmap' a report by the Office of the Assistant Secretary of Defense (Energy, Installations & Environment) Environment, Safety & Occupational Health Directorate, June 2014

"U.S. Navy Climate Change Roadmap" a report by Task Force Climate Change / Oceanographer of the Navy, April 2010

"Report of the Defense Science Board Task Force on Trends and Implications of Climate Change for National and International Security" a report by the Office of the Under Secretary of Defense for Acquisition, Technology, and Logistics, 2011

"$21 Trillion dollars is missing from the US government. That is $65,000 for every person in America. That is as much as our entire national debt!" an article appearing on *solari.com*

"Pentagon Racks Up $35 Trillion in Accounting Changes in One Year" an article by Anthony Capaccio, published by *Bloomberg*, January 22, 2020

Gladio: NATO's Dagger at the Heart of Europe: The Pentagon-Nazi-Mafia Terror Axis a book by Richard Cottrell, published by Progressive Press, 2012

Perilous Missions: Civil Air Transport and CIA Covert Operations in Asia a book by William M. Leary, published by Smithsonian Institution Press, 2002

The Killing of Uncle Sam: The Demise of the United States of America a book by Rodney Howard-Browne and Paul L. Williams, published by River Publishing, 2018

The Secret History of the CIA a book by Joseph J. Trento, published by Carrol & Graf, 2005

Cottrell: Samaritan of Science a book by Frank Cameron, published by Doubleday, 1952

"Frederick Gardner Cottrell 1877-1948" a National Academy of Sciences Biographical Memoir by Vannevar Bush, published by the National Academy of Sciences, 1952

100 Years of Supporting Science Innovation: Research Corporation for Science Advancement: 1912-2012 a book by the Research Corporation for Science Advancement, published by the Research Corporation for Science Advancement, 2012

US patent #1,948,384 "Method and Apparatus for the Acceleration of Ions" by Ernest O. Lawrence

An American Genius: The Life of Ernest Orlando Lawrence, Father of the Cyclotron a book by Herbert Childs, published by E.P. Dutton & Co., 1968

Research Corporation annual reports 1950-2018

"Former UCLA Physics Professor Agrees To Plead Guilty In Federal Contract Fraud Case And Pay Nearly $1.7 Million In Damages" a press release by the Department of Justice, U.S. Attorney's Office, Central District of California, published by the United States Department of Justice, May 9, 2013

The Research Corporation (1912-1952): Dedicated to Progress in Education and Science a booklet containing a transcript of a speech by Dr. Joseph Warren Barker, published by the Newcomen Society, 1952

Chapter 16
SOLUTIONS

As with every problem life throws our way, things can be done about chemtrails too. We are not defenseless. Let this chapter answer the skeptics who say that chemtrails would never be sprayed because the people responsible for the spraying would be exposed as well. Chemtrails *are* being sprayed. But people *can* exist on this planet and have little to no personal exposure. Having boatloads of money sure helps, but there are also less expensive options. We can limit our exposure. We can remediate our bodies and our environment with supplements, cleanses, and bioremediation. We can also investigate, legislate, and take legal action. These are the solutions to the New Manhattan Project.

Location

Chemtrails are not being reported in all areas of the world. As noted earlier, they are mostly appearing in NATO countries. Excluding NATO countries leaves about 85% of the Earth's surface, maybe more. Not all of that surface is land, but people can live on boats, too. Ultra-rich people have giant luxury yachts upon which they live like kings. Ultra-rich people also have fortified compounds and secret getaways in second and third world countries where chemtrails have rarely, if ever been seen. Although most of us would find it very difficult to move away from the Western world, the point is that it is possible - very difficult unless you are mega-rich, but not impossible. Not only that, but anecdotal evidence seems to indicate that certain areas of the contiguous United States see less chemtrails than others. People say that Wyoming and surrounding regions tend to have less chemtrails than the coastal population centers.

Air purification

At home, if one closes all the windows and doors and runs an air purifier, the air gets a lot better very quickly. Decent HEPA air purifiers are not very expensive. One can pick up a good one at Home Depot for about $100. Some air purifiers are incredibly expensive. Some work better than others. It takes some shopping around.

Really good air purifiers employ the electrostatic precipitation method of air purification described in chapter 6. They can get particles much smaller than the best HEPA purifiers can. Yes, they can even get nano-sized particles. Decently affordable electrostatic air purifiers are available on *Amazon.com*. Your author bought one there made by The Sharper Image and he is quite pleased with it.

The best building air purification systems are centralized like a central-heating system. Centralized air purification systems are standard in buildings like hospitals and high-tech manufacturing facilities where the air needs to be free of dust, molds, pathogens, and the like. They are called HVAC (heating, ventilation, and air conditioning) systems.

It is possible to be in a home environment completely free of any airborne impurities. Air-tight houses exist. They are called 'passive houses.' There is a *Wikipedia* listing. Now, if one were to couple a passive house with a centralized electrostatic air purification system, then the chemtrail jets could be dumping their megatons of toxic garbage overhead without one being exposed at all. Not everybody has enough money to achieve this, but again, it is possible.

Humidifiers probably help too. Being that the materials dispersed by the airplanes are hygroscopic in nature, replenishing the air with moisture is probably a good idea. Some air purifiers have humidifiers built-in.

Portable air purifiers can be utilized in automobiles as well. Maybe some high-end automobiles even offer built-in HEPA cabin air filtration. But, even if you don't have a high-end automobile with a HEPA cabin filter you can still reduce the amount of your air particulates. According to a Washington University study, if one rolls up all the windows of one's car, turns on the air conditioning and sets the air flow to recirculation, one can reduce the amount of pollutants in the car by 20 to 34 percent. And yes, the evaporator in your car's air-conditioning system even collects nano-sized particles.

One can also monitor one's air quality. Among more straight-forward air monitoring devices, an outfit called Awair is selling cool little gadgets like one that plugs into your wall socket, makes different colored lights, and turns air-purifying appliances on and off depending upon the air quality. Now finally here's a smart-home type of gadget that actually makes sense!

Supplements

Certain supplements can help keep your body functioning properly. Dr. Russell Blaylock (whose work is featured in chapter 9) recommends turmeric. On Dr. Blaylock's recommendation, your author has been taking turmeric for years. I believe that it is turmeric which enhanced my sense of smell. Daily chemtrail exposure clogs one's olfactory nerves; the nerves that allow one to smell. Dr. Blaylock says that turmeric in one's blood takes away the particles clogging up the olfactory system. My experience has been consistent. Many years ago, I could not smell chemtrails. Now I can.

Although I am not paid in any way, shape, or form to promote any products besides this book, your author is suggesting some potential remedies. The Original Himalayan Crystal Salt Inhaler is available on Amazon. It is a stout ceramic container about the size of a man's fist. It contains Himalayan pink salt rocks. The inhaler has a spout with an opening at the end. One places one's mouth at the opening on the spout and inhales. The air that is inhaled passes over the Himalayan salt and goes into one's lungs. Doctors say that this salty air is good for the lungs. I have had one of these inhalers for quite a while now and I find it soothing. After use, I usually experience better breathing.

Throat and nasal sprays can also help. I have found that, when they are dry, a nasal spray called Xlear makes my nasal passages feel a lot better. Xlear is not cheap, but it is available at local Whole Foods outlets. Another product I have found at Whole Foods is a throat spray called Singer's. It comes in a few different formulas. I like the Honey Lemon. I find that it makes me a little sleepy, but it does wonders for my throat. Dr. Group (who is introduced in the next section) sells a spray which is supposed to help one's lungs, but I tried it and found that it irritated my throat. It might work for other people. I might try it again.

Detoxification and cleansing

The toxic chemtrail spray we ingest on a daily basis can be removed from your body. That process is known as detoxification. This is how we go about restoring our health. For what is needed to accomplish this, your author suggests one source.

One of the doctors mentioned in chapter 11 as being chemtrail aware is Dr. Edward Group. At his website *GlobalHealing.com*, Dr.

Group sells a full line of supplements needed for detoxification as well as detoxifying cleanse kits. He also provides people with specific instructions as to which supplements to use and how to use them.

Dr. Edward Group

One of his products is called Zeotrex and it is especially relevant to chemtrail detoxification because it is described as a, "chemical & toxic metal body cleanser." He also offers a chemical and toxic metal cleanse kit which includes Zeotrex, Detox Foot Pads, and instructions. We will discuss Dr. Group's toxic metal cleanse kit shortly. The products made available by Dr. Group can provide you with a full range of chemtrail detoxification options for all the bodily organs most affected by chemtrails. Chemtrails or no, cleansing your body of toxins is generally good for one's health. Before doing any cleanses, though, please consult your doctor first.

The first Dr. Group cleanse I did many years ago was the colon cleanse. It is a six-day cleanse that consists of one sticking to a diet of fruit and taking supplements both in the evening and in the morning. During one of these colon cleanses, one might feel a little down because the cleanse causes your body to release stored toxins. When these toxins have been released into your system, before they are passed, they make one feel bad. But I have performed this cleanse on three different occasions now and every time, after completion, it has made me feel *great*.

Dr. Group's liver and gallbladder cleanse is quite interesting. It takes about six days and involves the consumption of his supplements along with a specific diet of food. The initial phase of the cleanse goes on for five days. In the evening before the last day of the cleanse, one is instructed to consume a tablespoon of epsom salt dissolved in 8 ounces of warm water. It's some of the worst stuff I have ever tasted, but it apparently opens up one's gallbladder. One hour after that, Dr. Group instructs the cleanser to consume 8 ounces of organic, extra-virgin olive oil. The olive oil, if taken straight is almost too much for one to keep down, but I find that if the olive oil is mixed with orange juice, it goes down much easier. The olive oil apparently lubricates the liver and the gallbladder stones, allowing the stones to pass into the digestive system. The next morning, the cleanser wakes up and takes a dose of Dr. Group's digestive system-clearing supplement known as Oxy-Powder along with lots of water. I have found that orange juice works well, too. Usually by mid-day, one should be passing gallbladder stones.

As someone who has performed Dr. Group's liver and gallbladder cleanse six times, your author can attest to its effectiveness as the cleanse has allowed me to pass easily hundreds of liver and gallbladder stones. Over the course of a few years, I have kept on doing Dr. Group's liver and gallbladder cleanse. The first time I did it, I passed only some neon-green, globulous type of material. I think that was cholesterol from my liver, but no gallbladder stones. The second time I did it, I passed about 20 small gallbladder stones along with more of the neon-green globulous material. The third time I did it, I passed quite a number of gallbladder stones, along with one that had the diameter of a quarter. One knows about the gallbladder stones one has passed because gallbladder stones float. The fourth time was the big gallbladder stone dump. On the fourth cleanse, I passed scores of pea-sized stones, maybe 10 with the diameter of about a penny, two with the diameter of a quarter, and one with the diameter of a half-dollar piece - I call that one the Rock of Gibraltar. There were so many other, smaller stones that they were too numerous to count.

I should warn the reader. On the night before this big gallbladder stone dump, I felt absolutely horrible. As the massive amount of stones and other waste from my gallbladder was making its way through my digestive system, I was tossing and turning in bed. I was sweating. I couldn't sleep and I had a headache. My stomach felt bad. It felt like I had drunk way too much alcohol. My malaise continued into the next morning. It was not until I passed the stones in the middle of the next day that I felt better. In fact, after I passed the stones, I felt *great*.

About two months after that, I did Dr. Group's liver and gallbladder cleanse one more time in order to make sure to get rid of all the gallbladder stones that I could. This fifth cleanse produced a handful of mostly pea-sized stones, with some that were a little bigger. These stones were very dark - as if they were very old and from the bottom of my gallbladder. I probably had these stones in my gallbladder since I was a little kid. I'm glad I got these out because it's apparently when old stones start to calcify that one begins to have real health problems.

I should also tell you about my experiences with Dr. Group's heavy metal detox cleanse. The cleanse involves taking his supplement called Zeotrex and applying detox foot pads to one's feet every night before bed. The cleanse is pretty easy to perform, but it goes on for 30 days, which is a little grueling. By about the second week, I was feeling somewhat down because the cleanse was releasing previously stored toxins into the rest of my body.

You see, the feet store a lot of toxins. It is this way because they are the lowest point in the body. That's where things tend to settle. When one applies these detoxifying foot pads every night, the toxins stored in one's feet are disturbed and much of it goes into one's blood stream. A lot of it is sucked up by the foot pads as well, but my feet were apparently so toxic that the foot pads were quickly filled up and I was left with lots of previously stored toxins circulating in my blood stream. I believe that this caused some residual effects such as a reduced quality of blood circulation which persisted for many months.

I figured out later, and this was corroborated by some of Dr. Group's information, that before one performs one of these heavy metal detoxes, one should apparently first make sure that they have very clean, highly-functioning kidneys, a clean, highly-functioning liver, and a clean, highly-functioning colon as well. If one has these three things going, one's body can deal with the onslaught of toxins released during a heavy metal detox. Apparently, at the time I did my heavy metal detox, my body's detoxification systems were not quite up to snuff. I plan to do Dr. Group's cleanses targeting these three systems, then try his heavy metal detox once more. Again, if you are contemplating doing any of this type of stuff, please consult your doctor first.

Microbes

When the enemy comes in like a flood, the Spirit of the Lord will lift up a standard against it. In this case, that standard is microbes. The larger area of study is known as 'bioremediation.' Microbes are microscopic organisms as small as a single cell which are absolutely everywhere. Earth's most accommodating and most inhospitable environs are teeming with them. They are inside and all over our bodies. They live floating around in the stratosphere, in the deepest depths of the oceans, and even in outer space. They are the oldest forms of life on this planet. Only a tiny fraction of them have ever been officially discovered and categorized.

In the context of the chemtrail phenomenon, we concern ourselves with microbes because they can remediate the environmental damage done. Microbes can consume and neutralize the toxins which have been so unceremoniously dumped upon us and our world. This may very well happen naturally. As the chemtrail spray proliferates, so will the microbes which feast on it. This may also happen by acts of man. We can produce and disperse beneficial microbes in a concerted effort to remediate our environment. However it happens, the best case scenario is that the spraying be stopped so that beneficial microbes are given a chance to work their magic without being overwhelmed.

Chemtrail activist Allan Buckmann (ch 5) has been working with microbes. He recently wrote a short piece which is presented here:

"I do not profess to be a microbial expert, however in the microbial business for soil, water, and plant production I found out a lot about these little creatures that make up most of our lives, and without them cause deficiencies.

"There are countless microbial species on this planet that fill every niche of the ecosystem. They are known to be present everywhere and

within their collective capacity eat everything... including radioactive materials.

"Some are known and used, but most (trillions) are still undiscovered for what they eat, defecate, or exchange. We are most familiar with those that cause disease, or are used for health, or agriculture. Those used for environmental treatments are all known beneficial species.

"The human body carries billions of microbes and most people don't know it. We have specialists in our armpits (generally a determining factor for body odor), in our glands, eyes, intestines that aid digestion, and in every other area. We have evolved with microbial modifications in our systems and they are an important part of us.

"They are also important for 'healthy' soil and agricultural or forestry production and are the difference between high value, colorful, and tasty fruits and vegetables like we can raise in our own gardens with good humus (rich with healthy microbes), and the bland crops grown with artificial chemicals. They are the difference between fertile and infertile soils. Mycorrhiza species were found to stimulate root development and increase nutrient uptake, and that the roots also interacted with other plants for nutrient and genetic exchange. From such studies it was found that combinations of species could enhance overall production and that they work together best in social mixes.

"So, yes I have been promoting the use of microbial products for bio-remediation of soil and plant deficiencies, and discovered that jet clouds and jet exhaust particles falling to the earth, and other particulates do not stop microbial actions as of this date.

"As the past president of Microbe Tech, I became familiar with and had many applications using a line of microbial products formulated by Hugh Muir (deceased) and still manufactured here in the United States, and I still use these products. We have successfully treated soil, water, and plants with great results. This is in spite of any accumulation of metals in the environment so it is possible to counteract the effects of pollution."

Investigations

It's all hands on deck. This is not a spectator sport. This is all of our lives that they are going about ruining. If we all do our parts, the chemtrails will end sooner and the people responsible will be brought to justice.

Collecting evidence is always good. Rainwater sample testing can add to the body of evidence. One can simply place a clean bowl, cup, or rain gauge at least 8 feet off the ground outside on a day when there is precipitation and visible chemtrails. Collect some rainwater and send it off to a local analytical testing laboratory. Ask the lab how much they need and what types of containers should be shipped. Analytical testing laboratories can be found all around the country. Soil and water testing is a common procedure. Frances Mangels uses Basic Laboratory in Redding, CA. Patrick Roddie uses McCampbell Analytical in Pittsburgh, CA.

You can test for the presence of any element you wish. Labs charge by the element, so if you want them to test for everything, it will be quite expensive. If your rainwater sample test results show high levels of aluminum, barium, and/or strontium, then you have created more evidence! One can then post these findings on such websites as: *GeoengineeringWatch.org*, *ChemtrailsProjectUK.com*, and/or *GlobalSkywatch.com*.

Videotaping chemtrail sprayers can be a productive activity. Getting a good shot of an airplane at 35,000 feet is not the easiest thing in the world, but it does sometimes yield some good evidence. You might be able to capture all white, unmarked jumbo jets. Those would be aircraft of the dedicated chemtrail fleet and, due to their relative rarity, the best ones to photograph.

There are many thousands of videos posted online, but one *YouTube* video in particular is impossible to refute. Dane Wigington at *GeoengineeringWatch.org* says he was sent this video titled "Undeniable Footage of Jet Aircraft Spraying." It depicts a chemtrail spraying aircraft with a constant dispersion coming from the wings and an *intermittent* dispersion coming from the fuselage. This video proves beyond a shadow of a doubt that this is not a contrail and that this plane is spraying something. Unless there were operations publicly disclosed, then this is not part of an overt silver iodide cloud seeding operation. This is most probably part of the covert New Manhattan Project fleets, caught on tape.

We the people can help, but the government investigators we pay for are the real heavy hitters. They have powers to uncover evidence which your average citizen cannot. The Government Accountability Office (GAO) is the right organization for the job. The GAO mission statement reads:

"*The Government Accountability Office, the audit, evaluation, and investigative arm of Congress, exists to support Congress in meeting its constitutional responsibilities and to help improve the performance and accountability of the federal government for the American people. GAO examines the use of public funds; evaluates federal programs and policies; and provides analyses, recommendations, and other assistance to help Congress make informed oversight, policy, and funding decisions. GAO's commitment to good government is reflected in its core values of accountability, integrity, and reliability.*"

Being that it appears that public funds have been used to develop and probably implement the New Manhattan Project (ch 15), we deserve a GAO investigation. In fact, the GAO has already been taking a look at SRM geoengineering. Although their many reports on the related issue of climate change indicate an adoption of the theory of man-made global warming, their few reports on climate mitigation measures such as SRM geoengineering express concern. Their 2010 report titled "Preliminary Observations on Geoengineering Science, Federal Efforts, and Governance Issues" was based on the congressional testimony of the director of their Natural Resources and Environment division. The first sentence of the report states, "Substantial uncertainties remain on the efficacy and potential environmental impacts of proposed geoengineering approaches..." The last sentence of the first paragraph reads, "Specifically, just as the effects of climate change in general are expected to vary by region, so would the effects of certain large-scale geoengineering efforts, therefore, potentially creating relative winners and losers and thus sowing the seeds of future conflict."

A congressional investigation is also in order. Although lots of credible evidence already exists, the proper progression here is probably first a GAO investigation, then a congressional investigation. It's best that our congress has rock solid evidence coming from professional government investigators upon which to proceed. It may take a little longer this way, but there's a better chance that it will be done properly.

Another way to gather rock-solid information would be through the formation of a federal grand jury. If a federal grand jury is convened to investigate geoengineering, the investigators would be imbued with tremendous powers such as the power of legal subpoenas which can force all types of organizations, public and private, to produce relevant information. Personally, I would love to be a part of something like that. Many others who have been leading this charge would probably be willing to participate as well. If such a grand jury is convened, the author has already compiled and is ready to produce an extensive list of persons and organizations of interest complete with background histories.

Legal action

Let this book serve as a preliminary investigation into violations resulting in the largest class-action lawsuit in history. Dane Wigington notes that the currently ongoing, large-scale SRM geoengineering program violates a raft of state and federal laws pertaining to the disposal of toxic substances, protection of the general public, and the accurate presentation of official environmental data.

Although the New Manhattan Project violates many laws, the strongest legal basis for suits against it probably lies with the principle of informed consent. This is the idea that people don't get to just go ahead and ruin our health without our being informed as to what is happening and without our approval. The most salient example of this concept is the Nuremberg Code. The trials of Nazi doctors which resulted in the Nuremberg Code were conducted by the United States. Since this code was established, it has served as a legal and ethical foundation for Human experimentations worldwide. The Nuremberg Code reads:

"The voluntary consent of the human subject is absolutely essential. This means that the person involved should have legal capacity to give consent; should be so situated as to be able to exercise free power of choice, without the intervention of any element of force, fraud, deceit, duress, overreaching, or other ulterior form of constraint or coercion; and should have sufficient knowledge and comprehension

of the elements of the subject matter involved as to enable him to make an understanding and enlightened decision. This latter element requires that before the acceptance of an affirmative decision by the experimental subject there should be made known to him the nature, duration, and purpose of the experiment; the method and means by which it is to be conducted; all inconveniences and hazards reasonably to be expected; and the effects upon his health or person which may possibly come from his participation in the experiment."

Now does it sound like our rights have been violated under the Nuremberg Code?

A U.S. law called the Common Rule for the Protection of Human Subjects is based on the Nuremberg Code and is part of the the U.S. Code of Federal Regulations. It reads:

"*Except as provided elsewhere in this policy, no investigator may involve a human being as a subject in research covered by this policy unless the investigator has obtained the legally effective informed consent of the subject or the subject's legally authorized representative. An investigator shall seek such consent only under circumstances that provide the prospective subject or the representative sufficient opportunity to consider whether or not to participate and that minimize the possibility of coercion or undue influence. The information that is given to the subject or the representative shall be in language understandable to the subject or the representative. No informed consent, whether oral or written, may include any exculpatory language through which the subject or the representative is made to waive or appear to waive any of the subject's legal rights, or releases or appears to release the investigator, the sponsor, the institution or its agents from liability for negligence.*"

In fact, the usual suspects have signed onto it. That means: the Department of Energy, NASA, the Department of Defense, and the National Science Foundation are all signatories. The CIA, by executive order, must comply with it.

Further, Title 50 Chapter 32 U.S. Code §1520a. Restrictions on use of human subjects for testing of chemical or biological agents states that the Secretary of Defense may not use any type of biological or chemical agent on Human subjects without first attaining the subject's informed consent AND Congressional consent.

They should be frightened. The legal resistance is already building. On March 23, 2016 Dane Wigington at *GeoengineeringWatch.org* reported that a Canadian class-action chemtrail lawsuit has been filed. The article is titled "Pushing Back Against Climate Engineering, Canadian Lawsuit Is Filed." The lawsuit, which was produced with the active participation of Mr. Wigington, claims that geoengineering activities have caused significant economic and environmental damage. The defendant is the Queen of England. Mr. Wigington is (as one might be able to discern from the material in this book) at the forefront of the battle against geoengineering. To keep up the legal fights, Dane is working with an organization called the Legal Alliance to Stop Geoengineering. They state that their purpose is to, "spearhead national and international legal efforts in order to fully expose and halt the ongoing illegal covert climate engineering programs." Through their website, one may donate to their organization.

Legislation

We can petition our government to write new laws in an effort to expose and stop chemtrails. It's not something that gets done overnight. But, like my grandmother used to say, nothing worthwhile is easy. Many milestones have already been achieved. As we saw in chapter 11, there has already been quite a bit of progress in this area. But we can use more. I offer you an amalgamation of Berkeley's "Resolution in Support of the Space Preservation Act," Maui's proposed "Clean Sky Ordinance" and my contributions gleaned from working with local activists and governments. I ask here and now that people go about getting a local council member to sponsor this:

A generic proposed resolution to declare your town or city a 'geoengineering/weather/climate modification-free zone'

BACKGROUND: A Multitude of extremely powerful organizations both foreign and domestic have produced voluminous amounts of documentation proposing that the Earth's biota be routinely sprayed with megatons of toxic materials from aircraft. This practice is commonly known as Solar Radiation Management geoengineering and also as climate modification. Some of the organizations suggesting such activities include, but are not limited to: The Council on Foreign

Relations, the Central Intelligence Agency, the United Nations, the National Academies of Science, Lawrence Livermore National Laboratory, the Trilateral Commission, the U.K. Royal Society, the U.S. Congressional Research Service, the Rand Corporation, the National Research Council, and the Carnegie Foundation.

A RESOLUTION OF *your council* TO DECLARE *your town/city* A 'GEOENGINEERING/WEATHER/CLIMATE MODIFICATION-FREE ZONE'

WHEREAS, the people of your town/city recognize that geoengineering, weather modification, and climate modification programs and activities are either known to employ or propose employing aircraft and other means to disperse substances in order to affect the weather and/or climate; and

WHEREAS, the people of your town/city have the right to be informed by full disclosure through an Environmental Impact Statement of the health and environmental implications caused by any geoengineering, weather modification, and/or climate modification program or activity; and

WHEREAS, any geoengineering, weather modification, and/or climate modification program or experimental program or activity with potentially adverse health and environmental implications for the people of your town/city needs the informed consent of the people of your town/city; and

NOW, THEREFORE, BE IT HEREBY RESOLVED:

I. That it is the will of your council to declare your town/city a 'Geoengineering/Weather/Climate Modification-Free Zone'; and

II. That your council hereby declares that the space above your town/city is a 'Geoengineering/Weather/Climate Modification-Free Zone'; and

III. That an official committee of your town/city be convened for the purpose of producing a report to your council and the people of your town/city in order to determine if your town/city and the residents of your town/city have been exposed to substances associated with geoengineering, weather modification, and/or climate modification activities; and

IV. That this resolution of your town/city is hereby recommended to be adopted by all Municipalities in the United States of America and worldwide.

~ ~ ~

If you decide to accept this mission, please learn from my mistakes. Don't just go to your local council meeting and float the idea during the public comment section. It's easy for them to dismiss somebody that way and they might not know what you're talking about. One goes about this the proper way by firstly contacting as many other interested community members as possible. Standing on a street corner with a petition and a sign would be effective. Form a group with these people and give yourselves a name. Making a website or a social media page doesn't hurt. Then gather your best information (copies of this book might fit the bill) and send it to your city council and mayor by email AND snail mail. Follow up with phone calls, letting them know that your group wants them to sponsor the proposed resolution. Hopefully by the time you address the council, you will have a council member or the mayor ready, willing, and able to officially sponsor your resolution. If your representatives don't want to sponsor it right now, ask them to form a sub-committee to study the issue before they do. When your sub-committee completes your report, then you can then ask the mayor or a council member for sponsorship. Once your proposed resolution is officially sponsored, your council will (hopefully) vote on it.

Free energy

As mentioned in chapter 10, free energy technologies have been around for over 100 years. They go back to Nikola Tesla. There is not another single issue more important to the liberty of Humanity or more indicative of our culturally-inverted society.

There are literally scores of examples. It doesn't take much web searching to find mind-blowing information. In the documentary *Thrive*, the best encapsulation of these technologies known to the author is presented. *Thrive* is available for free in its entirety on

YouTube. It's only got about 15 million views. The section on free energy begins at about 34 minutes in and runs to about the 43 minute mark.

While the geniuses at Lawrence Livermore Labs and CERN spend billions of dollars valiantly searching for the holy grail of cold fusion, a 13 year old kid recently made a free energy device for under $15. He made it out of a coffee can, some wire, two coils, and a spoon. He used it to light up a string of LED lights wrapped around his brother. You can read all about it in a May 18, 2016 article which appeared on the website *WakingTimes.com* titled "13-Year-Old Invents Tesla Inspired Free Energy Device for Under $15."

My science advisor tells me that a patent application released last year may describe a device that generates more energy than it consumes - a kind of free energy device. The patent application is titled "Plasma Compression Fusion Device," the inventor is one Salvatore Pais, and the assignee is the U.S. Navy. Shortly after the publication of this patent application, Jennifer Leman at *Popular Mechanics* wrote an article describing the new technology as something that has the potential to revolutionize the world. Salvatore Pais has a *Wikipedia* listing if one wishes to investigate further. My science advisor recommends two books on the subject: *Quest for Zero Point Energy-Engineering Principles* and *Tapping the Zero Point Energy - How Free Energy and Anti-gravity Might Be Possible with Today's Physics* both by Moray B. King.

Free energy is particularly important to the topic of this book because, as we know, the reason given for spraying toxic chemicals out of aircraft is that man's carbon emissions (specifically the burning of hydrocarbon fuels) is causing dangerous climate anomalies and that SRM geoengineering can avert a man-made climate disaster. If we have no more need for carbon-based fuels like oil and gas, and just use clean, free energy, then there is no more reason to spray stuff out of airplanes. You can bet that if free energy technologies are unleashed, though, geoengineers will start selling their activities as a necessary global weather modification project. But by that point, their agenda will be irreparably damaged.

Aside from derailing the chemtrail agenda, free energy would be the single most transformative development in Human history. It would set Humanity and our planet free. The problem is that the most powerful people in the world don't want Humanity and the Earth to be free. They would rather have us suffer under their Draconian New Manhattan Project. Are we going to allow this to continue? Their tyranny requires our consent.

References

"Comparing on-road real-time simultaneous in-cabin and outdoor particulate and gaseous concentrations for a range of ventilation scenarios" a study by Anna Leavey, Nathan Reed, Sameer Patel, Kevin Bradley, Pramod Kulkami, and Pratim Biswas, published in *Atmospheric Environment*, July 16, 2017

"Preliminary Observations on Geoengineering Science, Federal Efforts, and Governance Issues" a report by the Government Accountability Office, 2010

Geoengineering: Chronicles of Indictment: Exposing the Global Climate Engineering Cover-up a book by Dane Wigington, published by Geoengineering Watch Publishing and Media LLC, 2017

"60 Million USD Claim Against Government of Sweden for Failure to Respond to Citizens Request for Information on Chemtrails" an article by Dane Wiggington, published by *GeoengineeringWatch.org*, April 9, 2014

"Pushing Back Against Climate Engineering, Canadian Lawsuit Is Filed" an article by Dane Wiggington, published by *GeoengineeringWatch.org*, March 23, 2016

Thrive a documentary video by Foster and Kimberly Gamble, 2011

"13-Year-Old Invents Tesla Inspired Free Energy Device for Under $15" an article by Terence Newton, published by *WakingTimes.com*, May 18, 2016

U.S. patent application "Plasma Compression Fusion Device" published Sept. 26, 2019

“The Navy’s Patent for a Compact Nuclear Fusion Reactor Is Wild” an article by Jennifer Leman, published by *Popular Mechanics*, October 10, 2019

AFTERWORD

In the 2015 movie *The Big Short*, actor Steve Carell, playing the role of financial trader Mark Baum, characterizes our time as the 'era of fraud.' The movie is specifically about financial fraud that led to the housing and stock market crash of 2008, but this characterization of our current society as a whole is justified. There are myriad gigantic frauds currently being perpetrated against Americans right here at home - not the least of which being thousands of large aircraft routinely dumping tens of thousands of megatons of toxic waste into our biosphere.

In years past, in America at least, the major crimes of the establishment were hidden, and whenever they were exposed, there was hell to pay. Now, institutionalized, domestic crimes are on full display. Now it is all explained away with finely-crafted disinformation campaigns telling us that these things are helpful and good. Water fluoridation, GMOs, childhood vaccines, chemtrails, 5G and so much more are sold to the public as beneficial. The fake news Old Media has everything to do with this.

As Michael Ellner writes, "Just look at us. Everything is backwards, everything is upside down. Doctors destroy health, lawyers destroy justice, psychiatrists destroy minds, scientists destroy truth, media destroys information, religions destroy spirituality and governments destroy freedom." With the addition of 'teachers destroy education,' the New Manhattan Project is a testament to this statement.

America has been the epicenter of the fraud. America is supposed to be the guiding light to the rest of the world. It truly was for a long time. But, throughout the last 100 years plus, as Rockefeller minions have seized control of the levers of power at seemingly every level and in every area of our society, that guiding light has gotten dimmer and dimmer. All too often the American light of freedom and liberty has appeared as a voracious black hole, attempting to suck the life out of whatever comes into its contact. The New Manhattan Project requires this type of environment. It is the only type of environment in which it can exist. And as America goes, so goes the rest of the world.

There is still hope, though, and your author is quite optimistic about the future. But it requires that enough people stand up and say 'no' to this tyranny. As David Icke says, it's time for the Human race to get off its knees. Now that this global tyranny is out in the open, it is up to the citizenry to, once again, declare our independence and take a stand.

Remember: what the men at Morristown and Valley Forge did during the American Revolution was hard. What we do today, fighting in the information war, is easy. People sacrificed everything they had, including their lives, to create a space where we can have this discussion. I suggest that we proceed with the discourse that will forever free Humanity from the global tyranny we have been under all throughout history. America will lead the way. It starts with the individual and it is God's will that we succeed. Thank you.

INDEX

C

D

H

J

K

N

P

S

W

X

Made in the USA
Coppell, TX
04 June 2021

56880110R00356